T0185812

Power Systems

Electrical power has been the technological foundation of industrial societies for many years. Although the systems designed to provide and apply electrical energy have reached a high degree of maturity, unforeseen problems are constantly encountered, necessitating the design of more efficient and reliable systems based on novel technologies. The book series Power Systems is aimed at providing detailed, accurate and sound technical information about these new developments in electrical power engineering. It includes topics on power generation, storage and transmission as well as electrical machines. The monographs and advanced textbooks in this series address researchers, lecturers, industrial engineers and senior students in electrical engineering.

More information about this series at http://www.springer.com/series/4622

Hongming Zhang • Slaven Kincic
Sherrill Edwards

Advanced Power Applications for System Reliability Monitoring

 Springer

Hongming Zhang
Power System Engineer Center (PSEC)
National Renewable Energy Laboratory
Golden, CO, USA

Slaven Kincic
Pacific Northwest National Laboratory
Richland, WA, USA

Sherrill Edwards
MWConsulting
WA, USA

ISSN 1612-1287   ISSN 1860-4676   (electronic)
Power Systems
ISBN 978-3-030-44546-1   ISBN 978-3-030-44544-7   (eBook)
https://doi.org/10.1007/978-3-030-44544-7

This Springer imprint is published by the registered company Springer Nature Switzerland AG
The registered company address is: Gewerbestrasse 11, 6330 Cham, Switzerland

*To our fellow colleagues who were part of
Peak Reliability and/or WECC RC
for dedicated services to assure the wide
area view on the Western Interconnection
between January 1, 2009, and
December 3, 2019*

Foreword

The significance of this book is evident in several major ways. It is the first book that provides a complete description of the key functions within a major Reliability Coordinator (RC). It uniquely documents the practices and tools used in a real-world RC, many of which are more advanced by any measure than the state of the art in North America. The authors provide sufficient descriptions and references to many of the already realized new technologies that grew out of PEAK's R&D efforts within the last 10 years. Ultimately, this book will serve as a time capsule, documenting the legendary achievements by many forward-looking engineers and will be an important touchstone for years to come.

I have known and worked with the book's lead author Dr. Hongming Zhang professionally for more than a decade. His passion and dedication to the advancement of the power industry have no doubt greatly contributed to the development of many new technologies that are described in this book. The authors' commitment to share the knowledge and practical experience of their team at Peak RC and beyond is the main motivation of this book. Through the writing of this book, the authors have had the opportunity to pay tribute to many outstanding engineers at Peak RC, who have helped build a futuristic example of a modern reliability coordination entity. In addition, there had been a tremendous effort by this group to try to preserve proven engineering work including custom software and essential modeling data properly before Peak RC ceased to exist.

The need to ensure power system reliability has grown significantly in recent years. This is not only because of the increasing need to secure the supply of electricity, but to face new challenges brought by the increasing inverter-based resource and loads connected to the power grid. Modern theories and technologies are needed to enhance the reliability of the wide area bulk electric system (BES).

For more than a decade working as the RC for the Western Interconnection, Peak RC has conducted extensive R&D work in developing and implementing innovative system reliability enhancement tools and processes. Among all of these great achievements, I would like to highlight the following examples that have been used routinely or running in real time.

- They developed a model of the full Western Interconnection, i.e., the West-wide System Model (WSM). This model improved the real-time assessment accuracy and reliability coordination efficiency of the Energy Management System (EMS). WSM was also used by Peak RC's member utilities to improve their external models, hence the study scope and quality of the real-time and planning operation.
- A variety of new technologies were developed and applied in collaboration with software vendors, universities, national labs, and private consultants. These technologies are used in large system modeling, such as the network and dynamic modeling including remedial action schemes, and real-time system stability monitoring and analysis, such as voltage stability and inter-area oscillation modes.
- Tools and procedures were developed to bridge the EMS operation modes with Western Electric Coordination Council (WECC) planning models through data sharing and model conversion. This work leveraged the steady-state and dynamic system model validation and event analysis of the Western Interconnection system.

Even after Peak RC closed its doors, their work remains impactful on the industry and will no doubt become common practice in the future. For example, through new collaborations between the Western Utilities and U.S. national laboratories, Peak RC's legacy work in areas of the WSM data and real-time analytical tools is extended to the development and validation of the Western Interconnection Resilience Model, an essential part of the North American Energy Resilience Model, or NAERM envisioned by the Department of Energy (DOE).

Even though the book has Peak RC as the primary background, it provides a comprehensive view of the development of a modern reliability system and solutions to many practical engineering problems that can be easily adopted by others. As an educator, I see even more value in its detailed treatment of routine business and tools used in reliability coordination. I highly recommend this book to anyone in the power industry, especially those who would value having a complete picture of reliability coordination.

It has been a tremendous educational journey to read this book's chapters (though not necessarily in the order it is laid out now). I am confident it will be of great value to engineers, researchers, and students in the power system field.

UT/ORNL Governor's Chair Professor,   Yilu Liu
member of NAE, NAI
University of Tennessee
Knoxville, TN, USA

Oak Ridge National Laboratory
Oak Ridge, TN, USA

Preface

The story behind this book began on August 22, 2007, when I received a surprising email from Eric Whitley, who at the time was the EMS group Manager of MISO. Eric informed me in the email that he had accepted the new position of EMS Manager with WECC (Western Electricity Coordinating Council). He planned to build a startup team to create and manage a West-wide System Model (WSM) and to implement a new EMS tool suite, with Advanced Applications and SCADA as its main focus. Eric expected me to be onboard as one of the senior staff to kick off this fun but very challenging work.

There was no question that I was very excited about Eric's project, as this opportunity was something bigger than what I've done previously. Eric had gotten to know me and my expertise while I had supported his team on CAISO's EMS/DTS implementation project as ABB Sr. Applications Software Engineer between 2003 and 2005.

My family had just moved to the Greater Austin area from Houston, and I had started a new job with the Electricity Reliability Coordinating Council of Texas (ERCOT) about 10 months ago. Thus, my first reaction was "not ready to move" again since my family needed a break.

However, as Eric continuously updated me via email about the WSM project scope, progression, and significance, my heart was attracted to this fun but very challenging project. I clearly remember the day when WECC HR Director Melanie M. Frye (now WECC CEO) emailed me Eric's offer letter for my signature on December 12, 2007. It was my birthday, and I had just made the decision to go with Eric and join his team. Thanks to Eric's call, I jumped into this great journey of 11 years to work on the WSM project and EMS applications, until December 3, 2019, when Peak Reliability shut down the control centers upon completion of the RC transition.

In early 2019, I was contacted by Springer Nature with an invitation to submit an engineering book proposal. Once the proposal was accepted, two of my colleagues, Dr. Slaven Kincic and Mr. Sherrill Edwards, joined me, thankfully, to plan and draft this unique book about Peak RC's WSM modeling practice, advanced applications implementation, and lessons learned over a decade.

Professor Yilu Liu, University of Tennessee at Knoxville, has consistently encouraged me to wrap up Peak RC's experience on the WSM project and real-time tools for industry reference. Dr. Liu and her students and aides thankfully reviewed the entire book draft and provided valuable comments and edits. Moreover, she graciously wrote the foreword to this book.

Dr. Benjamin Kroposki, National Renewable Energy Laboratory, sponsored me to work on the North America Energy Resilience Model (NAERM) project regarding real-time use cases as a Consultant in late 2019. I was inspired by the project to recap Peak RC's proven work for planning the NAERM project implementation strategy. I would like to give special thanks to both Dr. Liu and Dr. Kroposki for their support and visions on the book.

The authors gratefully acknowledge the contributions of the RC management and the RC staff who had put forth significant effort in building the tools and processes described in this book. We would specifically acknowledge the following, our former colleagues, for their contributions to the work written in this book:

- EMS Modeling: Mr. Brett Wangen, Mr. Gareth Lim, Mr. Thomas Curtis, Mr. Krishna Karnamadakala, Ms. Celesani, Dr. Zhouxing Hu, etc.
- RTCA/RAS Modeling and Visualization: Mr. Matthew Veghte, Mr. Madhukar Gaddam, Mr. Ran Xu, Dr. Haoyu "Harry" Yuan, Ms. Mengjia "Jesse" Xiao, Mr. Ronald Evjen, Dr. NDR Sarma, Dr. Uttam Adhikari, etc.
- Real-Time Voltage Security Assessment (RT-VSA) Implementation: Mr. Saad Malik, Mr. James O'Brien, Mr. Ran Xu, Mr. Madhukar Gaddam, Dr. Jiawei "Alex" Ning, Mr. Matthew Veghte, etc.
- Online TSAT Implementation: Mr. Ronald Evjen, Dr. Haoyu "Harry" Yuan, Dr. Luoyang Fang, Dr. Xiaoyuan Fan, Dr. May Mahmoudi and Dr. Yidan Lu, etc.
- DTS Tool and Training: Zea Flores, Hari Ramana, Stephanie Conn, Patrick Olin, Arnold Schaff, etc.
- Oscillation Detection and Mode Meters: Mr. James O'Brien, Mr. Ran Xu, Dr. Jiawei "Alex" Ning, Dr. Tianying "Lily" Wu, Dr. Haoyu "Harry" Yuan, Dr. Ellen Wu, and Dr. Lakshmi Sundaresh.
- Peak RC Custom Software Development: Mr. Kirk Stewart, Mr. Arvind Kamath, Mr. Madhukar Gaddam, Mr. Kory Wilson, Mr. Todd Mccune, Mr. Jeff Shambaugh, etc.
- IT Applications Support: Mr. Murat Uludogan, Mr. Lon Kepler, Mr. Seong Chow, Mr. Peter Tang, Mr. Dayna Aronson, Mr. Mark Bowl, Mr. Joshua Harcourt, Mr. Vincent Devine, Mr. Andrew Esselman, Mr. Steve Pharo, etc.

The authors also gratefully acknowledge their major software vendors, Western Utilities, and innovation research partners. The RC's advanced real-time tools could not be implemented and operated successfully without their dedicated support and collaborations:

- GE/Alstom for EMS/DTS/WAMS Support: Mr. Srinivas M., Mr. Davis Huang, Mr. Suresh Prenus, Dr. Xingkang Wang, and Dr. Manu Parashar etc.

- PowerTech Lab for DSA Manager/TSAT Support: Mr. Frederic Howell, Dr. Lei Wang, and Dr. Lin Xi
- V&R Energy for POM/ROSE RT-VSA Support: Ms. Marianna Vaiman and Dr. Mike Vaiman
- MontanaTech for MAS Support: Prof. Dan Trudnowski and Dr. Matthew Donnelly
- Washington State University for OMS and FODSL support: Prof. V. Venkatasubramanian (Mani) and his PhD students including Addipour Farrokhifard and Yuan Zhi etc.
- WECC/Peak RC member utilities: AESO, APS, BPA, BC Hydro, CAISO, IPCO, PacifiCorp, PG&E, SCE, SDG&E, SRP, WAPA, etc.
- National Renewable Energy Laboratory (NREL)
- Pacific Northwest National Laboratory (PNNL)
- University of Tennessee, Knoxville, and CURENT Program

We thank the US Department of Energy for supporting our R&D work by co-funding the Western Interconnection Synchrophasor Program (WISP), Peak Reliability Synchrophasor Program (PRSP), and the Consortium for Electric Reliability Technology Solutions (CERTS).

We thank Springer Nature and associated editors for their efforts in publishing this book.

And of course, as almost always, last but not least, I would like to thank my family.

My wife, Xianghong Lily Liu, who had been constantly standing behind me every time I pursued new dreams. I wouldn't have gone through this journey without you.

My two children, Micah and Jeannie, who were always my sunshine. I am very proud of you. As Micah graduated from the University of Colorado Boulder with a major in Aerospace Engineering this May, Jeannie will step into Georgia Tech to study Mechanical Engineering this August, I hope this book could shed light on your own engineering dreams.

Respectfully,

Hongming Zhang

Fort Collins, CO, USA

Contents

About the Authors

Dr. Hongming Zhang is the Chief Energy Systems Operation Modeling Engineer of the Power Systems Engineering Center at the National Renewable Energy Laboratory (NREL), where he is responsible for Bulk Electric System (BES) modeling, next-generation control room capability development, and real-time grid operation simulator implementation.

Dr. Zhang received his B.S. and M.S. degrees in Electrical Engineering from Shanghai Jiao Tong University, Shanghai, China, and Ph.D. degree from the Texas A&M University, College Station, Texas. His expertise is focused on large-scale power system modeling for control room function, EMS/DTS software development, online voltage stability analysis and transient stability analysis implementation, and synchrophasor technology innovations.

Prior to his assignment with NREL, Dr. Zhang assumed the position of EMS/DTS group manager with WECC and Peak Reliability consecutively, where he took one of the critical roles in building and maintaining the first West-wide System Model (WSM) and advanced control room tools for Western Interconnection Reliability Coordination function. He had worked in ERCOT for EMS and Market System Applications support and ABB Automaton Inc. for advanced EMS/DTS software development.

Dr. Zhang is a Senior IEEE Member and holds an NERC Reliability Coordinator (RC) certificate. He has published more than 100 papers/technical reports in his domain areas. He serves as the Chair of WECC Joint Synchronized Information Subcommittee (JSIS) and a member of North American SynchroPhasor Initiative (NASPI) leadership team.

Slaven Kincic (IEEE Senior Member) currently works for PNNL. His previous work experience include Peak Reliability (Reliability Coordinator for Western Interconnection), where he led implementation and deployment of multiple control room applications, WECC, BPA-consulting for wind desk development (on behalf of PNNL), WECC Reliability Coordination startup of the new Reliability Coordination control centers, Modeling Lead for the Pacific Northwest for building the West-wide System Model, and EMS Analyst for Idaho Power Company.

Dr. Kincic obtained his B.Eng. degree from the Faculty of Electrical Engineering and Computing, Zagreb, Croatia, in 1996, and M.E. and Ph.D. degrees from École de technologie supérieure (ETS), University of Quebec, Montreal, Canada, in 2000 and 2006, respectively, where he worked as a researcher and lecturer. His main research interests are real-time application for control room, PMUs, FACTS, Renewables, Power Electronics, Power Systems, Power System Modeling, Magnetics, and State Estimation. He is a member of multiple WECC committees such as JSIS, MWVG, and TSS where he chaired multiple task forces.

Sherrill Edwards is a Consulting Engineer with electroMAGIC Tools where he supports various clients in EMS Modeling, Network Applications and Simulator Maintenance and training. He was the Senior DTS Engineer with Peak Reliability, Vancouver, WA. He used the Dispatcher Training System (DTS) for simulating real world events and training scenarios for Reliability Coordinator. Prior to Peak, he worked in the electrical utilities area with Nevada Power and Alstom Grid. He received a Bachelor degree in electrical engineering from the Georgia Institute of Technology, Atlanta, and an MS degree in electrical engineering from New Mexico State University, Las Cruces. He is currently certified as a NERC Certified System Operator-Reliability.

Disclaimer

This book is a work of the Authors and reflects the work that they, Peak Reliability, and their fellow co-workers at Peak Reliability accomplished before the company shutdown in December 2019. The Authors' current employers had no involvement in the book and thus this book doesn't reflect any opinion by those employers concerning the content of this book. The content was completed before the Authors started employment with those employers and only the final editing occurred after employment and then on the Authors' own time.

Chapter 1
Introduction

Hongming Zhang

Organization Acronyms

AESO	Alberta Electric System Operator
BPA	Bonneville Power Administration
CAISO	California Independent System Operator
CEN	Centro Nacional de Control de Energia
CFE	Comisión Federal de Electricidad
CMRC	California and Mexico (Northern Baja California) Reliability Coordinator
DOE	US Department of Energy
FERC	Federal Energy Regulatory Commission
IID	Imperial Irrigation District
NERC	North American Electric Reliability Corporation
NEVP	Sierra Pacific and Nevada Power
RDRC	Rocky Mountains and Desert Southwest Reliability Coordinator
SDG&E	San Diego Gas and Electric
SPP	Southwest Power Pool
PACE	PacifiCorp East
PACW	PacifiCorp West
PEAK	Peak Reliability or Peak RC
PNNL	Pacific Northwest National Laboratory
PNSC	Pacific Northern Security Coordinator
WAPA	Western Area Power Administration
WECC	Western Electricity Coordinating Council
WSCC	Western Systems Coordinating Council

© Springer Nature Switzerland AG 2021
H. Zhang et al., *Advanced Power Applications for System Reliability
Monitoring*, Power Systems, https://doi.org/10.1007/978-3-030-44544-7_1

NERC Acronyms

ACE	Area Control Error
AGC	Automatic Generation Control
BA	Balancing authority
BAAL	Balancing Authority ACE Limit
BES	Bulk Electric System
CIM	Common Information Model
DCS	Disturbance Control Standard
DTS	Dispatcher Training System
EHV	Extra High Voltage
EI	Eastern Interconnection
EMS	Energy Management System
ERCOT	Electric Reliability Council of Texas
FAL	Frequency Alarm Limit
FRL	Frequency Relay Limit
FTL	Frequency Threshold Limit
GOP	Generator Operator
ICCP	Inter-Control Center Communications Protocol
IROL	Interconnection Reliability Operating Limit
MSSC	Most severe single contingency
PMU	Phasor Measurement Unit
RAS	Remedial Action Scheme
RC	Reliability Coordinator
RTCA	Real-Time Contingency Analysis
SPS	Special Protection Scheme
WI	Western Interconnection

Other Acronyms

COI	California-Oregon Intertie
DPF	Decoupled Power Flow
ECC	Enhanced Curtailment Calculator
eLSE	enhanced Linear State Estimator
FOD	Forced Outage Detection
GSA	Grid Stability Assessment
KCL	Kirchhoff's Current Law
KVL	Kirchhoff's Voltage Law
QKNET	Quick Network Topology Processor
MAS	Model Authority Set
NETSENS	Network Sensitivity Calculator
NPF	Newton Power Flow
OAG	Open Access Gateway
PRSP	Peak Reliability Synchrophasor Program
PST	Phase-shifting transformer
RCO	Reliability Coordinator Office

RCSO	Reliability Coordinator System Operator
RMTNET	Real-time Multi-point of Time Network Analysis
ROE	Real-Time Operation Engineer
RSG	Reserve Sharing Group
RTDYN	Real-Time Dynamic Ratings
RTLODFC	Real-Time Line Outage Distribution Factor Calculator
RTNET	Real-Time Network Analysis
RT-VSA	Real-Time Voltage Stability Analysis
SMTNET	Study Network Analysis for Multi-Time of Point
TSAT	Transient Stability Analysis Tool
WISP	Western Interconnection Synchrophasor Program
WIT	WECC Interchange Tool
WSM	West-wide System Model

1.1 History

1.1.1 WECC vs. Peak Reliability

Western Electricity Coordinating Council (WECC) is geographically the largest and most diverse of the six regional entities given authority by the North American Electric Reliability Corporation (NERC) and the Federal Energy Regulatory Commission (FERC). The WECC region extends from Canada to Mexico and includes the provinces of Alberta and British Columbia, the northern portion of Baja California, Mexico, and all or portions of the 14 Western states between. Currently, WECC embraces 39 balancing authorities (BAs), including Southwest Power Pool (SPP). WECC BAs Map and NERC Region Map are given below (Fig. 1.1).

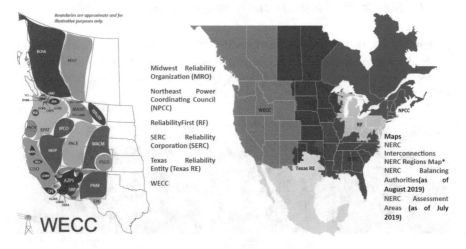

Fig. 1.1 WECC balancing authorities map and NERC regional entities map

WECC has a long history of assuring reliability in the west that began when it was originally formed in 1967 by 40 power systems and then known as the Western Systems Coordinating Council (WSCC). Thirty-five years later in 2002, the WSCC became WECC when three regional transmission associations merged. WECC was designated a Regional Entity for the Western Interconnection in 2007 after the North American Electric Reliability Corporation (NERC) delegated some of the authority it had received from the Federal Energy Regulatory Commission (FERC) to create, monitor, and enforce reliability standards.

Prior to 2014, WECC had the dual responsibilities: Reliability Coordinator and the Regional Entity responsible for compliance monitoring and enforcement across the whole Western Interconnection. This meant that WECC had responsibility to possibly cite itself for compliance violation as the Reliability Coordinator.

Therefore, in June 2013 it was decided that these two responsibilities should be separated. Peak Reliability was formed as a result of the bifurcation of WECC into a Regional Entity (WECC) and a Reliability Coordinator (Peak RC). The bifurcation of WECC received final approval from the Federal Energy Regulatory Commission (FERC) on February 12, 2011. Figure 1.2 outlines the history of WECC that has evolved since 1967 [1].

Peak Reliability, as NERC Reliability Coordinator, a company wholly independent of WECC, performed the Reliability Coordinator function in its RC Area in the Western Interconnection. Peak RC's Reliability Coordinator Area included all or parts of 14 western states, British Columbia, and the northern portion of Baja California, Mexico. Note that Alberta Electric System Operator (AESO) was not part of Peak's RC Area but is part of WECC. As of July 30, 2019, there were 39 balancing authorities (BAs), 57 transmission operators (TOPs), and 231 generator operators (GOPs) in the Western Interconnection (WI) [2].

1.1.2   Single RC Function Independence from Any Host

In January 1, 2009, the three previous WECC RC offices (PNSC, RDRC, and CMRC) were being consolidated into two WECC RC offices (RCO) to be located in Vancouver, WA and Loveland, CO, respectively. The monitoring footprints by individual WECC RCOs were changed as shown on Fig. 1.3.

Fig. 1.2 About WECC history. (Source: WECC)

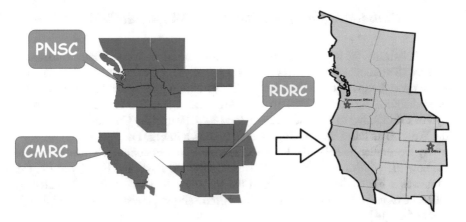

Fig. 1.3 Three WECC RCO map (prior to 2009) vs. Two WECC RCO map in 2009

The new WECC RC function demonstrated significant advantages over the old ones:

- New West-wide System Model and EMS system independent from any host utility.

  - 80,000+ ICCP points received from with 36+ BAs and TOPs.
  - Model representing the entire Western Interconnection BES.
  - Operator tools – OSI PI, Equinox Product CROW, COS, etc.

- Operate as one unit – overseeing security of Western Interconnection using common tools and a common view through the West-wide System Model.
- The two new RCOs are mirror images of each other.
- If a RCO needs to evacuate, the other RCO is already logged in and monitoring the Western Interconnection.
- No lag or gap in exchange of information or system status.

When the WSM was cutover for operation on January 1, 2009, the Loveland RCO was monitoring the areas of Rocky Mountain, Desert Southwest, Pac-East (PACE), Sierra Pacific, Nevada Power (NEVP), and Imperial Irrigation District (IID). The Vancouver RCO was monitoring the areas of Northwest – excluding Pac-East, British Columbia of Canada (BC Hydro), Alberta of Canada (AESO), California (CAISO), and Mexico (North Baja portion or CFE).

After the WECC RC function was transformed into Peak Reliability in February 2014, the Vancouver RCO was expanded to have two RC desks/groups: one for Northwest, BC Hydro, and another for CAISO and CFE. In the meantime, a new group of Real-Time Operation Engineers (ROE) were formed up to support Reliability Coordinator System Operator (RCSO) for watching and tuning real-time tools and performing operation studies as needed. Prior to July 1, 2019, Peak Reliability operated the power system of 1.6 million square miles and 110,129 miles of transmission and served a population of 74 millions.

1.1.3   Peak Reliability Wind Down and Western RCs Transition

Peak Reliability, the Reliability Coordinator for most of the Western Interconnection, on July 18 announced it would cease operation at the end of 2019. The decision was made upon feedback from Peak Reliability's funding parties that "clearly indicates overwhelming support for Peak to wind down the organization after a year of effort by Peak Reliability to provide a viable, long-term Reliability Coordinator option for the West". The latest WI RC Region map is shown in Fig. 1.4.

By December 3, 2019, the Peak RC service area was completely turned over to the new RC entities in the West: RC West/CAISO, SPP RC and BC Hydro RC [3]. This was a big challenge resulting in the transition of Western RCs, as stated by NERC CEO: the Western RC transition is the "single largest risk of reliability in front of us" [4].

- On July 1, 2019, Peak RC turned over RC responsibility for California and Mexico to RC West, an LLC for California ISO (CAISO).
- On September 1, 2019, Peak RC turned over RC Responsibility for the Province of British Columbia (BC) to BC Hydro.
- On November 3, 2019, Peak RC turned over RC Responsibility to CAISO RC West for a number of Western utilities outside California (see CAISO RC West

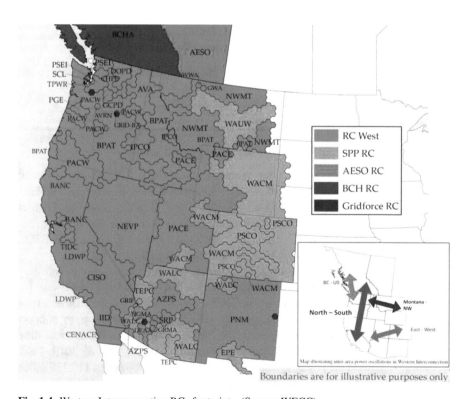

Fig. 1.4 Western Interconnection RCs footprints. (Source: WECC)

service areas in Fig. 1.4 for reference). Gridforce has turned to CAISO RC West for reliability service provider until the end of 2020, while it starts its own RC service then if applicable.

- On December 3, 2019, Peak RC turned over RC responsibility to SPP for the following Western entities:

  - Arizona Electric Power Cooperative, Inc.
  - Black Hills Energy's three electric utilities: Black Hills Power, Inc., Cheyenne Light, Fuel and Power Company, and Black Hills Colorado Electric, Inc.
  - City of Farmington, NM
  - Colorado Springs Utilities
  - El Paso Electric Company
  - Intermountain Rural Electric Association
  - Platte River Power Authority
  - Public Service Company of Colorado (Xcel Energy)
  - Tri-State Generation and Transmission Association
  - Tucson Electric Power
  - Western Area Power Administration (WAPA) Desert Southwest Region, WAPA Rocky Mountain Region, and WAPA Upper Great Plains – West

1.1.4   Reliability Definition and Reliability Coordinator's Role and Duties

Regardless of the continuous evolution of Reliability Coordinator(s) in the Western Interconnection, Reliability Coordination function remains the same in terms of Reliability Definitions and RC's primary roles and duties:

- Defines a reliable BES as one that is able to meet the electricity needs of end-use customers even when unexpected equipment failures reduce the amount of available electricity. In practice, a reliable BES can be measured by:

  - Adequacy – sufficient resources
  - Security – ability of system to withstand sudden and expected disturbances
  - Operational excellence – seasonal coordination, operation planning, and real-time operation

- The primary role of the RC function is to provide situational awareness, analysis, and coordination of the reliable operation of the BES for its RC Area in the operations planning horizon.
- The RC maintains real-time operating reliability by maintaining a wide-area view (including situational awareness of both transmission and balancing operations), analyzing and communicating pre- and post-contingency system conditions, and coordinating or directing actions to mitigate system issues. Peak

ensured that the BES was operated within specific limits and that system conditions were stable within its RC Area.

Over the last decade, to comply with NERC's reliability standards and guidelines and meet customer service requirements, WECC RC and Peak Reliability subsequently made great use of best practices and lessons learned from monitoring the western BPS for the assurance of system reliability, security, and the wide-area view under normal and stressed operating conditions. The rest of this chapter reviews WECC system reliability challenges/issues, development/implementation of the WSM, and the associated real-time tools used by WECC RC and Peak Reliability continuously.

1.2 Reliability Challenges in WECC System

As highest level of authority responsible for the reliable operation of the Bulk Electric System (BES), the RC has to perform three main duties:

- Authority to prevent or mitigate emergencies in the operations planning and next-day analysis/real time
- Assuring wide-area view of BES for the WECC footprint
- Seasonal coordination, operations planning, and real-time operation monitoring

Among a number of daily operation tasks, the RC needs to be ready to handle several major challenges, i.e., a Disturbance Control Standard (DCS) event, a frequency event, a System Operating Limits (SOL) violation, an Interconnection Reliability Operating Limits (IROL) violation, loop flow occurence, a Remedial Action Schemes (RAS) being triggered, and so on.

1.2.1 Disturbance Control Standard (DCS) Event

A DCS event occurs when a BA exceeds the ACE for their control area by an amount greater than a limit value due to a disturbance. Disturbance has the NERC definition:

1. An unplanned event that produces an abnormal system condition
2. Any perturbation to the electric system
3. The unexpected change in ACE that is caused by the sudden failure of generation or interruption of load

The RC need to monitor DC events in real time according to the NERC requirements:

- The BA ACE must return to pre-disturbance levels or zero within 15 min after the start of a reportable disturbance.

- A reportable disturbance is any event that causes an ACE change greater than or equal to 80% of a balancing authority's or Reserve Sharing Group's most severe single contingency.
- A most severe single contingency (MSSC) is a loss, due to a single contingency identified using system models maintained within the Reserve Sharing Group (RSG) or a balancing authority's area (if not participating in a RSG), which would result in the greatest loss (measured in MW) of resource output used by the RSG or a balancing authority (if not participating in a RSG) at the time of the event to meet firm demand and export obligation.
- BAs are required to have sufficient reserves to cover a disturbance equal to the MSSC. This can be done in a shared way by being part of a RSG.
- ½ of the reserves must be available within 10 min and respond to any frequency deviation due to the disturbance.
- A remaining 30% must be within the next 5 min to meet the DCS recovery time period
- If a BA won't make the 15-min time frame, the RC will direct the BA to shed sufficient load to meet the ACE recovery value.
- The BA reserves must recover to handle the MMSC (which might be different after the disturbance) value again within 60 min of the start of the disturbance.

To monitor the DCS events effectively, Peak RC has implemented custom DCS alarming logic in SCADA and DCS event monitoring displays in PI ProcessBook (see Fig. 1.5) to identify a DCS event accurately and track the event correctly.

1.2.2   Frequency Event

A frequency event is when the system frequency has deviated from 60 Hz by more than 0.068 Hz. The NERC defines three levels or types of frequency limits:

- Frequency Threshold Limit (FTL) is the limit set by deviating greater than 0.068 Hz.
- Frequency Alarm Limit (FAL) is a further limit set by deviating greater than 0.2 Hz.
- Frequency Relay Limit (FRL) is an even further limit set by deviating by greater than 0.5 Hz.

  - Any time the frequency exceed any of the limits, a frequency event has occurred.
  - A FAL event would include a FTL event.
  - A FRL event would include a FAL event and a FTL event.
  - Most frequency events are caused by either a sudden loss of generation or a sudden gain of significant load.

Peak RC developed a custom visualization display in PI ProcessBook in the above Fig. 1.5 to help RCSO obtain situational awareness of various system

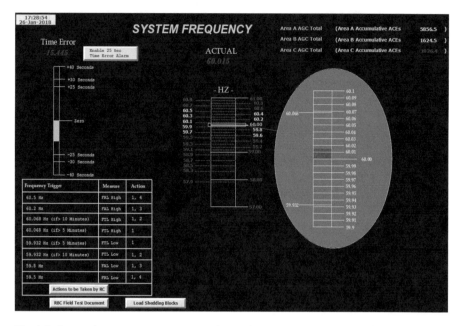

Fig. 1.5 System frequency and ACE monitoring PI display. (Source: Peak Reliability)

frequency events and take appropriate actions to correct the system frequency abnormalities.

1.2.3   System Operating Limits (SOL)

A SOL is defined as the value (such as MW, MVAR, amperes, frequency, or volts) that satisfies the most limiting of the prescribed operating criteria for a specified system configuration to ensure operation within acceptable reliability criteria. System Operating Limits are based upon certain operating criteria. These include, but are not limited to:

- Facility ratings (applicable pre- and post-contingency equipment ratings or Facility Ratings)
- Transient stability ratings (applicable pre- and post-contingency stability limits)
- Voltage stability ratings (applicable pre- and post-contingency voltage stability)
- System voltage limits (applicable pre- and post-contingency voltage limits)

Peak RC implemented the revised SOL methodologies to reinforce monitoring acceptable system performances and SOL exceedances in real-time and operation planning horizons. In particular,

- A SOL usually has a time frame for the exceedance of limit to be brought back under by.

- Acceptable system performance requires that BES elements are monitored for SOL violations for current pre- and post-contingency element conditions (voltage, thermal, or stability).
- Pre-contingency monitoring is done by checking the real-time SCADA and/or state-estimated values and ensuring that flows and voltages remain within Facility Ratings, voltage limits, and stability limits.
- Pre-contingency monitoring is done by checking the real-time SCADA and/or state-estimated values and ensuring that flows and voltages remain within Facility Ratings, voltage limits, and stability limits.
- Post-contingency monitoring is done in two ways:

    - Monitoring Real-Time Contingency Analysis (RTCA) to evaluate expected post-contingency conditions in the event of N-1 contingencies.
    - Monitoring contingency-based SOLs to ensure that applicable Facility Ratings, voltage limits, or stability limits are not exceeded upon occurrence of pre-identified contingencies. Contingency-based SOLs can be established and expressed in a number of ways.

1.2.4   Interconnection Reliability Operating Limits (IROL)

An IROL is a value (such as MW, MVAR, amperes, frequency, or volts) derived from, or a subset of the System Operating Limits, which if exceeded, could expose a widespread area of the Bulk Electric System to instability, uncontrolled separation(s), or cascading outages.

1.2.4.1   What Is a Potential IROL?

- Excessive branch exceedance: Single contingency (N-1) results in a branch exceedance (type BR) in excess of 125 percent of the emergency limit of the monitored facility.
- Multiple bus under voltage condition: Single contingency results in an undervoltage condition characterized by bus voltages of less than 90 percent across three or more related BES facilities.
- Unsolved contingency: An unsolved contingency is present in the RTCA.

1.2.4.2   Testing for IROLs: Evaluation and Validation

- Flag potential IROLs by examining RTCA results.
- Perform internal validation of RTCA results and test excessive branch exceedances for adverse system impacts and potential cascading.
- Contact TOP for external validation of limits and study results.

- If discussions with the TOP indicate that the condition is valid (i.e., the next N-1 contingency could result in instability, uncontrolled separation, or cascading), then TOPs must take action.

1.2.4.3 Known WECC IROLs

1. Northwest Washington Load Area Net Import
2. San Diego Gas and Electric (SDG&E) Area Summer (6/1 to 10/31) Import
3. SDG&E/CFE (or SDG&E/CEN) Non-Summer (11/1 to 5/31) Import
4. Oregon Net Export (also named Northwest Net Export)
5. Lugo-Victorville IROL (thermal overload cascading)

Peak RC implemented the Real-Time Voltage Stability Analysis (RT-VSA) tool to perform real-time assessment on most of the known WECC IROLs that are limited by voltage stability issues.

1.2.4.4 Loop Flow

The WECC system power flow is classified into two components: scheduled flow and unscheduled flow components, because electricity doesn't follow transaction schedules, it follows the path of least resistance, as conceptually illustrated on Fig. 1.6.

Scheduled Flow
- The WECC transmission network has schedules put in place to define the energy flow between utilities.

Unscheduled Flow
- The WECC transmission network is electrically strongest around the edges and weaker in the center. In fact, the network is often compared to a donut with a hole in the Utah-Nevada center.
- Even the edges have differing properties. The west side of the donut has a thick 500 kV side, and the east side of the donut has a thinner 345 kV/230 kV side.
- The Pacific Northwest has abundant hydro, while California has heavy loads. In years with abundant water, the Northwest has abundant surplus energy and schedules heavily to California. Unscheduled flow from those schedules flows east across Idaho, south through Utah and Colorado, and back west toward California.
- Many times, this clockwise unscheduled flow has loaded lines in Western Colorado and Northern Arizona such that systems in those areas were forced to make major cuts in their own schedules.
- Any single schedule is unlikely to cause a problem by itself. Each single schedule contributes only a small part to unscheduled flow at any other given point. Combining the effects of many schedules, however, can and often does cause difficulties at distant points on the network.

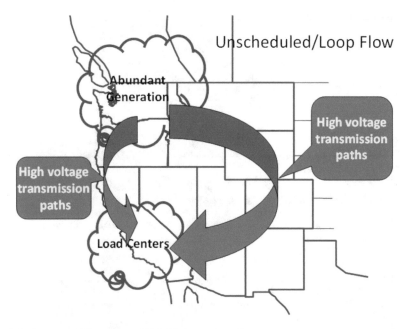

Fig. 1.6 Conceptual illustration of WECC system loop flow

- There are times California depends more on generation from the Utah-Colorado and Four Corners areas. This can cause counterclockwise unscheduled flow that travels westward across Idaho and south on the California-Oregon Intertie (COI). This unscheduled flow may cause systems in the Northwest to curtail schedules to California.
- The major problem regions in WECC have been Northern California, Southern California just north of Los Angeles, the Arizona-California ties, Northern Arizona, the Four Corners area, and Western Colorado.

1.2.4.5 Remedial Action Schemes (RAS)

A RAS is a scheme designed to detect predetermined system conditions and automatically take corrective actions that may include, but are not limited to, adjusting or tripping generation (MW and MVAR), tripping load, or reconfiguring a system(s). The WECC system is heavily deployed with various RAS. From the WECC RAS Distribution Map given in Fig. 1.7, there were 520 substations across the WECC region which have been operated with or protected by RAS protections. In 2018, Peak RC had modeled hundreds of WECC-approved RAS in the RTCA, Real-Time Voltage Stability Analysis, and online Transient Stability Analysis Tools, respectively.

Main purpose of RAS is to address stability, voltage, and/or overload issues in the system and maximize the transmission utilization (increase path rating) (i.e.,

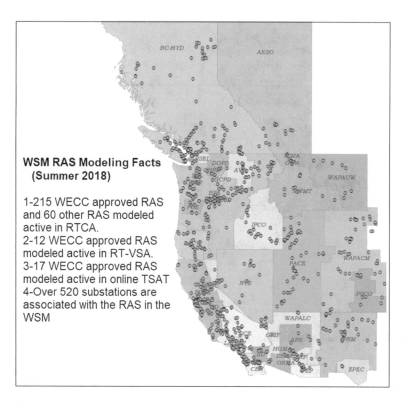

WSM RAS Modeling Facts (Summer 2018)

1-215 WECC approved RAS and 60 other RAS modeled active in RTCA.
2-12 WECC approved RAS modeled active in RT-VSA.
3-17 WECC approved RAS modeled active in online TSAT
4-Over 520 substations are associated with the RAS in the WSM

Fig. 1.7 WECC RAS distribution map in the WSM

cheaper than building new transmission) and quicker control measures to put in place.

Accurate and inclusive RAS modeling is critical for achieving high-quality real-time assessment and operations planning studies. RAS accomplish objectives as follows:

- Meet requirements identified in the NERC Reliability Standards.
- Maintain Bulk Electric System (BES) stability.
- Maintain acceptable BES voltages.
- Maintain acceptable BES power flows.
- Limit the impact of cascading or extreme events.

1.3   Building the West-wide System Model (WSM)

1.3.1   Overview of the WSM

The WSM was the first full Western bulk system model for real-time operation use. Generally, all the substations, transmission circuits, and DC ties above 100 kV voltage level are explicitly represented on the WSM. About a few thousands of sub-100 kV stations, typically associated with BA tie lines, units/SVCs, phase-shifting trans-formers (PST), and/or major loads, are contained in the WSM.

The initial version of the WSM was migrated from three existing operational models from BPA, CAISO, and WACM. The WSM was cut over in the WECC RC control rooms for operation use in January 1, 2009. By June 2009, the WSM already had

- 11,000 bus and 14,000 branch model (dynamic).
- 75% of buses are measurement observable.
- 19,200 degrees of freedom (# means – # variables).
- 6000 stations; 11,000 lines; and 2,400 generators.
- 80,000 measurements mapped to the State Estimator (SE) model.

SE solved the WSM on a 1-min periodic trigger. Over 99.5% SE solution avail-ability was achieved since January 1, 2009. As of September 2019, the WSM has reached a new record of modeling scales and coverages:

- 8938 substations
  - 6796 stations ≥100kV
- 15,658 buses (dynamic)
  - 8967 buses/115k Nodes ≥ 100 kV
- 14,062 transmission lines
  - 10433 AC Line Segments ≥100 kV
- 3830 generators
  - 295 generating units≥ 200 MW
  - 46 generating units≥ 500 MW
- 15 DC lines (full DC line representation)
- 285,302 MW total generation capacity
  - Gen fuel types are listed on Fig. 1.8.
- 5678 transformers with 45 phase shifters
- 8065 contingencies and 515 RAS modeled active in EMS/RTCA

The Peak RC network model has roughly 177,290 measurements mapped to the model. The ICCP measurements are broke down into three categories:

1. ICCP analog points: 75,613
2. ICCP status points: 101,677
3. SCADA calculation points: 26,657

The progression curves of ICCP points and mapping to SE models are presented in Fig. 1.9.

Peak RC met a modeling goal to have a redundancy ratio of > 2.53 for all voltage levels > 100 kV. The average redundancy ratio for BAs within the Peak RC network

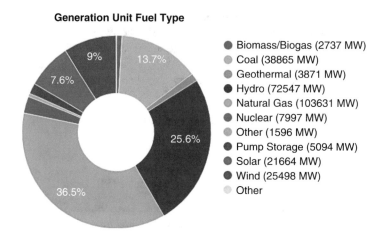

Fig. 1.8 WSM generation unit fuel types

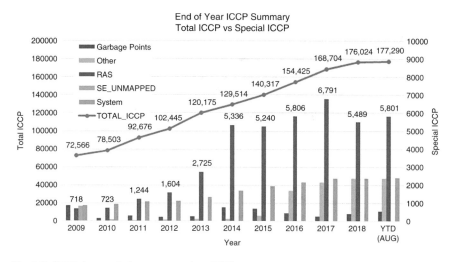

Fig. 1.9 ICCP data statistics summary since 2009

model is around 2.3. The system peak load record on the summer of 2018 was 161,000 MW per ICCP measurements and 153,000 MW from SE estimated, respectively.

1.3.2   WSM Model Update

Peak RC installed both production and Test EMS systems to enable performing software upgrade and EMS model update with no interruption to operations. Figure 1.10 illustrates the real-time ICCP data flow.

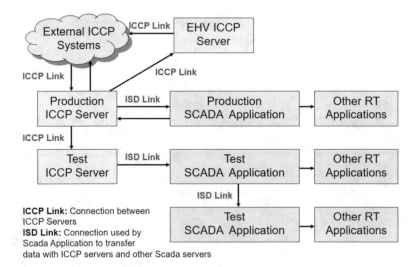

Fig. 1.10 Real-time ICCP data flow

1.3.2.1 WSM Model Update Overview

The WSM covers the footprints of 38 BAs, numerous TOPs, and other operating entities in the West. The RC depends on BAs and TOPs telling us when changes are going to be made to the power system. Typically, WECC's EMS model update is performed at a cycle of 4–6 weeks. The model update process consists of five main steps, which is briefly illustrated on Fig. 1.11 [5].

Basically, the network model change is the starting point and cornerstone for each model update process. The modelers usually take 2 weeks to complete various network change projects on an integrated modeling platform named **e-terra**source. The new modeling tool is set apart to the traditional ones by a few distinctive features:

- Import/export of both full and incremental network models in the CIM/XML file format.
- Import/export of a full network model and contingency database from/to save-case flat files.
- Offer a graphic interface for modelers to add/remove/update the network changes efficiently.

One key concept of **e-terra**source is the Model Authority Set (MAS), which is a complete model (or a model portion) at a given point in time. WECC creates a new MAS version for each cutoff new base case. The following is a typical workflow to build network model projects:

1. Modelers initialize their own modeling workspaces from the latest base model, i.e., the newest MAS version.

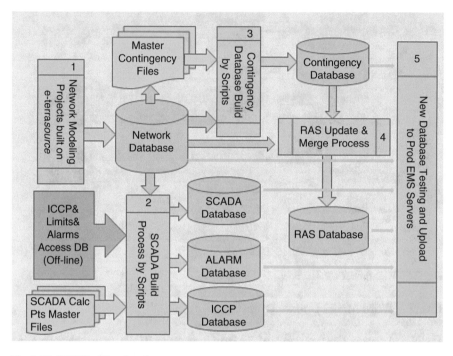

Fig. 1.11 WSM build and update process

2. Modelers make model change projects and validate and approve them on their workspaces.
3. Modeling coordinator merges all the approved projects to his workspace and export the model to a network model savecase file.
4. Move the case to test server and run first-pass validation.
5. Execute the custom scripts to patch up missing operational data and voltage regulation schedules, etc. and run second-pass validation.
6. Run power flow and validate the solution.
7. Map new network model to SCADA and load the mapped network model to test SE for confidence testing.

Peak RC use in-house automation scripts and an offline WSM database in access to build other application databases, such as ICCP or Open Access Gateway (OAG) database, SCADA and alarm databases, and contingency list database. The RAS models are updated by the NetApps team using a combination of RAS modeling UI and in-house scripts. The Voltage Stability and Transient Stability Analysis Tools have separate processes for checking and update upon EMS model changes.

1.3.2.2 WSM Model Build Flowchart

There are hundreds of models building subtasks and automation scripts developed by Peak RC EMS engineering personnel. A flowchart of the comprehensive WSM build process is described in Fig. 1.12 for reference. Although the automatic build process was developed in reference to specific EMS data structures defined in GE/Alstom EMS products, i.e., e-terrahabitat databases, overall framework could be useful for large-scale EMS model build and update.

1.3.2.3 Initial Model Integration Workflow

Figure 1.13 describes initial WSM model integration workflow with guidelines of steps needed to:

- Create the new initial Network Model (i.e., NETMOM) from **e-terra**source.
- Cross-validate and load the new initial Network Model (NETMOM) into the Test EMS system for initial model testing.
- Perform various Network Model Topological Verification Processes.
- Set up the **e-terra**source environment with the new Model Authority Set.
- Export the CIM/XML model and supplemental CSV files.

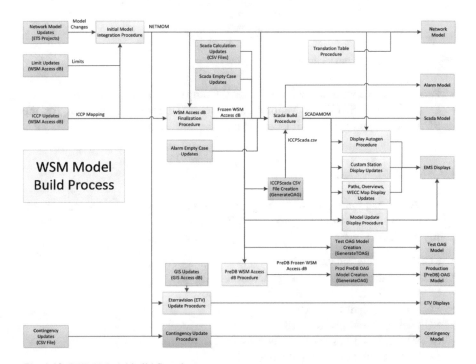

Fig. 1.12 WSM Model build flowchart

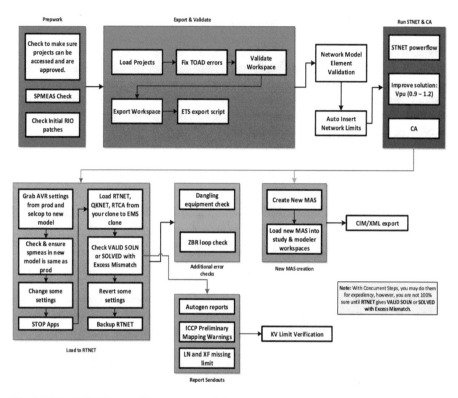

Fig. 1.13 Initial WSM model integration workflow

- Perform the Network Model (NETMOM) Comparison Autogen Process.
- Perform the WSM Data Maintenance Access Database verification for the SCADA_Mapping and limit tables against the new model release.

1.3.2.4 SCADA Calculations

To use ICCP measurements for aggregated information, overview summary, intelligent alarming, RAS modeling, etc., the WSM need be built with lots of calculation points in SCADA, such as,

- SCADA calculations for paths
- SCADA calculations for total generation
- SCADA calculations for miscellaneous purposes
- SCADA calculations for BAAL
- SCADA calculations for RAS
- SCADA calculations for boundary tie lines
- SCADA calculations for frequency
- SCADA calculations for PMU

In the EMS system at Peak Reliability, all paths (or sometimes called interfaces/corridors/cut planes), including WECC paths and balancing authority (BA)'s internal paths, are modeled for monitoring, alarming, and display purposes using GE/Alstom's EMS applications. The SCADA calculations for paths are currently developed and maintained in the *.csv* files named as **scada_calc_pathcalc** and **scada_calc_paths**. These files should cover all of the paths, including all presently defined WECC paths and some of the BA's internal paths, which are monitored by Peak Reliability, and provide guidelines to assist any modification of existing path calculations and/or generation of new path calculations. Next, we use an example of time error calculation to demonstrate the SCADA calculation work.

1.3.2.4.1   Time Error

Peak RC is responsible for monitoring the time error in its footprint and, when necessary, initiate time error correction requests to the BA systems operators.

- Power system time error describes the difference between the synchronous electric clock and the standard time resource, e.g., GPS time or atomic time.
- When the system frequency runs faster than the nominal or standard one, e.g., 60 Hz for the Western Interconnection, the synchronous electric clock would run faster than the standard time, generating a positive time error and vice versa.

1.3.2.4.2   Calculating Time Error

One of the time error calculations is to produce the hourly delta value of the time error to inform the RC how much time error has been added for the previous hour (see Fig. 1.14). The input, the instant value of time error (DTIM), is received from an external source. The time error of the previous hour (TERR) in Peak's system is calculated by setting a 3600 s delay timer to hold the value for the starting and ending points of the previous hour. This value varies only at the beginning of each hour

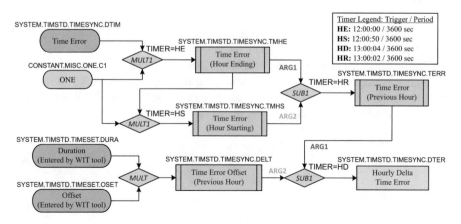

Fig. 1.14 SCADA calculations for time error values

and represents the amount of time error changes of the previous hour. When the RC initiates a request for time error correction, the WECC Interchange Tool (WIT) will enter duration and offset values into the SCADA. The production of these values is also delayed by 1 h to represent the time error offset target of the previous hour. The hourly delta time error (DTER) will be calculated by subtracting the targeted correction value (DELT) from the new error (TERR) for the past hour.

1.3.2.4.3   Calculating Time Error for Daily Report and Alarming

Another more complicated set of calculations, demonstrated in Fig. 1.15, **produces time error report to our BA/TOPs daily at 2 p.m. Pacific Prevailing Time (PPT)**. It is worth noting that with the current setup, **"0" should be entered for summer daylight saving and "1" for winter without daylight saving**.

Daylight saving: 0, summer; 1, winter without daylight saving

For example, during daylight saving summer time, "0" is entered for the "Daylight Saving." If the "Absolute Value of Time Error" is greater than 0 at 2 p.m., the "Not Daylight Saving" will be calculated as "1," which sets the "Time Error Alarm in Summer" to "1" and triggers the "Time Error Alarm" as well.

1.3.2.4.4   Calculating 25 Sec. Time Error Correction Alarm

Figures 1.4, 1.5, and 1.6 demonstrates the calculations of the "Terminate Time Error Correction" alarm, which **notifies the RC that time error correction should be terminated as time error falls into the predefined range**, i.e., −25 to 25 s.

The alarm would never be triggered if the "Enable 25 Sec Time Error Alarm" is not enabled (see Fig. 1.16).

1.3.2.4.5   Visualizing SCADA Calculations in PI Displays

As a large amount of SCADA calculation points are stored in PI historian, it's straightforward to visualized SCADA calculation values in PI displays. Peak RC developed hundreds of PI displays to enhance situational awareness of system operation. For example, the SCADA calculations for time error monitor and correction, system and BA AGC summary (reserves, frequency, ACE, total Geb/load, schedules, L10, BAAL, path limits, etc.) are displayed in Fig. 1.17.

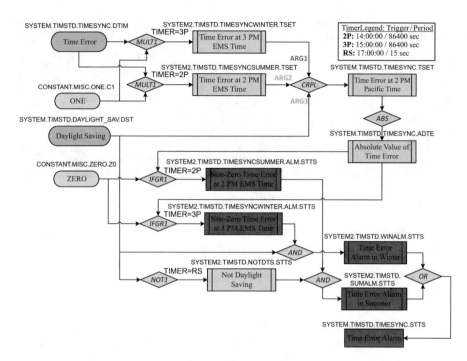

Fig. 1.15 SCADA calculation for time error report

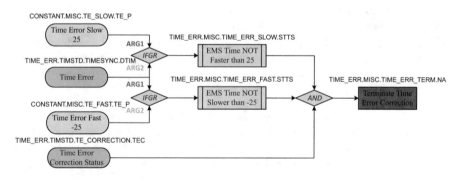

Fig. 1.16 SCADA calculations for 25s time error correction alarm

1.3.2.5 Model Export Process

The West-wide System Model (WSM) is an operational model used by the Peak Reliability Coordination offices to provide situational awareness and real-time supervision of the entire Western Interconnection. The WSM consists of a full node-breaker model which depicts breaker configuration details. It is integrated with real-time telemetry (via ICCP) providing situational awareness of the flows and loading of the Bulk Electric System (BES).

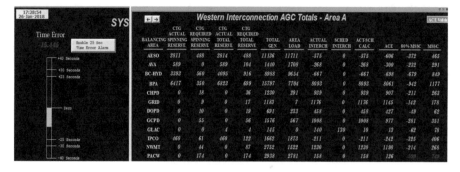

Fig. 1.17 Time error and AGC summary PI displays

Once the new network model is cut-off and released in production. The RC can share it with BAs or TOPs in multiple export file formats.

1.3.2.5.1   Preliminary

The WSM is built using the GE/Alstom **e-terra***source* modeling application. The WSM **e-terra***source* model is stored in an Oracle relational database, which is used to create the CIM model and a Preliminary Alstom Netmom Savecase.

The Preliminary Alstom Netmom Savecase is then tested and validated using:

- Alstom Network Model Validation code
- Alstom Study Power Flow application
- Alstom Real-Time State Estimation, using all available mapped real-time telemetry (received via ICCP), and Alstom Real-Time Contingency Analysis

Once tested and validated, the new WSM model is implemented on the Peak Reliability Production EMS system to be used by the Reliability Coordinators.

The final WSM Network Model is used to create the GE/Alstom Netmom Savecase, the PowerWorld Aux File, and the Microsoft Access Database Export File. The WSM Network Model is then converted to the PTI raw file. A Model Release Highlights document and EMS Substation SCADA Displays were also generated.

1.3.2.5.2   Model Export Formats

Below are additional details regarding the available Model Export formats:

1. Model Release Highlights Document
2. High-level description of the major changes performed in the latest WSM Network Model release – transmission > 230kV equipment additions/removals

(including major topological changes), generating unit additions/removals, large shunt reactive device additions/removals

3. EMS Substation SCADA Displays
4. Common Information Model (CIM) Version 15 File
5. GE/Alstom Network Model Savecase File
6. GE/Alstom SCADA Model Savecase File
7. WSM Export Network Model CSV Files
8. Microsoft Access Database Export File
9. PTI version 30 Raw Format File
10. Branch MVA Limits File
11. WSM-to-WECC Planning Basecase Mapping File
12. GE/Alstom One-lines Study Display Files
13. WSM-to-WECC Planning Dynamic Model Mapping Files
14. WSM Station Display Names (Long Names) File
15. WSM Contingency Definitions File

1.4 Implementing EMS Applications and Advanced Real-Time Tools

1.4.1 Outline of RC Real-Time Tools Implementation

As part of the WSM project aims, the RC implemented GE/Alstom EMS software tool suite and cut it over for real-time assessment and operation planning study on top of the WSM in January 1, 2009. Since then, the RC continuously enhanced the EMS features and implemented additional real-time tools for situational awareness [5, 6]. Figure 1.18 highlights the progression milestones on implementation of the advanced real-time tools for control room function:

- January 2009   GE/Alstom EMS EMP 2.5 software suite cutoff for RC operation use
- October 2012   GE/Alstom EMS migrated to EMP 2.6 with custom enhancements

  - RAS software enhancements
  - Network Sensitivity (NETSENS) calculator built in RTCA and study mode
  - Real-time Line Outage Distribution Factor (RTLODF) calculator

| EMS-EMP2.5 Online (Jan-2009) | EMS Upgrade to EMP2.6 (Oct-2012) | EMS/DTS SW Enhancements (2014-2017) | POM/RT-VSA Online (Jul-2015) | OATI/ECC Online (Jun-2016) | PowerTech DSA /TSAT Online (Oct-2018) |

Fig. 1.18 Implementation milestones for RC real-time tools

- July 2013   GE/Alstom EMP2.6 DTS cutover for RCSO operation training
- 2014~2017   EMS/DTS software enhancements
  - Dynamic Ratings application
  - Study Network Analysis for Multi-Time of Point (SMTNET)
  - Power flow trackers for DTS blackstart drills
  - Grid Stability Assessment (GSA) integrating mode meters into EMS
  - Real-time SMTNET (look-ahead RTCA)
  - Forced Outage Detection

- July 2015   V&R POM/ROSE Real-Time Voltage Stability Analysis (RT-VSA) was operational for real-time assessment of the Western IROLs.
- June 2016   OATI ECC tool cutover for real-time situational awareness
- October 2018   PowerTech DSA Manager/TSAT cutoff for RAS operation monitoring against contingencies.

1.4.2 System Architecture

The configuration of the WSM EMS applications and the advanced real-time tools integrated with EMS is shown on Fig. 1.19. Peak RC developed custom software for integrating different software vendor products, i.e., V&R, OATI, and PowerTech with GE/EMS. This "middleware" software provides the means for transferring the data files between the different products and the EMS so that the products can be used more effectively.

1.4.3 EMS SCADA Applications

- SCADA receives ICCP data from entities and downsampled PMU data from openPDC.
  - SCADA calculations and alarms
  - Station one-lines and overview displays
- RTDYN-Dynamic Ratings application runs every 5min.
  - Calculate dynamic thermal limits
- Fast topology processor (QKNET) runs every 10s.
  - Topology history update, compute delta changes on branch flows, generate unit output and busk voltage magnitudes, and issue an alarm if a change exceeds the threshold
  - Determine equipment in-service status or not for indication in system map, overview displays and station one-lines, etc.

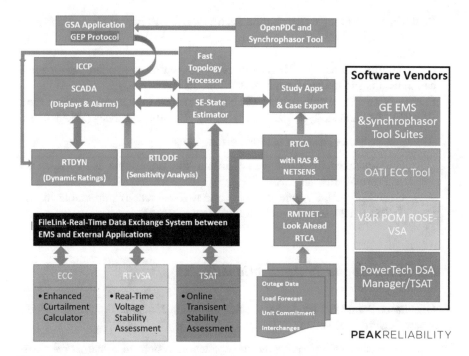

Fig. 1.19 Architecture of Peak RC Control Room Real-Time Tools

- Forced Outage Detection (FOD) runs in parallel to QKNET to manage network equipment outages in two categories: planned outages vs. forced outages. New forced outages are alarmed for operation awareness.

1.4.4 EMS Network Applications

- Real-Time Network Analysis (RTNET), i.e., State Estimator (SE) runs every 1 min.
- Real-Time Contingency Analysis (RTCA) runs every 5 min, including:

  - Screening of complete RAS models (500 RAS records modeled active in RTCA)
  - Network Sensitivity Analysis – NETSENS

- Real-time Multi-point of Time Network analysis (RMTNET), i.e., look-ahead RTCA runs every 15 min to predict 1-h and 2-h ahead CA violations.

  - The same RAS and contingencies as RTCA
  - Input data files for Outage, Load Forecast, Unit Commitment, and Interchange Schedules

- Real-Time Line Outage Distribution Factor (RTLODF) runs every 1 min to back up RTCA for a subset of transmission outages selected from the SOLs and IROLs of interest.
- Grid Stability Assessment (GSA) runs every 10s to receive synchrophasor application results and update them in EMS/GSA displays.
- Study Network Applications (STNET) include dispatch Power Flow (PWRFLOW) and study CA and study Voltage Stability Analysis (VSA)

  - Export snapshot cases for external offline tools

    Node-breaker model in ∗.csv format file importable to V&R POM, PowerWorld, and GE/PSLF
    GE/PSLF v29, 30, and 34 data format (bus-branch) importable to GE/PSLF, PTI, and PowerTech PSAT/VSAT/TSAT

1.4.5   Synchrophasor Applications [7]

- GE Phasor Point (PP) Product was integrated with Montana Tech MAS (Modal Analysis Software) version 1.0

  - Mode meters being configured in production system to monitor five to six Western system modes
  - Frequency events, islanding and system re-synchronization, and angle separation monitoring

- OpenPDC and GEP Applications of Grid Protection Alliance (GPA) product
- Washington State University (WSU) OMS-based Forced Oscillation Detection and Source Locating Tool (FODSL) were deployed in the test environment
- EPG's enhanced Linear State Estimator (eLSE) was deployed in the test environment

1.4.6   POM/ROSE RT-VSA Tool

Peak RC integrated the V&R Energy POM/ROSE Tool with the GE EMS system by a custom Filelink application for a Real-Time Voltage Stability Analysis (RT-VSA) tool. The tool imports a full topology model SE solution every 4–5 min. The tool was required to solve all the required IROLs transfer scenarios in that 5 min. Fast power flow solution and the competitive RAS modeling ability was the main business driver for Peak RC to select the V&R POM engine from several vendor products to develop Peak RC's RT-VSA tool for IROL assessment in near real-time conditions [8].

The RT-VSA tool has been used for real-time assessment of the known Western IROLs:

- SDGE Import (Summer)
- SDGE/CEN Import (Non-Summer)

- Northwest Washington Area Import
- Oregon Net Export (also called Northwest 500kV AC Net Export-NWNE)
- PNM Import Limit (one of the dynamic SOLs other than IROL)

The RT-VSA tool has the following important features:

- Solves multiple IROL scenarios in parallel using multi-threads computation.
- Applies RTCA contingencies and relevant RAS models by user-defined VB scripts.
- Performs both PV and QV analysis.
- Runs reverse transfer analysis as needed.
- Supports auto shunts switching schemes.
- Transfer output results to EMS for alarms and visualization.

1.4.7 PowerTech DSA Manager/TSAT

Peak RC has implemented online Transient Stability Analysis Tool (TSAT) from the PowerTech product. Peak RC developed custom application called XTSAT to manage EMS data transfer (including basecase raw file, unit D curve, and interface definition files) to TSAT servers for every 5–15 min [9]. Starting in 2014 Peak RC initiated a new project with PowerTech Labs to implement an online DSA manager/TSAT tool on the WSM for the underlying operation objectives/issues [9, 10]:

- Unacceptable frequency response that can result in under-frequency load shedding or tripping of thermal and nuclear units.
- Back up RTCA to monitor contingencies associated with a transient RAS.
- Monitoring transient or dynamic RAS operation and evaluating RAS gen drop impact to system frequency performance and any cascading outage scenarios.
- Loss of synchronism for a cluster of generators and consequently their removal from the grid on a fault.
- Real-time assessment of a stability limited SOL/IROL.
- Fast voltage collapse due to induction motor instability.

Peak RC's primary use case of the TSAT is to monitor a subset of RTCA contingencies that require dynamic or transient RAS models to mimic post-contingency conditions accurately. Otherwise, RTCA may report false CTG violations frequently, e.g.:

- False SOL exceedances
- Falsely unsolved [N-2] or [N-1-1] CTGs
- False unsolved islanding cases

1.4.8   OATI ECC Tool

Peak RC implemented the Enhanced Curtailment Calculator (ECC) to improve upon OATI's webSAS program and provide automation and centralization of congestion management in the Western Interconnection. The architecture of the ECC integration with EMS and main functions is illustrated on Fig. 1.20.

The primary objective of the ECC is to:

1. Forecast transmission flows that may cause SOL exceedances.
2. Determine the impact of generation, load, and interchange schedules contributing to such exceedances.
3. Provide users with the ability to take appropriate actions, such as curtailing schedules and adjusting generation, in a fair and equitable manner to mitigate SOL exceedances.

Over the past years, Peak RC has utilized ECC to cover the following use cases in operations:

- **Real-Time Situational Awareness** – provides situational awareness for real-time flows and congestion on defined elements. The release also provides visualization of impact "layers."
- **WebSAS integration** – integrates webSAS functionality into the ECC and replaces the webSAS tool.
- **Future-hour situational awareness** – integrates generation and load forecast inputs and introduces visualization that allow the user to view forecast conditions up to 3 h into the future (i.e., current OH+3).

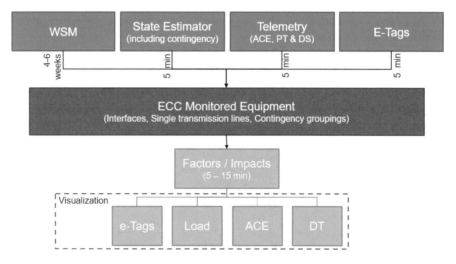

Fig. 1.20 Architecture of Peak RC ECC tool integration with EMS. (Source: Peak RC)

- **Expanded congestion management (pending)** – address the expansion of ECC functionality to allow for management of other defined elements in addition to qualified paths.

1.5   Engineering Lab for Technology Validation

In 2015, Peak RC built an Engineering Lab as a software test bed to validate new technology in a near real-time environment. The architecture is given in Fig. 1.21.

All RC real-time software tools except ECC were integrated into a single lab server (the server hardware and accessory parts costed less than $100k in total) to facilitate new software and model development and testing. Many synchrophasor new technologies were validated in the lab, such as,

- Washington State University Oscillation Monitoring and Forced Oscillation Source Locating Tools
- EPG's enhanced Linear State Estimator (eLSE) and RTDMS products
- GPA's open-source software tools
- PNNL Synchrophasor Analytical Tools

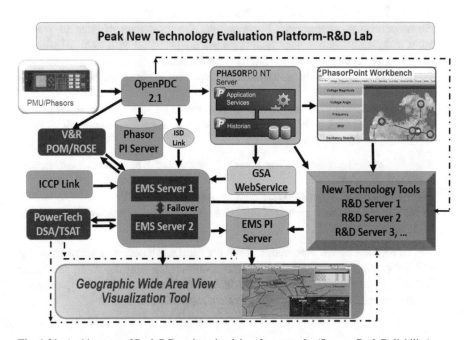

Fig. 1.21 Architecture of Peak RC engineering lab software tools. (Source: Peak Reliability)

1.6   Tool Tuning and Performance Metrics of Advanced Network Applications

1.6.1   State Estimator Tuning

In addition to SE software algorithm and measurement redundancy, there are a number of other factors affecting the SE solution quality. Here are top three issues found from SE tuning practice [5]:

- SE weights or accuracy class assignment.
- BA-BA boundary modeling issues, such as substations, units, loads, and tie lines being assigned to inappropriate BAs.
- Load modeling issues, such as splitting load, dynamic load or non-conforming load, low kV load reduction, inaccurate load distribution factors, etc.

To ensure WSM SE solution credibility, a number of modeling improvement projects were completed. First, a project for finely tuning SE weights was to incorporate the following criteria:

1. SE weight assignments for analog measurements are broken down by device types and/or voltage levels. For instance, phase shifters and regular transformers are assigned with different measurement weights; 230 kV and below branch measurements are less weighted than 345 kV and above branch measurements; etc.
2. Increased priority level for a network level device and a tighter accuracy class for the associated measurements. For example, the measurements tied to WECC paths and 345 kV above backbone branches are assigned with tight SE weights to have more accurate flow estimations. In addition, pseudo bus injection measurements on units, loads, and shunts are typically assigned with loose weights to minimize the impact.
3. Justify the threshold for anomalous measurements of various devices by setting different breakpoint numbers. The formulation to identify an anomalous measurement is given as follows:

Anomalous measurement when Wres> Break Point Where Wres = Residual/ STDV.
STDV means calculated standard deviation (=Base*STDEV).
STDEV means manual standard deviation (P.U.) Base is base MVA applied to network. devices.

From the above equation, a small unit and a big unit may share the same STDV but have different anomalous measurement thresholds by assigning separate breakpoint numbers. This feature will secure SE not dropping good measurements for less capacity devices improperly.

4. Optimize the SE accuracy setting on KCL equations to achieve an equilibrium between the bus and branch residual minimizing and also not downgrade robustness of SE solution iterations.
5. Introduce an accurate class called BOND to address certain measurements located in an unobservable or less observable pocket/area. Below is such an example.

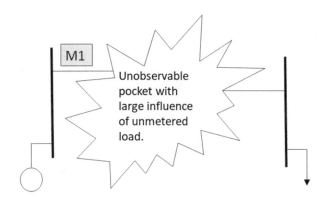

Measurement M1 is a valid measurement, yet due to influence of unobservable area, M1 is often declared anomalous when a standard weight is applied. Through the BOND accuracy class, the measurements in sensitive locations are weighted in SE and help secure a reasonable SE solution on the local pocket

Second, the engineering group conducted a thorough review on BA-BA boundary modeling in the WSM. The goal was to make sure our modeling of tie line, substation, unit, and load ownership/operatorship was in line with BA's real-time AGC model definition. An Excel format PI metrics tool (see Fig. 1.22) was developed to evaluate the gap between SCADA and SE solved values on area interchange/generation/load for each BA.

Third, improved load modeling, such as refining load distribution factors and area load schedules from SE solutions on desired peak hours, identifying splitting loads and non-conforming/dynamic loads, and modeling them properly in the WSM.

1.6.2   RTCA Tuning

RTCA uses the fast-decoupled power flow (DPF) algorithms as the primary engine. The DPF will diverge in the case of a Jacobian matrix singularity. A RCSO closely watches RTCA results to obtain the system security awareness. For a newly detected unsolved or islanding contingency, the on-shift RCSO will evaluate the credibility of the CA results by executing a study power flow. If confirmed by the study, RCSO will contact BA's or TOP's operation desk to discuss the contingency. Usually BA or TOP will sensibly check the contingency by their EMS study tools and get back

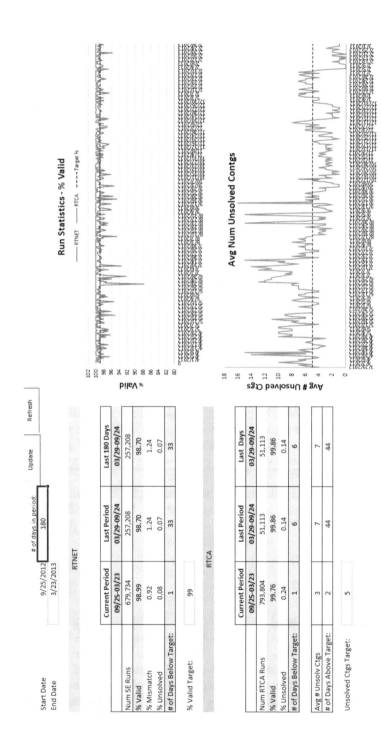

Fig. 1.22 RTNET (SE) and RTCA performance metrics tool

to RCSO with their study results. If the contingency violation is confirmed true, a mitigation plan shall be jointly developed by the RCSO and BA or TOP. Below are some of typical issues affecting RTCA solution [6]:

- Poorly solved SE basecases.
- Basecase Jacobian matrix-B' near to singular.
- Missing lines/transformers and/or substations.
- Incorrect contingency definition.
- Missing RAS or inappropriate RAS modeling.
- Wrong network limits cause false contingency violations of CA results.
- Unable to replicate RTCA results by study CA or outage simulation in power flow.

To resolve or mitigate these problems, WECC-RC EMS engineers took actions to improve CA's solution quality as follows:

- Polished settings on CA and DPF control parameters. Consolidate real-time CA and study CA setting parameters to make two results identical or closer.
- Corrected topology errors and improve SE solutions.
- Cleaned up network and contingency modeling errors.
- Added RAS measurements and missing RAS modeling.
- Assigned resources to monitor RTCA solution quality.

Meanwhile, a number of RTCA software bugs were identified on the EMS base product, e.g.:

- DPF Jacobian matrix became singular sometimes when series capacitors are switched in/out.
- CA screening process causes falsely unsolved, harmful, and islanding contingencies.
- Q-V iteration may oscillate or diverge sometimes, when unit reactive output is at the reactive limit.
- Parallel mode or CA workers issues, i.e., databases are out of sync between CA workers and CA boss, etc.

  - RAS modeling and screening software issues. Relatively speaking, the RAS software defects impacted RTCA solutions the most. Most RAS software defects were addressed by the EMS vendor.

After SE and RTCA tuning and software bugs fixing, SE and RTCA achieved high performance overall (see Table 1.1).

Table 1.1 WSM SE and RTCA annual performance summary

SE metrics	2009	2010	2011	2012	2014	2015	2016
%Valid	98.7	98.37	98.29	99.10	99.15	99.51	99.90
%Mismatch	1.14	1.52	1.61	0.87	0.8	0.42	0.09
%Unsolved	0.17	0.11	0.08	0.04	0.05	0.07	0.02
RTCA %valid	99.67	99.85	99.92	99.91	99.82	99.87	99.94

1.6.3   Operational Excellence and Daily Performance Metrics

Since 2017 Peak RC set a new bar to achieve operational excellence and maintain high standards on performance of critical real-time tools. The on-shift Real-Time Operation Engineer (ROE) watched the dashboard of critical real-time tools closely and addressed tool performance issues on a timely basis.

RTNET/SE tool performance settings are as follows:

- Threshold: 99.75%
- Target: 99.90%
- Excellent: 99.98%, and no RTNET 30 min continuous failure incidence for a whole year

RTCA tool availability goals are set as follows:

- Threshold: 99.9%
- Target: 99.95%
- Excellent: 99.98% and RTCA 30 min continuous failure incidences < 1 per year

A daily performance metrics tool was developed and implemented to track the performance is shown in Table 1.2.

1.6.4   RC Tool Tuning and Monitoring by ROE

The ROE monitors tool and system health status using PI dashboard in Fig. 1.23. A ROE would tune the SE to meet flow and voltage metrics for selected branches and buses as needed for the changing operational conditions. The monitored RC tools include:

- RTNET and RTCA
- RT-VSA
- Online TSAT
- Daily metrics spreadsheet
- CROW (coordinated outage management system)
- RMTNET (look-ahead RTCA)

The ROE leveraged the SE Tuning tools in RC Workbook (see Fig. 1.24) and iGRID Dashboard (see Fig. 1.25) to keep SE solution quality in good shape through out the 7x24 rotation schedules. A ROE used the tools to resolve most frequently found issues:

1. Branch/load/unit error (MW/MVAR)
2. Topology incoherency
3. Daily review for not-in-service measurements, etc.

Table 1.2 RC tool daily performance metrics

State Estimator	Max aggregate score = 3	Score	Weight	Weighted Score (0-3)	Aggregate Score
SE Convergence	Percentage of time SE converges; calculated based on number of 1 min periods in a 24 h period (max number of attempted solutions = 1440). Any continuous 30 min or greater outage of SE constitutes a 0 for the day. < 99.75%, 0; 99 75% – 99.9%, 1; 99 9% – 99 98%. 2; > 99 98%, 3	**100.00%**	1	1	3
SE Branch MW Solutions	Percentage of lines and transformers that solve within designated residual target (<=10 MW residual when SE flow <= 500 MW; <= 2% of SCADA MW flow when SE flow > 500 MW) < 88%. 0; 88% – 92%. 1; 92% · 96%. 2; >= 96%, 3	**97.57%**	1	1	
SE Bus Voltage Solutions	Percentage of bus voltages that solve within designated residual target (7 kV for 500 kV buses. <4 kV for 230/345 kV buses) < 95%. 0; 95% · 97.5%. 1; 97.5% · 99.0%. 2; >= 99.0%. 3	**99.39%**	1	1	

Next Day Study	Max aggregate score = 3	Generation	Load	Score	Weight	Weighted Score (0–3)	Aggregate Score
Inputs/Case Setup	Compare area load and generation in RTNET with STNET; measure difference as % of RTNET load and generation; average error > 8.02%. 0; >=7.15–8.02%. 1; 6.28–7.15%. 2; <= 6.28%. 3	**9.48%**	**5.62%**	**7.55%**	1	0.33	1.65
Study Process and Issue Resolution	Number of issues identified in real-time caused by schedule outage. Input data applied correctly (1), sufficient mitigation plan developed (1), issue and plan communicated to operators (1).	**1**	**2**	2	1.32		

(continued)

Table 1.2 (continued)

RTCA	Max aggregate score = 3		Score	Weight	Weighted Score	Aggregate Score
Availability	RTCA Availability means that we have a converged basecase solution. Availability in terms of percentage of completed solutions as compared to number of five minute periods in a day (288). Any continuous 30 minute or greater outage of RTCA constitutes a 0 for the day. < 99.90%. 0; 99 90% – 99 95%. 1; 99.95% – 99.98%. 2; >= 99.98%, 3		**100.00%**	1	1	3
Results Accuracy	RTCA results (Branch, Node Pair, Interface and Unsolved) are accurate and actionable as determined by the number of SharePoint tickets created for potentially inaccurate RTCA results; verified by after the fact review of RTCA results. 0 issues =3; 1-2 issues=2; 3 issues=1; >3 issues=0. Started calculating July 1st. 2017.	0	3	2	2

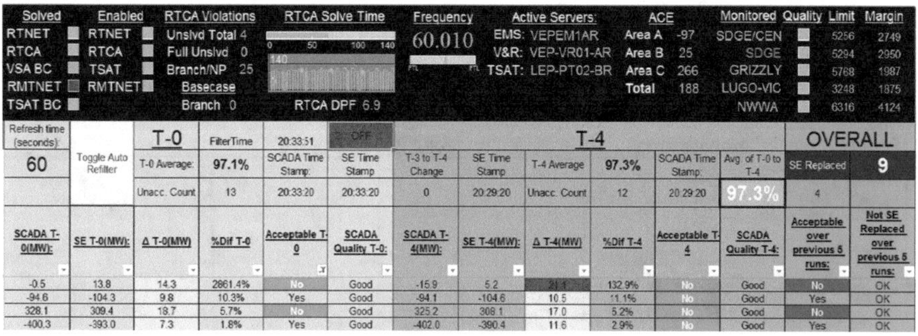

Fig. 1.23 RC tools health status dashboard

Fig. 1.24 SE tuning Tool in RC Workbook

Fig. 1.25 SE tuning Tool-iGrid view

1.7   Synchrophasor Application Implementation at Peak RC

The Western Interconnection Synchrophasor Program (WISP) began with a collaborative project to improve reliability of the Bulk Electric System (BES) in the Western Interconnection.

In 2010, the Western Electricity Coordinating Council (WECC) received $53.9 million in funding from the US Department of Energy (DOE) for WISP. The funding, awarded under the American Recovery and Reinvestment Act's Smart Grid Investment Grant initiative, matched dollars committed by nine WISP partners to extend and deploy synchrophasor technologies within their western electrical systems. The total funding for WISP was $107.8 million.

Between 2010 and March 2014, WISP installed more than 480 new or upgraded Phasor Measurement Units (PMUs) in the Western Interconnection. PMUs deliver high-speed synchronized measurements using a secure communications network and software in order to better manage the power grid. Under the WISP project, WECC Reliability Coordinator (RC) installed many state-of-the-art synchrophasor applications and tools, such as:

- OSI soft PI for PMU historian
- V&R Energy ROSE-Real-Time Voltage Stability Analysis (RT-VSA) software
- GE/Alstom PhasorPoint platform integrated with Montana Tech developed Modal Analysis Software-MAS 1.0
- GE/Alstom Grid Stability Analysis (GSA) application to transfer synchrophasor application results to EMS
- GE/Alstom openPDC and **e-terra**_vision_ (replaced by other products in 2016)

The new synchrophasor infrastructures and advanced applications enabled WECC RC and then Peak RC to play a leadership role in promoting new technologies for visualizing power system disturbances and minimizing the risk of these disturbances from evolving into a major, system-wide event on the Western BES.

Since WISP's phase-1 completion in March 2014, Peak RC continued to improve the quality and use of the synchrophasor data it receives. Peak Reliability applied for an additional grant from the DOE for this next phase of WISP, subsequent to the bifurcation of WECC into a Regional Entity (WECC) and a Regional Coordinator (Peak Reliability). The new DOE grant project was named Peak Reliability Synchrophasor Program (PRSP), which was widely considered WISP 2.0. The PRSP project was budgeted at $9 million. It consisted of nine partner entities: BPA, CAISO, Idaho Power, NV Energy, PacifiCorp, PG&E, Salt River Project, Southern California Edison, and Peak Reliability.

Through the PRSP project, Peak Reliability played an important role in the program by housing the PMU Registry, MAS 2.0 software enhancement, historical data archives, PMU data accuracy and event data export tools, disturbance reports, linear state estimation, inter-area oscillation mode meters/forced oscillation detection, as well as the wide-area visualization tools.

Peak Reliability received the inbound 6000 **PMU** signals of nearly 440 PMU from **18** WECC utilities. Most of the PMU were downsampled and sent out to SCADA. Hundreds of downsampled PMU voltage phasors were enabled in Peak RC EMS State Estimation solution.

1.8   Advanced Dispatcher Training Simulator

Peak RC utilizes a GE/Alstom **e-terra***platform* Energy Management System (EMS) for operating the Western Interconnection. Within the **e-terra***platform* system is the option for the **e-terra***simulator* Dispatcher Training System (DTS). After multiple-year efforts and large-scale training drills as shown on Fig. 1.26, Peak RC fielded a mature WSM-based DTS tool for various operation and tool training programs.

Notes
- 2010 – Simple RCSO training simulation
- 2011 – Initial black start drill (1 DTS support)
- 2013 – Migration to EMP 2.6
- Restoration drill simulation yearly since 2013
- Train the Trainer (TtT) 2016 held for areas B and C
- Restorations: B March 2016, C April 2016
- TtT for area A in September 2016
- Restoration area A drills in November 2016

The DTS tool was able to support large-scale system restoration drills with high quality. The Peak RC EMS Application was fully replicated. A Cloud Service was used that allowed up to 300 people participating in the drills either locally or remotely. The DTS handled island frequency controlling properly and simulate complicated system conditions including up to 50 islands. It has DTS software features for power flow tracking and Peak RC custom displays for Island Reserve Status and SBO control functionality.

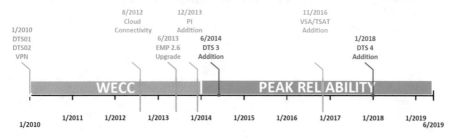

Fig. 1.26 Peak RC DTS training system development milestones

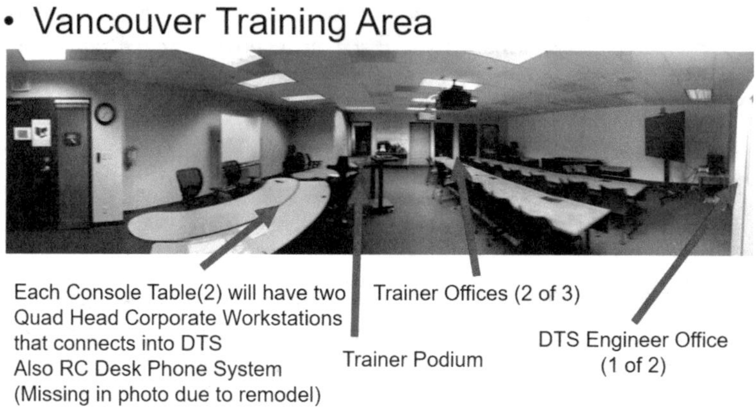

• Vancouver Training Area

Each Console Table(2) will have two   Trainer Offices (2 of 3)
Quad Head Corporate Workstations
that connects into DTS   DTS Engineer Office
Also RC Desk Phone System   Trainer Podium   (1 of 2)
(Missing in photo due to remodel)

Fig. 1.27 Peak RC training room settings

There were two RC training rooms within the Peak RC control centers. Figure 1.27 describes the settings of Vancouver RC training room.

1.9   Summary

Over one decade, WECC RC and Peak Reliability successively made efforts on development of the Western Interconnection BES operational model WSM and advanced real-time tools for wide-area view reliability monitoring and operational situational awareness and trainings. The remaining chapters of the book will discuss implementation experience for most innovative real-time tools, complex system modeling, and new synchrophasor applications. Lessons learned from operations and system events will be presented with specific study cases by practical examples.

In the following seven chapters, the authors will present Peak RC's implementation experience and lessons learned in detail on individual topics. In Chap. 9, we will review the RC's achievements on implementing the WSM and advanced EMS Network Applications to secure a wide-area view over the WECC system and provide several recommendations for the Western RCs to overcome emerging operation challenges on monitoring and managing renewable energy resources across the WI.

References

1. WECC History [Online] Available: https://www.wecc.org/Pages/AboutWECC.aspx
2. Peak Reliability History [Online] Available: https://www.peakrc.com/aboutus/Pages/default.aspx
3. WECC Reliability Coordinator Forum [Online] Available: https://www.wecc.org/Reliability/RC%20Web%20Ad.pdf
4. NERC CEO: Western RC transition is 'single largest risk of reliability in front of us' [Online] Available: https://www.utilitydive.com/news/nerc-ceo-western-rc-transition-is-single-largest-risk-of-reliability-in-f/557719/
5. Zhang, H., & Wangen, B. (2012, July 22–26). Implementation of a full western bulk system operational model for reliability monitoring. *Proceedings of IEEE PES General Meeting.*
6. Wangen, B., & Zhang, H. (2012, July 22–26). Monitoring for post-contingency system operating limit exceedance in the Western Interconnection. *Proceedings of IEEE PES General Meeting.*
7. Peak Reliability Synchrophasor Technology Implementation Roadmap and Current Progress in 2016 NASPI Meeting [Online] Available: https://www.naspi.org/sites/default/files/2017-03/01_pr_zhang_synchrophasor_roadmap_20161019.pdf
8. Zhang, H., Vaiman, M. et al. (2019, August 5–8). Implementing the Real-Tune voltage stability analysis tool on a large scale system model for IROL assessment in the Western Interconnection. *Proceedings of IEEE PES General Meeting.*
9. Mahmoudi, M., Kincic, S., Zhang, H., & Tomsovic, K. (2017). Implementation and testing of remedial action schemes for Real-Time transient stability studies. *PES General Meeting.*
10. Mahmoudi, M., Kincic, S., Zhang, H., & Tomsovic, K. (2018). *Model enhancements for Real-time transient stability assessment in western interconnection.* T&D conference 2018 Denver, CO

Chapter 2
Real-Time Contingency Analysis for Post-contingency SOL Exceedance Monitoring

Hongming Zhang

Organichzation Acronyms

AESO	Alberta Electric System Operator
APS	Arizona Public System
BC Hydro	British Columbia Hydro Electric Service
BPA	Bonneville Power Administration
CAISO	California Independent System Operator
CFE	Comisión Federal de Electricidad
FERC	Federal Energy Regulatory Commission
IID	Imperial Irrigation District
NERC	North American Electric Reliability Corporation
NWMT	Northwestern Energy
PAC	PacifiCorp
PG&E	Pacific Gas and Electric
SCE	Southern California Edison
SDG&E	San Diego Gas and Electric
SPP	Southwest Power Pool
WALC	Western Area Power Administration Lower Colorado Region
WECC	Western Electricity Coordinating Council

NERC/WECC Acronyms

ACE	Area control error
AGC	Automatic generation control
ATC	Available transfer capability
AVR	Automatic voltage regulation
AWR	Automatic MW regulation
BA	Balancing authority
BAAL	Balancing Authority ACE Limit
BES	Bulk Electric System
COI	California-Oregon Intertie

© Springer Nature Switzerland AG 2021
H. Zhang et al., *Advanced Power Applications for System Reliability Monitoring*, Power Systems, https://doi.org/10.1007/978-3-030-44544-7_2

DCS   Disturbance Control Standard
EI   Eastern Interconnection
EMS   Energy management system
ERCOT   Electric Reliability Council of Texas
FACTS   Flexible Alternating Current Transmission System
FAL   Frequency alarm limit
HVDC   High voltage direct current
ICCP   Inter-Control Center Communications Protocol
IROL   Interconnection Reliability Operating Limit
LAPS   Local Area Protection Scheme
LF   Load forecast
LTC   Load tap-changing transformers
NIS   Net interchange schedule
OC   NERC Operating Committee
PSS   Power system stabilizers
PST   Phase shifting transformer
RAS   Remedial Action Scheme
RASRS   (WECC) Remedial Action Scheme Review Subcommittee
RC   Reliability Coordinator
RCSO   Reliability Coordinator System Operator
RTCA   Real-Time Contingency Analysis
SCADA   Supervisory Control and Data Acquisition
SN   Safety Net
SPS   Special Protection Scheme
SSR   Sub-synchronous resonance
SVC   Static VAR compensator
UC   Unit commitment
UFLS   Under frequency load shedding
UFMP   Unscheduled Flow Mitigation Procedure
UVLS   Under voltage load shedding
WAPS   Wide Area Protection Scheme
WI   Western Interconnection
WSM   West-wide System Model

Other Acronyms

CAMS   IEEE PES Computer Analytical Methods Subcommittee
CFTA   IEEE PES Cascading Failures Task Force
DPF   Decoupled power flow
LDC   Line-drop compensation
MUC   Multiple outages contingency
NETSENS   (GE's EMS) Network Sensitivity Calculator
NPF   Newton power flow
PI   (OSIsoft's) plant information platform
QKNET   (GE's EMS) Quick Network Topology Processor
RMTNET   (GE's EMS) Real-Time Multi-Point of Time Network Analysis

ROE   Real-Time Operation Engineer
RTDYN   (GE's EMS) Real-Time Dynamic Ratings
RTNET   (GE's EMS) Real-Time Network Analysis (i.e., SE)
RT-VSA   (Peak RCs' ROSE) Real-Time Voltage Stability Analysis
SMTNET   (GE's EMS) Study Network Analysis for Multi-Time of Point
STNET   (GE's EMS) Study Network Analysis
TSAT   (PowerTech's) Transient Stability Analysis Tool

2.1 Introduction

As one of largest Reliability Coordinator Entities of the North American Electric Reliability Corporation (NERC), Peak Reliability (i.e., Peak RC) was responsible for coordinating and promoting the reliability of the Bulk Electric System (BES) in the entire Western Interconnection (WI). Until July 12, 2019, Peak RC consists of 38 balancing authorities (BA) and 336 member entities. Peak RC wound down operations by mid-December 2019, transitioning its RC service role to California Independent System Operator (CAISO), BC Hydro, and Southwest Power Pool (SPP) et al. entities, respectively.

The WI is a unique operating environment with long high-voltage transmission lines. Typical flow patterns are north to south or east to west. An abundance of hydroelectric power in the north and large coal and nuclear plants in the east are a source of power for the load centers in the west. WECC transfer paths are groupings of transmission facilities that transport power across multiple areas. Transfer paths provide the primary means for transmitting large amounts of power across the Western Interconnection. Due to the significant length of the transmission lines and the relative sparseness of the load pockets, the Western Interconnection is more apt to experience voltage or transient stability issues than other interconnections. Peak RC (formerly WECC RC prior to 2014) monitored these transfer paths for both actual and post-contingent exceedance of the path SOLs and directed corrective action when the path flows exceed their SOLs.

One major challenge for the WI is managing unscheduled flows, also widely called loop flows in many IEEE literature. Unscheduled flow results from the law of physics that causes power from a given source to flow over all possible paths to its destination. Since schedules between entities are made over specific contract paths, that part of the schedule flowing on other than the contract path is unscheduled on the path it uses. The unscheduled flows use up transmission capacity and may prevent path owners from scheduling as much as they would like on their own paths. The Peak RC System Operators monitor phase-shifting transformer (PST) tap positions to identify the effect on pre-contingency flows and evaluate available mitigation for post-contingent transmission constraints, as shown on the following WECC map in the EMS (Fig. 2.1).

The Peak RC System Operators (RCSOs) coordinated operation of PSTs based on the results of the Peak RC phase-shifting transformer program tool calculations.

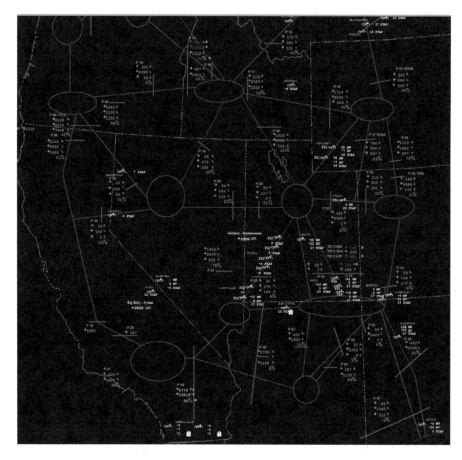

Fig. 2.1 WECC system map showing major phase-shifting transformers used for loop flow miti-gation and path flow overload relief. (Source: Peak Reliability)

In addition, RCSOs needed to know when the PST owners would use their devices in independent operation and who the owners are. The WECC Unscheduled Flow Mitigation Procedure (UFMP) is an interconnection-wide transmission load relief procedure used to mitigate overloads attributed to Unscheduled Flow on Qualified Transfer Paths. To monitor Western Interconnection operation, Peak RC mainly relied on the EMS advanced applications, i.e., State Estimator and RTCA to per-form real-time assessment for basecase and post-contingency conditions. The diagram given in Fig. 2.2 describes overall architecture of the EMS software suite deployed in Peak RC, where State Estimator (SE) and RTCA are two center-pieces [1, 2].

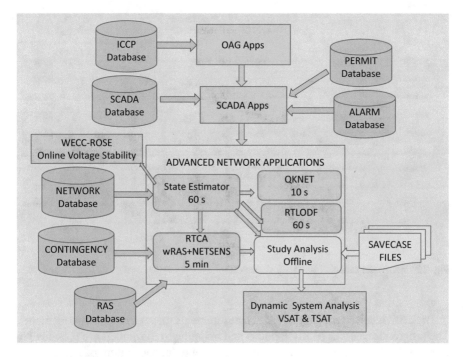

Fig. 2.2 Peak RC EMS applications system architecture

2.1.1 About the Peak RC Network Model

The main input to any RTCA application is the SE solution. Without a good SE solution, it is not practical to expect that contingency analysis results will be reliable and credible. Most importantly, it is extremely important that the RCSOs have accurate, actionable information available if they are expected to use contingency analysis information. There is a confidence level in the tools that needs to exist, and it starts with a quality SE solution.

Peak RC was the single Reliability Coordinator for the entire Western Interconnection (with the exception of the AESO area). From a modeling perspective, this was a big positive. In the Peak RC network model, there was no external model. For the eastern boundary of the network model, which connects to the Eastern Interconnection, there are eight DC ties that are modeled as injections (generation). Further details about the last Peak RC network model are as follows:

- 8938 substations
- 15,658 buses (dynamic)
- 14,062 transmission lines
- 5678 transformers
- 3830 generators
- 15 DC lines (full DC line representation)

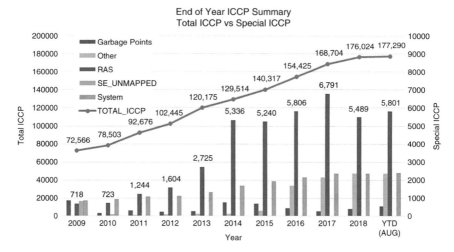

Fig. 2.3 ICCP data statistics summary since 2009

Peak RC received roughly 177,290 measurements mapped to the network model. The progression curves of ICCP points and mapping to SE models are presented in Fig. 2.3. Peak RC had a modeling goal to have a redundancy ratio of >2.53 for all voltage levels >100 kV. Peak RC tracked redundancy ratios for each balancing authority in the model. The average redundancy ratio for BAs within the Peak RC network model was 2.27.

As expected, the lower the voltage level, the lower the redundancy ratio. The downside to this was it contirbuted to poor SE solution quality at the lower voltage levels. This lead to false contingency violations and possible actual limit violations that go undetected in the contingency analysis solution. For these reasons, it was imperative to have high measurement redundancy in the lower voltage areas of the model. Significant progress was made toward improving the measurement redundancy ratios through the ongoing focus on adding new measurements to the Peak RC measurement set.

2.1.2 Monitoring SE Solution

The Peak RC SE runs once every minute. Within the control room, there was two RCSOs who were responsible for maintaining the solution quality of the state estimator. One of these RCSOs was responsible for the Loveland RC footprint, while the other was responsible for the Vancouver RC footprint. When the state estimator does not converge or if it solves with a MW or MVAR mismatch greater than defined tolerances, the RCSOs got an alarm indicating this condition. Generally, the RCSOs relied on these two key indicators to determine whether there is action needed to improve or correct the state estimator solution. Initially, there were many errors in the SE solutions that were undetected using these methods. Peak RC had a project completed in 2012 that improved visibility of the SE solution quality. This new tool helped RCSOs to visualize various types of errors in the state estimator solution in addition to bus mismatch or non-convergence:

- Measurement residuals and anomalies – often represent a topology error rather than an erroneous analog measurement.
- Bus voltage solution and residuals – it is important to have a quality voltage solution to minimize false post-contingent voltage violations.
- Area load, interchange, and generation errors – this is important to ensure overall solution quality and is an important metric to ensure a quality solution for the Peak RC next-day study process.

Peak RC had started to enhance the monitoring of the SE solution by building an on-shift rotation of operations engineers in 2013. Generally, the Peak RC's SE solution had been quite robust, with valid solution percentages around 99.5% since the EMS cutover in 2009. Excessive mismatch solutions occurred 1.3% of the time, while unsolved state estimator solutions occurred less than 0.05% of the time.

2.2 Post-contingency Power Flow Solution

Once receiving a valid SE basecase solution, the RTCA engine would solve for basecase power flow and post-contingency power flow for a handful of contingencies flagged as potentially harmful by the sensitivity analysis-based screening process, using two known algorithms: Newton power flow (NPF) or fast decoupled power flow (DPF). This section provides a brief overview of some aspects of EMS steady-state post-contingency power flow solution and overall configurations of RTCA for post-contingency studies. The descriptions might be useful in a comparison between the EMS tools used at Peak RC with those used by transmission operators of utilities and sub-regional study groups.

2.2.1 Post-contingency Generation Reallocation Principle

The reallocation of generation among generating units following a contingency depends on the following criteria.

2.2.1.1 Problem Definition

The network model typically consists of a single topological island and multiple operating areas/balancing authorities within that island.

For operating areas that transact a pre-defined interchange (i.e., the "area interchange" flag is checked, thereby enforcing interchange for that balancing authority (BA)), generating units within such areas contribute toward a power balance within those areas, i.e., they regulate to meet load, losses, and interchange requirements for their area. A unit-regulating MW will move up or down to lower mismatch as long as its MW output is within its P_{Max}/P_{Min} limits.

Generating units within operating areas that do not enforce area interchange transactions contribute toward the power balance for the rest of the topological island, i.e., they regulate to meet island load and losses within their MW limits.

In RTCA enforcement of interchange transactions is a user-selectable flag on a BA area basis.

2.2.1.2   Generation Participation

Irrespective of whether a generating unit contributes toward area power balance or island power balance, the proportion in which it picks up MW mismatch in the post-contingency power flow depends on a pre-defined participation factor.

The participation factor for a unit represents the proportion in which a generating unit picks up MW mismatch with respect to generating units in its operating area or the rest of the island, depending on whether or not area interchange is enforced.

For each generating unit, three possible values of participation factors can be specified. These three values model the following types of responses:

1. Governor/transient response
2. Automatic generation control (AGC) response
3. Economic dispatch response

2.2.1.3   Peak RC Configuration

- Peak RC implemented EMS software uses a hardcoded setting of AGC response for its base case power flow. For post-contingency power flow, out of the three types of generation participation in MW allocation, Peak RC has set up its software to use AGC response.
- Peak RC has set up the participation factors to be proportional to the unit P_{max} rating. Upon a simulated unit contingency, all units online with a non-zero participation factor will pick up those lost MW based on this proportion.
- Most generating units in the Peak RC WSM (excluding nuclear power plants, wind farms, and solar energy resources) are enabled with AGC response for post-contingency power flow.
- All existing static VAR compensator (SVC) devices are simply modeled as generating units (with $P_{max}/P_{min} = 0$) in the WSM. These devices only participate in bus voltage/Var regulation.

2.2.1.4   Special Setting for AESO

- Peak RC sets up only the Alberta Electric System Operator (AESO) balancing authority to enforce area interchange in the post-contingency power flow. Area scheduled interchange is enforced for AESO in order to ensure that post-contingency power flow does not produce false violations or PATH 1 and PATH 2 exceedances due to generators in the USA picking up slack in proportion to their participation factors for loss of an AESO unit.

2.2.2   Post-contingency Steady-State Controller Actions

The steady-state post-contingency behavior of control devices such as switched shunt capacitors or reactors, load tap-changing transformers (LTC), phase shifters, unit reactive capability, SVC devices, and series capacitors/reactors is described in this section.

2.2.2.1   Switched Shunts

Capacitors and reactors are switched based on the following criteria in the post-contingency power flow solution:

- Shunts switch in or out in order to bring the regulated bus voltage within the acceptable range. The target-regulated bus voltage in RTCA is set to the state estimator solved voltage at the bus. Shunts are switched when the voltage deviates from the target voltage by more than a modeled deviation threshold. Peak RC has this threshold set at 3.5%.
- In the event of multiple switched shunts regulating the voltage for the same bus, they are banked together in order to coordinate their switching.
- Shunts can be switched in or out based on selecting one of two global options: (i) modeled static sensitivity or (ii) user-specified priority.

  - **Option (i)** or the sensitivity option for switching shunts uses a modeled static sensitivity (a value specified in units of delta per unit (PU) voltage/MVAR) to decide how many shunts to switch in order to achieve the desired voltage regulation.
  - **Option (ii)** or the priority option relies on the user specifying a switching order for multiple shunts regulating voltage at the same bus. For example, if two shunts regulate bus voltage together and they have a priority number of "1" and "2" specified, the shunt that has priority 1 assigned will switch first, and if needed, the shunt with priority 2 switch second, etc., to bring the regulated bus voltage within the desired range. If the global switching-by-priority option is selected and if switching priorities are unspecified for individual shunts that are banked together, the shunts are switched in descending order of their nominal MVAR capacity.

- – During the iterative post-contingency power flow process, the switching of shunts is performed only after the MW/MVAR mismatch at buses has settled down to be below a user-definable threshold.
- – The user can designate individual shunts to be switchable or not in the post-contingency solution.

- **Peak RC configuration** – Peak RC has set up its model to switch shunt capacitors and reactors using priority. Wherever shunt capacitors/reactors are in a bank, the nominal MVAR of shunts are used to prioritize shunts for switching, and individual shunt devices are not modeled with a static priority value.

2.2.2.2   LTC Transformers

Peak RC has not enabled movement of taps for LTC transformers in its post-contingency solution.

2.2.2.3   Phase Shifters

Peak RC currently does not enable movement of taps for phase shifters in its post-contingency solution.

2.2.2.4   Reactive Control Devices

- **Unit reactive capability** – Most generating units in the WSM are modeled with piecewise linear reactive capability curves. The Qmax and Qmin are dynamically updated upon post-contingency unit MW output values.
- **SVC devices** – There are dozens of SVC devices being modeled as generating units ($P_{max}/P_{min} = 0$) in the WSM. The devices are pre-specified with static reactive capability limits-Qmax/Qmin.
- **Series capacitors/reactors** – No separate scheme for series capacitor/reactors switch is directly applied to post-contingency power flow. However, series capacitors/reactors can be defined as a RAS protection action to allow for switching for post-contingency power flow.

2.2.2.5   Post-contingency Generator Regulation

The EMS post-contingency power flow supports regulation of the generator terminal bus, the step-up transformer high side bus, or even remote regulation of a bus that is outside the generating station within the generator's reactive capability. During the iterative post-contingency power flow solution, when the MW injections of the units are updated to reallocate MW mismatch, the unit reactive capability curves are reevaluated. When a generator that is regulating bus voltage hits a high/

low reactive limit or a generator that was previously at a reactive limit is now able to move MVAR to regulate bus, type switching is done, and the generator moves from being a PV device to a PQ device or vice versa.

The EMS software supports **line-drop compensation** (LDC). A remote node is specified for LDC that is typically different from the node whose voltage is being regulated by the generator. The generator voltage regulation target, when LDC is enabled for a generator, is adjusted to compensate for the voltage difference between the regulated bus and the remote LDC node using a user-specified compensation % as follows:

$$VtargetLDC = Vtarget + (1 - LDCFraction) * (Vreg - Vldc\,bus)$$

where,

VtargetLDC: Revised generator voltage regulation target taking LDC into account
Vtarget: Generator voltage regulation target before LDC, typically the state estimator solved voltage at the regulated bus
LDCFraction: A value between 0.0 and 1.0 that is used to model the LDC as a fraction Vreg: Actual bus voltage corresponding to the modeled regulated node
Vldc bus: Actual bus voltage corresponding to the modeled LDC node

Peak RC configuration: Peak RC currently **does not** use LDC in modeling generator voltage regulation.

2.2.3 Load Modeling

Peak RC uses constant P/Q loads. Individual loads may have a base or time-invariant portion and a time-dependent portion. The post-contingency solution uses load MW/MVAR injection values from the SE basecase. In the SE, BA loads are telemetered directly from the respective operating entity and distributed down to individual loads using allocation factors for those loads that are not individually telemetered.

The SE also has the ability to calculate the injections of non-telemetered loads from measurements at nearby buses in order to come up with a better injection estimate. At the end of its solution, the state estimator might even redistribute bus mismatches to loads within their modeled high/low limits. Once estimated in this way, the constant P/Q loads in the basecase are passed onto contingency analysis for the post-contingency solution.

2.3 RAS Modeling in RTCA

Peak RC was the single Reliability Coordinator (RC) of the Western Interconnection monitoring for actual and/or potential System Operating Limit (SOL) and Interconnection System Operating Limit (IROL) to ensure the reliability of the

WECC system. The main software used to monitor the reliability of the system is GE-Alstom Real-Time Contingency Analysis (RTCA). RTCA used to run over 8000+ active contingencies every 5 min with 515 RAS modeled active. In this section we summarize main RAS modeling procedures and practice that are implemented in Peak RC's RTCA and contingency analysis in study mode.

2.3.1 NERC Definition of SPS/RAS

2.3.1.1 Definition of Special Protection Scheme (SPS)

North American Electric Reliability Corporation (NERC) defined SPS as "An automatic protection system designed to detect abnormal or predetermined system conditions, and take corrective actions other than and/or in addition to the isolation of faulted components to maintain system reliability. Such action may include changes in demand, generation (MW and Mvar), or system configuration to maintain system stability, acceptable voltage, or power flows. An SPS does not include (a) under frequency or under voltage load shedding or (b) fault conditions that must be isolated or (c) out-of-step relaying (not designed as an integral part of an SPS). Also called Remedial Action Scheme."

2.3.1.2 Definition of Remedial Action Scheme (RAS)

NERC defines RAS as "A scheme designed to detect predetermined System conditions and automatically take corrective actions that may include, but are not limited to, adjusting or tripping generation (MW and Mvar), tripping load, or reconfiguring a System(s). RAS accomplish objectives such as:

- Meet requirements identified in the NERC Reliability Standards,
- Maintain Bulk Electric System (BES) stability,
- Maintain acceptable BES voltages,
- Maintain acceptable BES power flows,
- Limit the impact of Cascading or extreme events."

2.3.1.3 Protection Schemes of Non-RAS SPS

Based on NERC definition, not all SPSs are RASs. The following SPSs do not constitute as a RAS:

- Protection systems installed for the purpose of detecting faults on BES elements and isolating the faulted elements.
- Schemes for automatic under-frequency load shedding (UFLS) and automatic undervoltage load shedding (UVLS) comprised of only distributed relays.

- Out-of-step tripping and power swing blocking.
- Automatic reclosing schemes.
- Schemes applied on an element for non-fault conditions, such as, but not limited to, generator loss of field, transformer top-oil temperature, overvoltage, or overload to protect the element against damage by removing it from service.
- Controllers that switch or regulate one or more of the following, series or shunt reactive devices, flexible alternating current transmission system (FACTS) devices, phase-shifting transformers, variable-frequency transformers, or tap-changing transformers, and that are located at and monitor quantities solely at the same station as the element being switched or regulated.
- FACTS controllers that remotely switch static shunt reactive devices located at other stations to regulate the output of a single FACTS device.
- Schemes or controllers that remotely switch shunt reactors and shunt capacitors for voltage regulation that would otherwise be manually switched.
- Schemes that automatically de-energize a line for a non-fault operation when one end of the line is open.
- Schemes that provide anti-islanding protection (e.g., protect load from effects of being isolated with generation that may not be capable of maintaining acceptable frequency and voltage).
- Automatic sequences that proceed when manually initiated solely by a system operator.
- Modulation of HVDC or FACTS via supplementary controls, such as angle damping or frequency damping applied to damp local or inter-area oscillations.
- Sub-synchronous resonance (SSR) protection schemes that directly detect sub-synchronous quantities (e.g., currents or torsional oscillations).
- Generator controls such as, but not limited to, automatic generation control (AGC), generation excitation [e.g., automatic voltage regulation (AVR) and power system stabilizers (PSS)], fast valving, and speed governing.

2.3.2 WECC RAS Type Classification

All RAS must be assessed for operation, coordination, and effectiveness at least every 5 years in order to comply with the requirement of NERC standard PRC-014-0. WECC Remedial Action Scheme Review Subcommittee (RASRS) has specific guideline on how to submit a RAS for assessment. WECC uses the same NERC RAS/SPS definitions to classify individual RAS, as described in Fig. 2.4.

If a SPS scheme is deemed as a RAS, WECC has three (3) types of RAS which are Local Area Protection Scheme (LAPS), Wide-Area Protection Scheme (WAPS), and Safety Net (SN).

A RAS that is classified as LAPS if whose failure to operate would not result in any of the following:

- Violations of TPL-001-WECC-RBP-2 System Performance RBP
- Maximum load loss $\geq$300 MW,

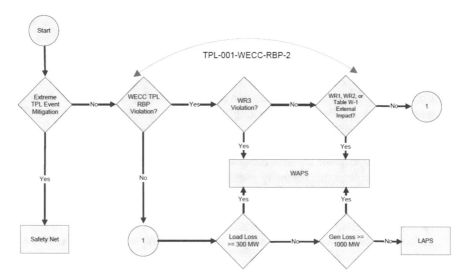

Fig. 2.4 WECC Remedial Action Scheme type definitions

- Maximum generation loss ≥1000 MW.

A RAS that is classified as WAPS if whose failure to operate would result in any of the following:

- Violations of TPL-001-WECC-RBP-2 System Performance RBP.
- Maximum load loss ≥300 MW,
- Maximum generation loss ≥1000 MW.

SN is a type of RAS designed to remediate NERC TPL-004-0 standard (System Performance Following Extreme Events Resulting in the Loss of Two or More Bulk Electric System Elements (Category D), http://www.nerc.com/files/tpl-004-0.pdf) or other extreme events.

2.3.3   RAS Modeling Practice at Peak RC

Figure 2.5 describes Peak RC's RAS modeling workflow.

All required RAS are supposed to be properly modeled in Peak RC's RTCA to perform real-time assessment. However, the RTCA software has limitation to model/monitor contingencies triggering transient or dynamic RAS by nature. For example, the RTCA software can't model certain transient RAS, rate of change, and out-of-step relay-based RAS. Those RAS need be modeled in the transient stability analysis tool (TSAT). Peak RC has also modeled associated RAS in the real-time voltage stability analysis (RT-VSA) tool for IROL/SOL assessment for every 5 min.

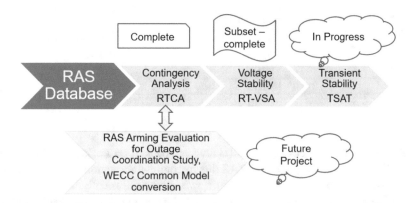

Fig. 2.5 A practical RAS modeling workflow

In principle, RAS measurements received from TO and TOP via ICCP need be used for RAS modeling in real-time assessment tools, such as RTCA, RT-VSA, and online TSAT. For operation analysis study in planning horizon, those real-time RAS measurements from a SE basecase may not be accurate or appropriate to future time of point operation planning studies. It's ideal to reevaluate RAS arming points by user-specified calculation in offline study tools, such as PowerWorld and GE/PSLF, etc.

2.3.3.1 Special RAS Modeling in RTCA

2.3.3.1.1 Basecase RAS Screening

Due to the system topology changes, the trigger condition of a RAS can be met in the basecase level that means the RAS will fire at the basecase level in RTCA. It is not the case in the actual power system operation. The RAS will be deactivated before taking the power system elements out of service to prevent the false operation of the RAS. To avoid RAS being fired in basecase, the RTCA software incorporates a global check for "Screening Basecase RAS." User shall check this global option in the RTCA display to achieve basecase RAS screening.

2.3.3.1.2 RAS First Logic

The RAS first logic is intended to work as follows: while processing a contingency, the topology changes of the contingency are applied first; before solving the power flow with the contingency applied, it is checked whether any RAS conditions would be triggered. If a RAS condition were to be triggered, appropriate protective actions are applied, and then power flow is solved. The protective action thus applied "might" enable the otherwise unsolved contingency, to be solved.

Table 2.1 ITB final output determination logic

ITB basecase status	ITB contingency status	ITB final output
FALSE	FALSE	FALSE
FALSE	**TRUE**	**TRUE**
TRUE	FALSE	FALSE
TRUE	TRUE	FALSE

The "RAS first" checkbox need be always checked on Peak RC's RTCA displays. It provides a way to model RAS due to contingency actions, and appropriate RAS actions can be taken before power flow solutions. Basically, when the RTCA is running a particular contingency, it applies the contingency and check the RAS trigger conditions, if the conditions are met, then apply the protection actions and then solve the power flow. The RAS first logic could mitigate some issues of RAS unfired due to invalid post-contingency power flow solution.

2.3.3.1.3   Independent Trigger Block

The purpose of independent trigger block (ITB) is to determine how RAS is supposed to work in the field when power system equipment that serves as inputs of the triggering condition is out of service for maintenance.

In Peak RC's EMS RAS database, each ITB definition is represented by a trigger condition (TRIGCON) and marked with a change of state considered status to distinguish it from the normal trigger condition in the current RAS scheme.

ITB is evaluated in both contingency analysis (CA) basecase and contingent states. The output of each ITB serves as input to **OR** gate. If ITBs are serving as inputs to an OR gate, the determination of the final output status sent to the gate input will depend upon their evaluated triggering condition statuses in the basecase and contingent states, as shown in Table 2.1. Use of ITB in RAS modeling enables simulation of some complex RAS logic and protective actions in RTCA to be compliance with actual RAS operation in the field.

2.3.3.1.4   RAS Processing in RTCA

During RTCA full processing, the sequence corresponding to RAS processing proceeds as follows:

1. Apply topology changes corresponding to a contingency.
2. Check for RAS first for all applicable RAS's.
3. Apply any protective actions for RAS's triggered during RAS first.
4. After the RAS first logic is complete, set up the PROCESS RAS flags for all RAS's. It relies on RasCheckTrig method to skip any RAS's or Stage's that need to be skipped.
5. Complete the power flow solution for the contingency.

6. Set up RAS's for processing in the post-power flow state. Commenting out the post RAS first cleanup.
7. Evaluate the RAS's in the post-power flow state.

Notes: there is a pointer for each RAS to link last processed stage of a RAS. RasCheckTrig method is "If the last processed stage was not triggered, continue processing the stage again. Do NOT skip processing the current RAS. Also, do not skip processing the RAS if the pointer directs to the last stage of the RAS."

2.3.3.2 RTCA RAS Modeling Procedure

All the WECC RASRS-approved RAS should be modeled in Peak RAS's EMS RAS model database except Safety Net schemes (Safety Net schemes are used to protect extreme events and need to be evaluated whether to model on a case-by-case basis).

2.3.3.2.1 Model a New RAS

When a new RAS needs to be modeled in the current model update cycle, below are the basic steps.

1. Identify and obtain the proper RAS-operating procedure and documentation.

   (a) BC Hydro sent RASMOM to Peak RC which could be translated using Peak naming convention and merge it to Peak's current RASMOM.

      (i) Performed a RASMOM compare to see which RAS needs to be updated or modeled.
      (ii) Translated the new RAS and merge to the master RASMOM.

   (b) BPA sent RASMOM to Peak RC which could be directly merged to Peak's RASMOM.

2. Read the documents and understand how the RAS works and build a logic diagram of how the RAS can be modeled.
3. Identify and request new ICCP points needed to model the RAS (overall arming statuses, real-time arming statuses for the trigger conditions and protection actions).
4. Once having the ICCP points, update them in the SCADA (SCADA mapping) and NETMOM (Special Measurement table (SPMEAS)) using the Access DB tool located at TEMS Non-Replicated Shared Drive\Access_Tools\QueryWSM_Access.xlsm.
5. If SCADA calculation is needed, follow the RAS SCADA calculation procedure. If new ICCP points cannot be obtained on time for the current model update cycle, an approximated model may be needed.
6. Identify which contingency(es) could trigger the RAS.

7. Model the actual RAS based on step 2 logic diagram.
8. Check the "Enable RAS" box for the identified CTGs in step 7 in RTCA (send email to operation engineer who is responsible of updating CTG_DEF database).
9. Once finishing modeling the RAS, make screenshots of trigger conditions and protective actions and save them in EMS RAS folder.
10. The engineer models the RAS and must test the RAS following the RAS testing template.
11. Move the RAS-related procedures and documents to the RAS folder under the same TOP/BA which owns/operates the RAS.
12. Update the RAS tracking spreadsheet.
13. Update the RAS display if any.
14. Second engineer review the RAS and make any necessary changes.
15. Save new RAS model in WECC_MASTER_RAS and WECC_MASTER_RAS_DBXX to STNET.
16. Load WECC_MASTER_RAS to RTCA on Test EMS server.

2.3.3.2.2 Modify Existing RAS

For any existing RAS, if a new operating procedure is received, follow the steps below:

Review the update section in the new procedure to see if any RAS modification is needed.
If not, save the new procedure in the RAS folder under the same TOP/BA which owns/operates the RAS.
If yes, make necessary changes to the RAS following "Adding a new RAS" steps 3–16.

2.3.3.2.3 Deactivate Existing RAS

For seasonal-dependent RAS, make sure the one appropriate to the current model cycle is enabled, and the other one is disabled if not using real-time arming status 0.

If receiving email for TOP to temporary deactivate a particular RAS, make sure to deactivate the RAS in real time and the current model update cycle. Also, create a RAS SharePoint ticket includes when the RAS will be back in service.

2.3.3.2.4 Delete Existing RAS

When receiving information of RAS decommissioning, Peak RC engineers delete the RAS from the current master WECC RAS list. Also, update the RAS tracking spreadsheet with the date of decommission.

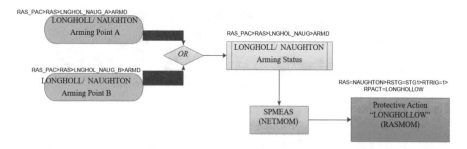

Fig. 2.6 Sample of RAS calculation logic diagram

2.3.3.2.5 SCADA Calculation

Although RAS application in EMS is capable of performing logic and analog calculations under IPIN/RPACT records, some simple ones are still preferred to (or must) be added into SCADA for easier maintenance. Those calculations are currently maintained in **scada_calc_ras.csv** file. RAS application can only link special measurements as input or output; thus all of the RAS-related records need to be added into the SPMEAS table in NETMOM in addition to the creation of the SCADA calculations. Refer to the below document for more detailed information about SCADA calculation function.

In the csv file, all records should have RAS as DEVTYP (ID_DEVTPY). The substation name of RAS records could either be RAS [BA] or any WSM substation name. For each SCADA calculation point, the first record under column "#Record Type (C*6)" must be "MEAS."

The RAS calculations are currently maintained by Network Applications Team. There is no generic structure for RAS calculations as seen in other calculations. A simple example is given in Fig. 2.6. The logic is to enable the protective action "LONGHOLLOW" if the arming status of either LONGHOL_NAUG_A or LONGHOL_NAUG_B is armed. It is worth noting again that both input arming points A and B and the output arming status should be specified in the SPMEAS table of the WSM access database.

The output of a POINT is either true or false (the result of a Boolean logic – AND, OR, NOT, IFGR, etc.). It is often used when several arming points need to be combined into one to have the final arming point (RAS A or B -> overall RAS arming point).

As shown on the example in Fig. 2.7, the four points under "PNTREF" are the input (ARG1) to the OR gate (the record type must be PNTREF for a point), and the final results that can be used by RAS (ARG2) are the last row with a protective action.

Figure 2.8 shows how the point is calculated under SCADA substation topology (scada/subtop).

#	Record Type	ID_SUBSTN	ID_DEVTYP	ID_DEVICE	(ID_MEAS	TYPE_MEAS	FLAGS_MEAS	ID	AREA	REF SUBSTN	REF DEVTYP	REF DEVICE	REF ID
1	#Record Type	ID_SUBSTN	ID_DEVTYP	ID_DEVICE	(ID_MEAS	TYPE_MEAS	FLAGS_MEAS	ID	AREA	REF SUBSTN	REF DEVTYP	REF DEVICE	REF ID
2	MEAS	RAS_SPPC	RAS	05_RAS	ARMD	OR	CALC						
3	PNTREF	RAS_SPPC	RAS	05_RAS	ARMD			STS1	WECC	RAS_SPPC	RAS	SPP_LEVEL_1	ARMD
4	PNTREF	RAS_SPPC	RAS	05_RAS	ARMD			STS2	WECC	RAS_SPPC	RAS	SPP_LEVEL_2	ARMD
5	PNTREF	RAS_SPPC	RAS	05_RAS	ARMD			STS3	WECC	RAS_SPPC	RAS	SPP_LEVEL_3	ARMD
6	PNTREF	RAS_SPPC	RAS	05_RAS	ARMD			STS4	WECC	RAS_SPPC	RAS	SPP_LEVEL_4	ARMD
7	POINT	RAS_SPPC	RAS	05_RAS	ARMD			ARMD	WECC				

Fig. 2.7 Sample of RAS arming points

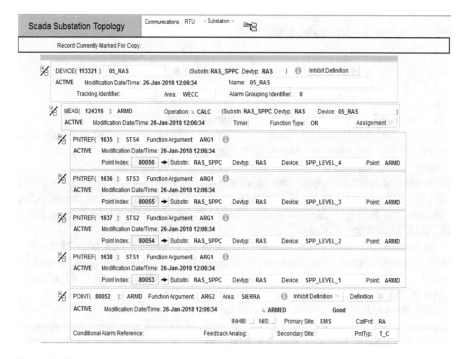

Fig. 2.8 SCADA substation topology for point in Fig. 2.7

2.3.3.2.6   RASMOM Validation

After modeling all the RAS for the current model update cycle, Network Applications Engineer must validate the new RASMOM in EMS to ensure no error. If there is, errors need to be fixed before saving as the master RASMOM case.

2.3.3.3   Peak RC RTCA RAS Modeling Statistics

Remedial Action Scheme/Special Protection System (RAS/SPS) modeling has a significant impact on the WECC RC's real-time contingency analysis solution quality. Roughly 50% of the unsolved or islanding contingencies that are detected by real-time contingency analysis are due to a RAS not being modeled or a RAS not being triggered properly. In accordance with Peak RC RAS_PRC-013, there were about 272 RASs, Local Area Protection Schemes, and Safety Net Schemes that may have inter-BA or inter-TOP impacts. The RAS model definition is primarily driven by the RAS-operating procedures written by WECC Remedial Action Scheme Reliability Subcommittee (RASRS) or individual RAS-operating TOPs. Below were the statistics of the Peak RC RAS models:

- 5801 RAS measurements (arming/triggering status points and analog points) were received from TOPs via ICCP.
- Most of RAS measurements available in SCADA were mapped to state estimator and transferred to contingency analysis for processing.
- 515 RASs/SPSs were finally modeled in the WSM.
- All of them had been validated and activated for contingency analysis processing.
- 9200 contingencies were flagged out for RAS screening.

2.4 Monitoring SOL and IROL Post-contingent Exceedances

Peak RC utilizes an in-house developed tool known as the RC Workbook to monitor real-time contingency analysis results. The EMS software suite tools provided by most standard vendors are not practical for monitoring large amounts of contingency analysis results. The Peak RC EMS contingency analysis software is tabular in nature with limited ability to sort, filter, acknowledge, or track information associated with the contingency results. It is a suitable secondary means for monitoring, but is not practical for daily usage. Peak RC developed custom tool called "RC Workbook" to visualize EMS Network Application solution results, including RTCA. Fig. 2.9 illustrates the RC tool's RTCA summary page. It was designed with the following key features:

- Flexible sorting and filtering to allow the RC to focus on the limit exceedances that are of a concern. Filtering allows the RC to remove items from the list that are outside of the RC's area of responsibility, for example, a Loveland RC does not need to address Vancouver area issues. Sorting is primarily setup to sort by monitored element voltage level or severity in terms of percentage outside limit.
- A notes feature provides a means for the RC to document findings. This may be a mitigation plan if the contingency were to occur or a temporary note addressing a unique condition caused by an existing outage or abnormal grid condition. Notes can be permanently or temporarily associated with the contingency to document the findings. Notes that are temporarily assigned will disappear once the limit exceedance is no longer detected by real-time contingency analysis.
- Allow for acknowledgment of a post-contingent limit exceedance. This feature is a combination of standard alarm monitoring and contingency analysis monitoring. When a new limit exceedance is presented to the RC, the RC can review the

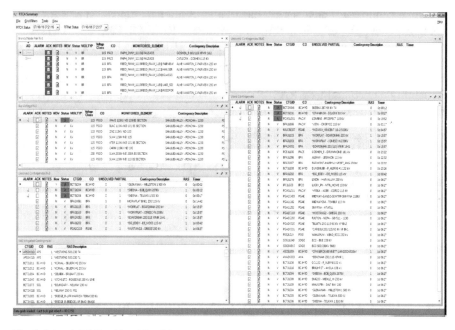

Fig. 2.9 Peak RC Workbook/RTCA summary display

results, determine the reason for the problem and a mitigation plan, and document those in the notes. After complete review and documentation, the limit exceedance can be acknowledged, removing that item from the RC's contingency analysis results view. If the RC needs all of the acknowledged, just click Acknowledge All button.

- Use of color to provide heightened awareness of new violations or violations that have not been acknowledged by the Peak RC.
- Summary information about the most recent real- time contingency analysis execution: number of limit exceedances detected, number of unsolved contingencies, number of islanding contingencies, etc.
- Alarms with the enhancement feature of timer to track the time duration of post-contingent SOL and IROL exceedances.
- Special view tables for monitoring multi-equipment outages contingency violations.

2.4.1   Real-Time Detection of Potential IROLs

The NERC definition of an IROL is "A System Operating Limit that, if violated, could lead to instability, uncontrolled separation, or Cascading Outages that may adversely impact the reliability of the Bulk Electric System." NERC Standards, specifically NERC FAC-011-2, identify the Reliability Coordinator as the responsible entity for identifying the subset of SOLs that qualify as IROLs. Known IROLs, if exceeded, need to be mitigated without delay, and there is very little question about what needs to be done to remedy the situation. IROLs that are known about in advance have clear procedures in place for how to act if the IROL is exceeded. However, what happens when the system finds itself in an unstudied state? If there are unexpected outages, unusually high loading, or other abnormal operating conditions, the Western Interconnection could find itself operating with IROLs that were not identified in the planning or operating horizons.

The current method for the Peak RC to identify potential IROLs is to monitor for unexpected and unsolved contingencies or islanding contingencies in real-time contingency analysis. The RC Workbook is designed to provide immediate information to the RCSOs about both of these conditions. Figure 2.9 illustrates how the RCSO is made aware of unsolved or islanding contingencies detected in real-time contingency analysis.

When a new unsolved contingency appears in real-time contingency analysis, the RCSO performs the following steps [1]:

1. Review the state estimator solution quality in the general area of the unsolved contingency. Ensure the state estimator is accurately representing system conditions.
2. Move the state estimator solution into the study power flow environment and simulate the unsolved contingency. The Peak RC contingency analysis engine is a fast, decoupled power flow algorithm, while the study power flow runs a Newton-Raphson algorithm. This makes the contingency analysis application a bit sensitive to conditions such as a poor voltage solution or transmission with a high R/X ratio. Issues such as these can lead to unsolved contingencies in contingency analysis that can easily be solved by the study power flow application.
3. Validate the contingency definition to be sure the simulation is correctly removing the desired facilities.
4. Contact the TOP to discuss the operating conditions that could lead to this unsolved contingency.
5. Contact engineering support if the issue cannot be addressed with the initial set of steps.

Once the unsolved contingency is determined to not be a potential IROL, the condition is documented in the RC Log and in the RC Workbook so that if the condition occurs again in the future, the RCSO will immediately have better awareness of what caused the issue and will not have to do the assessment from the beginning.

If the condition is determined to be an IROL, per IRO-009-1, the RCSO will take immediate action to mitigate the IROL. This mitigation could include generation adjustment, transmission adjustment, load shedding, or other actions deemed effective by the RCSO and the impacted TOP. The RC Workbook is again used to document the IROL so that if the condition appears again, the RCSO will immediately have better awareness of the IROL condition.

2.4.2   SOL Monitoring Methodology

2.4.2.1   Monitoring Responsibilities in the Western Interconnection

The NERC IRO standards require the Reliability Coordinators to monitor the BES. Generally, this means that Peak RC should monitor any facility greater than 100 kV or other lower voltage facilities that are known to have an impact to the BES. The transmission operator (TOP) responsibilities are to monitor their own transmission systems and to monitor for System Operating Limit (SOL) exceedance within their operating boundaries. Many smaller TOPs have limited tools to perform this function and generally transfer much of the monitoring burden to the Peak RC.

NERC and FERC are somewhat inconsistent in their message to the industry about their expectations of the Reliability Coordinator and TOP roles. FERC has come out with the message that the Reliability Coordinators exist to monitor for IROLs, while the TOPs are primarily responsible for monitoring SOLs. However, application of NERC standards generally does not support this message, and this makes it very difficult for the industry to understand what the expected behaviors of the entities responsible for monitoring the BES are.

The monitoring of facilities within the Peak RC network model is maintained by the setting of a flag associated with the network model facility. If the Peak RC determines in the operations horizon that there is a need to monitor additional equipment due to unexpected operating conditions, this can be done on the fly. This assessment is typically done in the day-ahead study but can also be determined as part of real-time operations.

2.4.2.2   Definition of Credible Contingencies

By August 2019, Peak RC modeled a contingency list with 9267 contingencies and 515 RAS in the RTCA. 8065 of the defined contingencies are activated for post-contingent assessment and RAS screening.

- 92 kV and greater transmission lines
- 92 kV and greater transformers
- Sub-100 kV lines and transformers identified to impactful for BES
- Generators >50 MW output
- DC pole outages
- Common mode or common corridor multiple facility outages (credible N-2 contingencies)

All of the Peak RC contingencies are defined as breaker-to-breaker contingencies to ensure accurate topology of the network following the simulation of the contingencies.

2.4.2.3 Real-Time Detection of Cascading Overloads

2.4.2.3.1 Problem Definition

A cascading outage is a sequence of events in which an initial disturbance, or set of disturbances, triggers a sequence of one or more dependent component outages:

- In some cases they halt before the sequence results in the interruptions of electricity service.
- In many case, cascading outages have resulted in massive disruptions to electricity service.

Cascading failures continue to contribute significantly to blackouts [3, 4], such as the August 14, 2003 New York blackout, November 12, 2006 Europe blackout, November 10, 2009 Brazil blackout, September 8, 2011 Pacific Southwest blackout, etc. Interconnected power grids throughout the world are usually reliable but irregularly suffer massive blackouts with multi-billion dollar loss to society. Cascading failures present severe threats to power grid reliability, and thus reducing their likelihood, mitigation, and prevention is of significant importance.

Since 2008 the Cascading Failures Task Force (CFTA), under the IEEE PES Computer Analytical Methods Subcommittee (CAMS), have been primarily focusing on understanding, prediction, mitigation, and restoration of cascading phenomena and preventive measures against cascading outages. Among many mechanisms by which subsequent outages can propagate beyond the initial outages. Transmission overload and voltage collapse due to addition vegetation contact faults and/or overcurrent/under-voltage relay tripping are the main contributing factors to a cascading event. The CFTA proposed methodologies are concerned with estimation of risk needed to consider as follows:

- The probability of initiating events
- The probability distribution of cascading failure sizes
- The impacts of the blackout

Several IEEE papers were written to present the basic methodologies for mitigation, summarizes currently deployed special protection schemes, and lists cases of successful and unsuccessful mitigation of cascading outages and lessons learned, as well as discuss future developments and challenges in the area of mitigating cascading outages [5, 6].

2.4.2.3.2 Relevant NERC Standards

There are a few NERC standards defining specific operation requirements against cascading failures.

- NERC PRC-023 requires that protective relays reliably detect all fault conditions but are not set over conservatively so that they restrict the system power transfer capability.

- FAC-011-2 requires that System Operating Limits be established such that all single contingencies and certain multiple contingencies do not result in cascading outages.
- TPL-001–1 requires analysis of extreme events, and if the extreme events analysis concludes there are cascading outages caused by the occurrence of extreme events, an evaluation of implementing a change designed to reduce or mitigate the likelihood of such consequences shall be conducted.

According to the NERC standards, utilities need to implement real-time analytical tools for detection of potential cascading failure and development of viable mitigation plans in an effective and timely manner.

2.4.2.3.3   Best Practice at Peak RC

Peak RC was responsible for maintaining the RC SOL methodology for the Peak RC coordination area. This is a document that defines how the Western Interconnection TOPs are to determine SOLs and how the RC will identify IROLs, both for the operating horizon. This methodology is required under NERC standard FAC-011-2. Occasionally, real-time contingency analysis will detect a significant percent overload for a transmission line or transformer. At what point will a significant limit exceedance turn into a cascading overload? In other words, if a significantly overloaded transmission line was to trip due to relay protection, line sag, or any other issue, what would the impact be to neighboring facilities? Would additional facilities experience a similar or more significant overload that could lead to additional cascading outages?

To address these questions, Peak RC updated the SOL methodology to define a process for detecting potential cascading outage conditions. If a thermal overload is detected by real-time contingency analysis, that is greater than 125% of the emergency or highest facility limit, further study needs to be performed. The study would include the following [3]:

- Move state estimator solution into study power flow.
- Simulate the loss of both the contingent element and the monitored element that saw a relay tripping limit (if it is available in EMS) exceedance or an exceedance greater than 125% of the highest operating limit.
- Identify any further overloads that occur upon loss of both of these facilities. A growing number of significant overloads could be an indication of potential cascading outages.
- Check if there is cascading outages initiated by RAS interactions, i.e., firing of one RAS changes the system conditions, hence the changes result in firing of other RAS subsequently and so on. If so, identify and validate the impact of cascading events.
- Analyze any missing or inadequate RAS models and their impact to specific excessive transmission overload. For example, transient RAS with large gen drop actions may not be modeled accurately in RTCA. Missing of RAS actions

may lead to false transmission overloads or low bus voltages in post-contingent power flows.

• Provide comprehensive and practical trainings to RCSO and real-time operation engineer (ROE) and get them familiar with potential IROL violation and cascading failure test procedure and workflow.

2.4.2.4 Roadblocks to Accurate Contingency Analysis

2.4.2.4.1 Dynamic Facility and Transmission Path Limits

The Peak RC real-time contingency analysis monitors thousands of lines, transformers, buses, and other network equipment. Transmission paths are groupings of transmission lines and/or transformers that often have dynamic ratings that are in place due to a transmission path's simultaneous interaction with other transmission paths. Nomograms are developed that restrict the amount of MWs that can flow on a transmission path to ensure reliable operation of the transmission path. Peak RC does not have these nomograms modeled in contingency analysis; however, it does receive real-time indication of the path limit that is sent to Peak RC via Inter-control Center Communications Protocol (ICCP). In some cases, a path limit will automatically change (increase or decrease) due to the loss of a transmission facility. Peak RC does not simulate the automatic change in path rating due to the loss of the transmission facility in real-time contingency analysis.

Facility limits present unique challenges to monitoring because of the impact from ambient temperature. Peak RC currently does not have the ability to monitor temperature and adjust limits according to limit tables nor is Peak RC able to receive transmission line or transformer limits via ICCP. This leads to false limit exceedances detected by real-time contingency analysis. The RC's first step in reviewing a post-contingent limit exceedance is to contact the transmission operator to review system conditions and verify the facility limits. If limits are incorrect, the RC can document that information in the RC Workbook and pass the information on to the engineering team for potential resolution.

2.4.2.4.2 RTCA RAS Software Limitations

When adding more complex RAS models and reviewing contingency analysis solution quality issues, Peak RC has identified several RAS software problems or limitations. Some high-impact issues identified are:

• Generation/load reduction is not executed properly under certain circumstances. First, the RAS protective action raises generator MW rather than drops generator MW, when the current MW output is less than the RAS target MW. Second, the station reduction is executed by removing units or loads, instead of by defined shares. Third, it does not allow multiple RAS protective actions to reduce the same unit or load with different shares.

- The RAS could not be fired as needed when its triggering condition is to check against a MW level or a voltage value if the contingency simulation does not solve.
- The RAS basecase outage screening logic has a limitation to handle complicated RAS definitions where each RAS combines dozens of triggering conditions and protective actions. This causes a RAS to be triggered for many irrelevant contingencies.
- The RAS software does not include advanced calculation gates such as summation, multiplication, division, and node pair. Without such user-defined calculations, it is impossible to automatically update RAS arming/triggering points from a modified basecase in the next day-study process.
- It's difficult to model those RAS based on two-dimensional or three-dimentional nomograms accurately.
- RTCA RAS logic is unable to model transient RAS or dynamic RAS that include time-dependent components such as frequency rate of change, out-of-step relay zone, and so on.

Since 2014, Peak RC has continuously been working with the EMS vendor to enhance the RAS software capabilities to meet our RAS modeling requirements. New RAS triggers and protective actions were introduced as follows over the past years:

- Basecase RAS screening, ITB, and RAS first software enhancements.
- Load area can be defined as both protective actions and arming/triggering points.
- Node pair phase angles are allowed for a triggering point (for out-of-step relay).
- DC MW schedules can be changed by a MW amount or to a MW amount.
- Phase-shifting transformers (PST) Automatic MW Regulation (AWR) mode and target MW can be defined as protective actions.
- New calculation gates such as summation, multiplication, division, and square root are available.
- Interface (or WECC Path) is defined as gate input.
- Interface, lines, and transformers in/out-service status are defined as gate input.
- Enable assigning a basecase solution value to RAS triggering gates.

After the RAS enhancements were cutoff in production EMS before and in 2015, Peak RC improved RAS model accuracy and RTCA solution quality significantly.

2.4.2.4.3   Node-Pair Monitoring

Phase-angle monitoring will become a big part of power system monitoring with the advent of PMUs. In addition to real-time monitoring of key phase-angle pairs in the Western Interconnection, Peak RC has modeled these node pairs in real-time contingency analysis. There is much work yet to be done to determine appropriate thresholds for alarming or monitoring wide-area separation phase angles. Once these thresholds have been determined through a robust study process and collaboration with the transmission operators within the Western Interconnection, the Peak

RC should have the ability to predict phase-angle separation that exceeds defined thresholds in real-time contingency analysis.

2.4.2.4.4   Enhanced Contingency Analysis Visualization

Peak RC dedicated significant resources (both human and financial) to improve the way the RCSOs view real-time information, including contingency analysis results. Peak RC planed to build new geographically correct (based on GPS coordinates) overviews that have the ability to show detailed real-time actual data but also will have the ability to geographically illustrate post-contingent violations. The enhanced visualization would provide a new means for RCSOs to be aware of [N-1] contingency conditions in the Western Interconnection. This concept is one that should be pursued by other RCs, vendors, and governing authorities.

2.4.2.4.5   Online Stability Assessment

Peak RC recognized that steady-state contingency analysis is not going to identify many significant problems that exist in the Western Interconnection. Voltage and transient stability are significant concerns due to the large amount of power transfer that occurs over very long lines. Peak RC has implemented real-time voltage and transient stability in the control rooms to improve real-time operation monitoring beyond the RTCA scope. These tools were built successfully with significant collaboration between the Peak RC and WECC transmission operators (TOPs) to ensure consistency in how we implement the tools.

2.4.2.4.6   Remarks

The RTCA is one of critical real-time tools for Peak RC. Its predictive capabilities are only limited by the commitment that is made to ensure that the models and state estimator solutions are of highest possible quality. The Peak RC has moved forward on many fronts to improve our contingency analysis capabilities, which stressed the skills of RCSOs in the control room. In the following sections, we will focus on important lessons learned from the Pacific Southwest blackout and other system events and discuss new software features enabling RTCA to provide intuitive, accurate, and actionable information to the RCSOs when the information is needed.

2.5   Lessons Learned from September 8, 2011 Blackout

This section reviews the September 8, 2011 Pacific Southwest blackout event from a perspective of WECC RC (i.e., the predecessor of Peak RC) SE and RTCA results. The real-time snapshots created by SE during the disturbance period were used to validate the results of advanced network applications against the sequence of system events. Impact of the cascading outages to the blackout is reviewed, and the cause and effect of the blackout are analyzed.

Nomenclature

APS – Arizona public system
BA – Balancing authority
CFE – Comision Federal de Electricidad of Mexico
IID – Imperial Irrigation District
SCE – Southern California Edison
SDG&E – San Diego Gas & Electric
SONGS – San Onofre Nuclear Generating Station
TOP – Transmission operator
WALC – Western Area Power Administration Lower Colorado Region

2.5.1   Introduction

At around 3:38 PM on Thursday, September 8, 2011, a major power outage knocked out electricity to almost five million people in California, Arizona, and Mexico, bringing San Diego and Tijuana to a standstill leaving area workers and residents sweltering in the late-summer heat. How could a single line outage due to a blunder on maintenance work provoke such an infamous blackout across multi-areas? Was there any chance to avoid the blackout by pre-disturbance contingency screening and during the disturbance enact fast countermeasures? What improvements can be made to prevent a new blackout? The EMS in the WECC RC control centers started operational on a full West-wide System Model (WSM) since January 12, 2009. The WSM-based SE and RTCA solutions contained more intact disturbance data than any of WECC's member entities for event diagnosis [6].

To gain an understanding on the blackout, one can look into the California Bulk System Map shown on Fig. 2.10 about the network structure of impacted power systems. The inter-area power transfer capability of the 500 KV transmission network will be reduced significantly under one of [N-1] 500 kV transmission outage contingencies. Specifically,

- No strong 500 KV network loop is formed up in the area. Two 500 KV transmission corridors are shaped like two opened arms, with no connection on the far ends (close to load centers). With loss of North Gila to Hassayampa line, two other 500 KV lines on the same corridor will carry no loads. The power flows previously carried on the North Gila line are shifted to the other transmission corridor between Palo Verde and Devers and associated 345 and 230 KV networks.
- As the loading level of the rest 500 KV lines goes up sharply, positive reactive losses of the lines increase rapidly, which make relative bus voltages drop progressively.

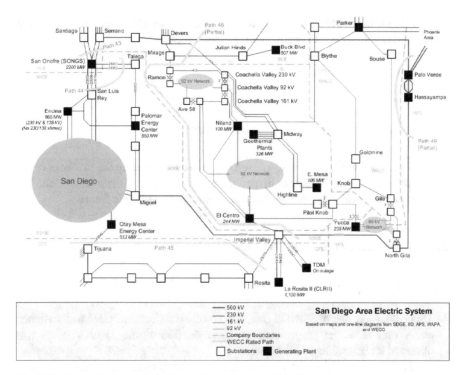

Fig. 2.10 Partial California Bulk System Map. (Source: CAISO)

- Under a normal operation paradigm, 161KV and 92 KV load substations in the area are supplied by nearby 500 KV substations. Once the North Gila-Hassayampa line is out, a massive flow shift occurs and aggregates the 161 KV and 92 KV low-voltage network. Overflows on the branches and rising reactive losses quickly decrease the reactive reserves on the local Var resources. The bus voltage drop inversely increases the relevant inductive loads.
- Cascading outages may follow up upon the limiting branch and generation equipment. A voltage instability phenomenon can be resulted from the new outages.

In addition, a Safety Net scheme-"SONGS separation scheme" was designed to separate the SONGS substation into two sections and trip two large nuclear units with 2300 MW generations in seconds once SONGS 230 kV bus voltage falls below 218 kV at minimum. When the SDG&E system was operated in a stressed condition on a hot summer day with massive air-conditioning loads, loss of any major 500 KV line, and 2 large nuclear units at SONGS close to the load centers could push the system in an insecure operation condition quickly.

As a matter of fact, the blackout event was initially triggered by loss of North Gila 500 kV line, followed by a continuation of 11 min of voltage decline, primarily on WALC and IID 161 and 92 KV buses. A sequence of the cascading outages pushed down the SONGS 230 KV bus voltages below the threshold of 218 kV, which initiated SONGS separation scheme and dropped two nuclear units in a few seconds. It took only 12 s to collapse a big portion of the SDG&E system after the SONGS units were tripped.

2.5.2   Real-Time EMS Application Results

The RC's EMS applications RTCA results were archived during the blackout event. The historical data and savecases helped the event investigation and learned the lessons to prevent similar system disturbance from occurring in the future.

2.5.2.1   Sequence of Events from SCADA Alarms

The SCADA had alarmed and archived a set of disturbance events. The relative sequence of events was recorded as follows:

At 15:17:00, AVR flag of SONGS (i.e., San Onofre) nuclear unit-GN3 was switched on. A very slow, yet progressive voltage reduction on the 230 KV buses was observed an hour ahead of the time.

At 15:27:43, 500 KV line between North Gila and Hassayampa was mistakenly opened at North Gila end. After 21 s, the line was disconnected at the other end. The line outage resulted in an overload on Path 45 at 15:28:04. The following were CFE BAAL exceeding and IID ACE excursion at 15:28:13 and 15:28:21, respectively.

At 15:28:24, two 230/92 KV and one 161/92 KV transformer at Coachella Valley were relay opened due to emergency overloading. The cascading outages instantly caused voltage drops on certain 161 and 92 KV substations within IID and WALC areas.

At 15:32:22, two 161 KV lines associated with IID Niland substation was opened. The units at North Gila and Coachella Valley were relay tripped. The events dropped 100 MW generations and caused 161 and 92 KV bus voltages down to a danger level.

At 15:36:59 and 15:37:59, Pilot Knob-Yucca 161 KV line and Yucca 161/92 KV transformer were sequentially out of service. The events momentarily dropped SONGS 230 KV main bus voltage to 217.9 KV at **15:38:04**, below the emergency limit of 218 KV.

At 15:38:07, the line of Cenroa-CCM230 was out. The 230 KV line between Imperial Valley and El Centro was opened at Imperial Valley end.

At 15:38:19, Imperial Valley-El Centro 230 KV line became disconnected at both ends. 75 MW generations of two units at El Centro were dropped.

At 15:38:24, the SONGS separation scheme was initiated. **At 15:38:26**, Imperial Valley-North Gila 500 KV line was opened end at North Gila substation.

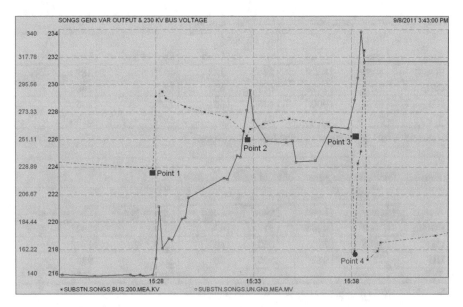

Fig. 2.11 SONGS unit reactive generation output and 230 KV main bus voltage PI trend. (Source: Peak Reliability)

Following these two events, the system of SDGE, plus portions of IID and CFE systems, collapsed within 12 s.

To illustrate the impact of these outages on development of the blackout in 11 min (i.e., **15:27:43~15:38:41**), we use a plot of PI trend on SONGS' 230 KV bus voltages and GN3 unit Mvar output in Fig. 2.11.

The PI plot clearly indicates:

- 230 KV bus voltage sharply changed at points 1, 2, 3, and 4. For each turning point, Var demand was increased, while the reactive reserve margin at SONGS 230 KV bus was decreased. We'll make use of VSAT to quantitatively explain the point later.
- Point 1 timed at 15:27:43 corresponds to North Gila-Hassayampa 500KV line outage.
- Point 2 is stamped on 15:32:22. The trigger events were two 161KV line outages and loss of two units at Niland and Coachella Valley.
- Point 3 is around 15:38:00, when the line of Imperial Valley-El Centro was fully disconnected and two units of El Centro were out of service. 75 MW generations were lost.
- Point 4 is stamped on 15:38:24 when separation of SONGS substation took place. Starting from this point, SDGE system collapsed in 15 s.

To conclude, SONGS 230 KV bus voltage is a key indicator on development of the blackout disturbances. The instability of SONGS 230 KV bus voltage triggered

Table 2.2 Estimated area generations and loads (unit: MW)

Area	At 15:23:53		At 15:38:55		Loss of gen	Loss of load
	Gen	Load	Gen	Load		
SDGE	4052[a]	4008	34	138	4018	3870
CFE	2164	2215	647	654	1517	1461
IID	681	900	446	65	235	835
Σ					**5770**	**6166**

[a]Include SONGS Generations of 2254 MW (SCE Shares)

Table 2.3 Telemetry area generations and loads (unit: MW)

Area	At 15:23:50		At 15:38:50		Loss of gen	Loss of load
	Gen	Load	Gen	Load		
SDGE	4090[a]	4263	517	436	3573	3827
CFE	1965	2205	660	883	1305	1322
IID	721	952	506	26	215	926
Σ					**5093**	**6075**

[a]Include SONGS Generation of 2284 MW (SCE Shares)

the autoseparation scheme and nuclear unit shutdown. The SDG&E system collapsed in 12 s after loss of two SONGS units.

2.5.2.2   State Estimation Solution

The RC's EMS Real-Time Network Analysis (RTNET, i.e., SE) was configured to execute for every minute, yet the SE snapshot case was archived for every 5 min. For the blackout event, fortunately the autosave SE case captured the pre-disturbance snapshot at 15:23:53 and the post-blackout snapshot at 15:38:55, immediately after the SDG&E system collapsed.

At 15:38:55, SE accurately detected the abnormal islanding condition and issued an alarm properly. SE reported three islands split in the WECC footprint, i.e., the main island, CFE island with 52 substations, and SDG&E island with 147 substations plus SONGS substation. Both the SDG&E and CFE islands became diverged on the basecase solution. As the RC's primary goal is to monitor and coordinate the reliability of Bulk Electric System (BES), typically the distribution stations under 100 KV level are simply modeled as equivalent units and loads tied to a higher KV substation.

To measure the SE solution accuracy under the major disturbance, the solved MW generations and MW loads before and after the blackout are compared in Table 2.2. The differences between two results are counted as loss of generations or loss of loads for the impacted areas. The underlying Table 2.2 accounts for SE solved losses of area generations and area loads.

Without loss of generality, historical SCADA AGC measurements received from entities via ICCP were pulled up from PI and summarized in Table 2.3. Note the

AGC measurements in use were sent from BAs at a 10 s interval, which provide a benchmark for evaluating the SE solutions.

From the above tables, the total losses of MW loads from the WSM SE solutions match closely with the values computed from real-time measurements. On the other hand, it appears SE over estimated MW generation loss by 677 MW or 13.3%, largely on SDG&E and CFE areas. The gap is likely due to the topology errors and bad SCADA measurements.

2.5.2.3 RTCA Results

The RTCA was configured to execute per 5 min. Unlike the SE, the RTCA snapshot is not automatically archived. Usually the RCSO have two means to look into historical RTCA results: PI system and RC Workbook. As an in-house tool, RC Workbook allows users to retrieve the last run real-time CA and study CA solutions and exhibit the results with flexible and user-friendly interfaces. Beyond these features, the tool is capable of storing historical harmful contingencies or contingency violations of RTCA for ad hoc diagnosis. What the RC Workbook is missing is the records of unsolved and islanding contingencies. The PI system stores a high-level RTCA results, such as total unsolved or islanding or harmful contingency numbers, as well as RTCA task execution time and status, etc.

During the blackout disturbance, as expected, the RTCA also perceived multiple islands by loading SE snapshot/basecase. The RTCA alarms of so-called Partial Valid showed up, while the areas of SDG&E, IID, and CFE were isolated and collapsed at 15:38:39. The alarms were triggered by two non-solvable islands (i.e., SDG&E and CFE). Unfortunately, the full-scale RTCA snapshots under the disturbance were not manually captured. Nevertheless, the real-time SE snapshots and PI trend enable users to replicate RTCA results by a post-disturbance study. The variant of unsolved contingencies between the pre-disturbance and post-blackout is plotted in Fig. 2.12 as follows.

In Fig. 2.12, the, solid line symbolizes the unsolved contingency number-NUNSOLV; dot line stands for RTCA completion status-SET_STATES, i.e., 0 for "Completed" or Valid; and 1 for "Partial Valid" (at least one island unsolved). Three checkpoints are marked on the screen. Here are the observations:

- **At 15:25:34** (Point 1), RTCA reported four unsolved contingencies.
- **At 15:35:26** (Point 2), RTCA reported 11 contingencies not solved. The basecase is under the disturbances.
- **At 15:40:39** (Point 3), RTCA unsolved contingencies number dropped to four.
- **At 15:45** (Point 4), RTCA turned to "Partial Valid" solution status, because two collapsed islands could not converge on the basecase.

Note that there was a couples of minutes delay for RTCA to catch the event sequence of SCADA, because of the intervals of RTNET and RTCA execution. With that being said, RTCA on Point 2 can only take the network outages that actually occurred prior to 15:30.

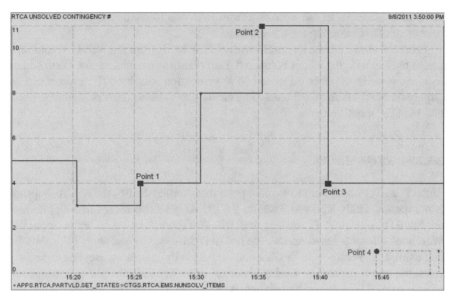

Fig. 2.12 PI trend of RTCA unsolved contingency no. variations. (Source: Peak Reliability)

An unsolved contingency signals the system may be unsecured when the contingency actually takes place. As such, a sudden rising of RTCA unsolved contingencies at 15:35 warned Western Interconnection was getting into an insecure system situation. To uncover which contingencies not solved then, a study power flow and CA run need be performed.

2.5.2.4   Post-disturbance Study Results

To begin with the pre-disturbance SE snapshot at 15:23:53, the outages to study can be applied in multiple steps. See Table 2.4 in details.

The study power flow and contingency analysis (CA) are performed against the outages on each step. When the basecase fails to solve with the specific outage, the outage is considered a critical event for the blackout progress. To proceed with the remaining outages, the corrective actions need be taken to make the study power flow solve.

The basecase power flow and CA solution are obtained under Step 1 and 2 outages. Removing North Gila 500KV line does not cause abnormal voltages but incurs MVA limit violations on Path 44 and emergency overloads on the transformers of Coachella. After the overloaded transformers are removed, the 161 KV bus voltages of the border substations between IID and WALC drop significantly. The unsolved contingencies after Step 1 and 2 outages are displayed in Fig. 2.13.

Regarding the number of unsolved contingencies, study CA results under Step 2 roughly agree with the PI trend of RTCA solutions. Ten contingencies within SDGE

Table 2.4 Event outage sequences

Study step	Event time	Outage equipment
1	15:27:43	North Gila-Hassayampa 500KV line
2	15:28:24	Two 230/92 transformers and one 161/92 KV transformer at Coachella Valley Coachella Valley-Ramonz 230 KV line
3	15:32:25	Blythe-Niland 161 KV line Coachella Valley-Niland 161 KV line Unit-GT2 at Niland Unit-comic at Coachella Valley
4	15:36:59~15:38:09	Pilot knob-Yucca 161 KV line Pilot knob 161/92 KV transformer Cenroa-CCM230 230 KV line El Centro-Imperial Valley 230 KV line El Centro-pilot knob 161 KV line

ID	Contingency Description	ID	Contingency Description
PGA2X020	TBLMTN 230-115-60 KV T/F BN2	PGA2X020	TBLMTN 230-115-60 KV T/F BN2
SCE1L003	CONTRL-INYO24 #1 115 KV	SCE0U027	SONGS GEN 2
		SCE0U028	SONGS GEN 3
		SCE1L003	CONTRL-INYO24 #1 115 KV
N GILA-HASSYYAMPA 500 LINE OPENED		SCE24201	MIRAGE_RAMON_1230
		SCE2C019	DEVERS_MIRAGE_1230
		SDG2C007	IVALLY_CENROA2_2230
COACH HELLA VALLEY TRANSFORMERS + COACH HELLA-RAMONZ 230 LINE OUTAGES		SDG2C008	IVALLY_CENROA2_1230
		SDG2C027	IVALLY_ELCENTRO_1230
		SDG2C061	PALOMR ALL GEN
		SDG5L002	IMVALLY_N.GILA_1500

Fig. 2.13 CA study results: Unsolved contingencies under the outages at Steps 1 and 2. (Source: Peak Reliability)

and SCE become newly unsolved following the cascading outages at Coachella Valley. The outcome correctly indicates the system security is at high risk under the contingencies associated with the disturbance impacted areas. Unfortunately, several of CA-detected abnormal contingencies, for instance, the events defined as SDG2C027, SDG5L002, SCE0U27, and SCE0U28 contingencies, occurred in reality.

The basecase power flow still solves with the 161 KV line outages defined in Step 3. It becomes diverged after two units are tripped, due to a shortage of Var and voltage support on the local 161KV network. It appears the outages of Step 3 worsen the system voltage instability situation. To proceed with the study analysis, two adjustments on the basecase are made:

ID	Contingency Description	ID	Contingency Description
SDG5L001	* IMVALLY_MIGUEL_1500	IID1L004	ELCENTRO_NILAND_1161
SDG5X002	* MIGUEL T80 500-230	PGA2X020	TBLMTN 230-115-60 KV T/F BN2
SDG5X003	* MIGUEL T81 500-230	PGE2L017	CARLTON-SHERWOOD 230 KV
CFE0U004	CCM230 UNIT STG	SCE0U027	SONGS GEN 2
IID1L003	AVENUE 58 - EL CENTRO 161	SCE0U028	SONGS GEN 3
IID1L004	ELCENTRO_NILAND_1161		
IID2X001	ELCENTRO T230-160	SCE24201	MIRAGE_RAMON_1230
PGA2X020	TBLMTN 230-115-60 KV T/F BN2	SDG2C027	IVALLY_ELCENTRO_1230
SCE0U027	SONGS GEN 2	SDG5L002	IMVALLY_N.GILA_1500
SCE0U028	SONGS GEN 3		
SCE24201	MIRAGE_RAMON_1230	STEP-3 OUTAGES + 179 MW LOAD SHEDDING AT COACHELL SWITCH ST	
SCE2C019	DEVERS_MIRAGE_1230		
· SCE5L004	DEVERS_VALLEY_1500	STEP-3 OUTAGES W/O UNIT TRIPPING	
SDG2C007	IVALLY_CENROA2_2230		
SDG2C008	IVALLY_CENROA2_1230		
SDG2C027	IVALLY_ELCENTRO_1230		
SDG2C047	ENCINA_SANLUS_2230		
SDG2C061	PALOMR ALL GEN		

Fig. 2.14 CA study results: Unsolved contingencies under the outages at Step 3. (Source: Peak Reliability)

1. Keep two units in service.
2. Shed 179 MW load at Coachella Switch Station.

Two sets of CA study results under Outage Step 3 and under Outage Step 3 plus the corrections are compared side by side on Fig. 2.14.

From the load impact sensitivity study, it appears a load shedding at IID 92 KV substation effectively improves the system security level under the contingencies. The risk level of the system voltage instability can be measured by the existence of abnormal bus voltages resulting from the outages. The solved low bus voltages under Steps 2 and 3 are summarized in Fig. 2.15.

The study results reveal the following findings:

1. WALC/IID 161 KV bus voltages tank after loss of Coachella transformers.
2. IID 92 KV bus voltages fall significantly after two 161KV lines removed. Most of those buses are located at the substations modeled with large equivalent loads. The voltage decaying will further incur inductive load demands.
3. Load shedding at IID Coachella Switching Station effectively recovers the falling bus voltages on the 92 KV network.

To verify further which outage is the last straw to collapse the SDG&E system, we continue to apply the outage events described at Step 4, on top of Step 3 base-case with load shedding. The power flow remains solved until El Centro-Imperil Valley 230 line is forced out of service.

ABNORMAL BUS VOLTAGES W STEP-2 OUTAGES

| Island | Area | Station | - Low Voltage - | | Voltage(pu) |
			KV	BS#	
1	WALC	BLYTHE	161	11862	0.888
1	WALC	BOUSE	161	11865	0.888
1	SCE	BLYTHECM	161	9587	0.888
1	APS	YUCCA	161	952	0.891
1	IID	RAMON_Z	92	3790	0.893
1	IID	PILOTKNO	161	3786	0.893
1	WALC	KNOB	161	11945	0.894
1	IID	AVENUE58	161	3753	0.894
1	WALC	DOME_TAP	161	11884	0.898
1	WALC	KOFA	161	11946	0.898

ABNORMAL BUS VOLTAGES W STEP-3 161KV LINE OUTAGES

| Island | Area | Station | - Low Voltage - | | Voltage(pu) |
			KV	BS#	
1	IID	AVENUE42	92	3752	0.647
1	IID	AVENUE58	92	3754	0.664
1	IID	COACH_SS	92	3759	0.668
1	IID	COACHELL	92	3757	0.672
1	IID	AVENUE58	161	3753	0.712
1	IID	RAMON_Z	92	3790	0.720
1	IID	COACHELL	161	3756	0.721
1	APS	YUCCA	69	953	0.822
1	WALC	GOLDMINE	161	11913	0.826
1	APS	YUCCA	161	952	0.835
1	IID	PILOTKNO	161	3786	0.837

ABNORMAL BUS VOLTAGES W STEP-3 OUTAGES AND 179 LOAD SHEDDING AT COACHELLA SWITCH STATION

| Island | Area | Station | - Low Voltage - | | Voltage(pu) |
			KV	BS#	
1	IID	PILOTKNO	161	3786	0.888
1	WALC	DOME_TAP	161	11884	0.892
1	WALC	BOUSE	161	11865	0.896
1	WALC	WLTNMK	161	12014	0.897
1	IID	PILOTKNO	92	3787	0.898
1	WALC	KOFA	161	11946	0.899

Fig. 2.15 Abnormal bus voltage summary under three scenarios. (Source: Peak Reliability)

In reality, this line was overloaded and relay tripped. Following the line tripping, SONGS 230 KV bus voltage fell below 218 KV, which triggered the autoseparation scheme and resulted in the loss of 2300 MW generations at SONGS. The El Centro-Imperil 230 line was one of critical contingencies (outages) causing the collapse. A sequence of cascading outages that developed the blackout is described in Fig. 2.16. It's consistent with the FERC and NERC joint investigation report as shown on Fig. 2.17.

Between 15:36:59 and 15:38:07, multiple lines were tripped by the relay, as the real-time SCADA measurements indicate the overloads. One important fact to mention here is these overloads resulted from the outage events at 15:32:26, and then the overcurrent relays tripped the overloaded 230 kV lines a few minutes later. Historical line flow measurements and PI trend on El Centro-Imperial Valley 230 KV line loading are provided in Table 2.5 and Fig. 2.18, respectively.

Should fast-responsive load shedding be able to prevent the overloads of the lines?

The answer could be yes. As previously stated, the AC power flow fails on the low 92KV bus voltages at IID area after all of Step 3 outages are applied.

Regardless of reactive power constraints, let us run the DC flow only for two scenarios on Step 3 outages: (1) no corrective action and (2) shed 179 MW loads at Coachella Switch Station. It's noted the solved MW flow on El Centro-Imperil Valley 230 KV line is significantly mitigated from 205 MW (with no load shedding) to 84 MW (with 179 MW load shedding). The DC flow results confirm that if there were a undervoltage load shedding scheme in place, and timely initiated, the critical line tripping might not happen.

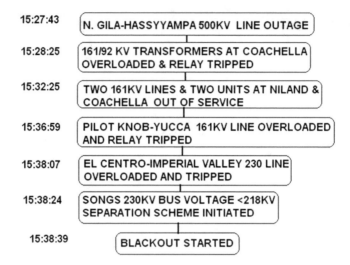

Fig. 2.16 Sequence of the cascading outages from Peak RC WSM-EMS

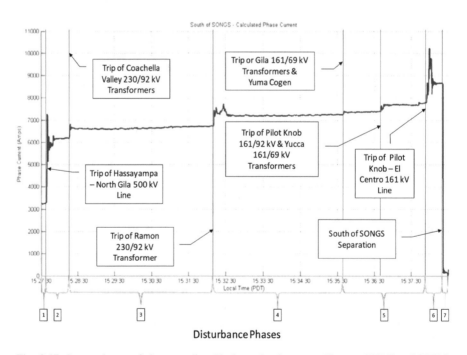

Fig. 2.17 Seven phases of the cascading blackout development. (Source: FERC and NERC Report)

Table 2.5 SOL exceedances on relay tripped lines from the CA study

Outage time	Line device ID	MW	MVAR	MVA	NORM limit	EMER limit
15:36:59	PILO_YUCC_1161	99	100	141	135	149
15:38:07	ELCE_IVAL_1230	−235	−126	267	239	258
15:38:09	ELCE_PILO_1161	−168	172	241	165	182

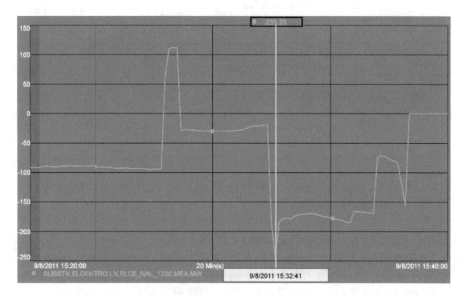

Fig. 2.18 El Centro-Imperial Valley 230 KV line flow in MW. (Source: Peak Reliability)

2.5.2.5   Post-disturbance Analysis by VSAT Simulations

To verify the system voltage stability margin under the disturbances, we used PowerTech VSAT to conduct post-disturbance analysis. In accordance with California Bulk System Map in Fig. 2.10, a study scenario of power transfer and voltage stability margin is created as:

- Composite source group: APS, SCE (with SONGS units), and CFE
- Load group: SDG&E
- Transfer interface: Path 44
- Monitoring bus voltage: SONGS 230 KV bus
- Monitoring criteria is 0.9 pu for basecase and 0.85 pu under contingency
- MW transfer target: 500 MW
- Unit governor response and AGC actions: Default

Starting from the SE snapshot at 15:23:53, the following conditions are assessed by VSAT:

1. Basecase + North Gila 500KV line outage at 15:27:43.
2. Condition 1 + cascading outages at 15:28:24.

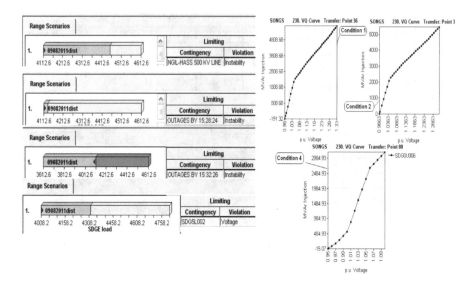

Fig. 2.19 VSAT calculated power transfer limits and VQ curves on September 8, 2011 disturbance

3. Condition 2 + cascading outages at 15:32:26.
4. Condition 3 + load shedding (179 MW) at Coachella Switch Station. One hundred of SDGE, IID, and SCE contingencies are applied to the stability analysis on this condition. The transfer limits calculated by VSAT are compared with the simulation results on other conditions.

Without loss of the generality, we run VSAT voltage stability margin assessment tool to compute both the maximum power transfer on Path 44 (P-V curve) and the reactive power margins at SONGS 230 KV bus (VQ curve) against each of four-study conditions. The analysis results are summarized in Fig. 2.19, respectively. Note the VQ curve is not plotted out on Condition 3, where the system is not secured with the outages.

The VSAT study results imply:

- The transfer capability of Path 44 is largely affected by the cascading events at 15:28:24. The basecase could not be secured against the outages at 15:32:26. The results re-confirm EMS application study results.
- Reactive reserve at SONGS 230 KV bus declines dramatically. There is no VAR margin left after 15:28:24 outages applied. This implies a voltage instability situation is resulted from the outages.
- Load shedding at IID 92 KV load substation significantly improves both the transfer capability and Q-reserve margin at SONGS 230 KV bus.
- SDG5L002, a limiting contingency for Condition 4, is identified as a CA unsolved case in Table 2.6.

Table 2.6 Bus participation factor of Mode 1

```
                           Right  Eigenvectors
----------------------------------------------------------------------------------
No.   Bus No., Name                   Area No.Name  Part.Fac. Voltage (Pu)
----------------------------------------------------------------------------------
  1   22536   NGILA          500.   2    APS      1.00000,    0.7032
  2   84836   NGILA          69.0   2    APS      0.99935,    0.6697
  3   19049   GILA           69.0  34    WALC     0.98316,    0.6526
  4   84841   SONORA         69.0  34    WALC     0.96817,    0.6618
  5   19050   GILA          161.   34    WALC     0.89582,    0.6871
  6   19063   WLTNMK        161.   34    WALC     0.89257,    0.6835
  7   19070   DOME_TAP      161.   34    WALC     0.86930,    0.6859
  8   19100   KOFA          161.   34    WALC     0.66230,    0.7331
  9   19051   KNOB          161.   34    WALC     0.56411,    0.7561
 10    4939   YUCCA          69.0   2    APS      0.51641,    0.7630
 11   19105   GOLDMINE      161.   34    WALC     0.50780,    0.7620
 12   21059   PILOTKNO      161.   12    IID      0.47667,    0.7801
```

- Modal analysis is applied on the instability point of Condition 4. The smallest Eigenvalue λ associated with one critical mode is 0.017461. The relative bus participation factors on right eigenvectors are computed in Table 2.6. The buses in the table are most relative locations to the instability mode [5].

It's noted that the VSAT falls into an insecure basecase after one of SONGS nuclear units is removed, no matter which disturbance condition is applied. This confirms loss of both the North Gila 500 KV line, and one nuclear unit of SONGS could cause the system to be instable or even collapse.

2.5.3   Discussions

The post-disturbance steady-state analysis and computations were conducted by RTCA and VSAT from the full WSM real-time SE snapshot cases. The following observations and facts are derived from the study results:

- The blackout started with a sequence of cascading failures and ended up in a voltage collapse. The impacted systems experienced about 11 min of voltage instability phenomena.
- 161 KV and 92 KV low-voltage network across the borders of WALC, IID, and APS is the bottleneck of system voltage stability under the a major 500 kV line outage. To secure the voltage stability margins at the critical buses such as SONGS, fast-responsive undervoltage load shedding scheme should be appropriately installed at IID 92 KV substations.
- Implementation of online voltage security monitoring system and cascading outage evaluation tool is highly important for RCSOs to be aware of the security situation and respond to the disturbance. Voltage stability analysis shall be incorporated into next-day study and real-time study process as necessary to assess SOL and IROL in Western Interconnections.

- In addition to the established WECC paths within the areas affected by the black-out, new cut planes to measure inter-area power transfer stress need be intro-duced and implemented for real-time monitoring. The limit calculation of the new cut planes should take voltage stability margins into account.
- The entire SDG&E and part of the IID/CFE areas collapsed after the SONGS separation scheme was initiated by a voltage drop at the 230 KV bus. It is crucial to compute the reactive security margin of SONGS 230 KV bus against 218 KV by a real-time reactive monitoring group. The correlation between the separation action and the developed voltage collapse could not be fully revealed by steady-state analysis. A time domain stability simulation is required on this course.
- IID 92 KV network suffered voltage instability with loss of a few 161KV lines and small units. There was no adequate reactive reserve under a stressful and disturbed condition. Both PI trend and study analysis indicate shortage of VAR reserves on IID 161 or 92 KV networks impacted the voltage stability margin of SONGS 230 KV buses. Hence IID reactive reserve groups need be accurately defined and online monitored.
- The definition of the RTCA contingency list for IID and its neighborhood need be modified to include those sensitive sub-100 kV facilities, such as Coachella 230/92 KV and 161/92 KV transformers et al. A thorough VSAT offline study shall be utilized to determine the voltage weak buses and associated voltage-sensitive devices.
- Some critical local protection schemes, such as SONGS separation plan, need be modeled as RAS/RAP in WSM model for real-time CA screening and processing.
- The accuracy of equivalent load modeling on WALC, IID, and APS et al. areas need be improved to ensure a more accurate RTCA solution on the branch flows under the contingency.

2.5.4   Lessons Learned About RTCA

From the Pacific Southwest blackout, we identified a few major modeling gaps that negatively impacted the credibility of RTCA reported violations:

First, disabled real-time monitoring for all sub-100 kV elements, including the sen-sitive ones, such as Coachella 230/92 KV and 161/92 KV transformers, because we didn't have right understanding on impact of sub-100 kV networks to BES. Otherwise, the RTCA would have detected the major risk a couples of hours ahead, i.e., Coachella 230/92 kV transformer exceeds the relay tripping limit under [N-1] 500 kV line outage contingency.

Second, the EMS tool suite had software features to monitor bus voltage angle pair exceedances in both SE and RTCA. However, before the blackout occurred, none of the bus voltage angle pairs were modeled in the WSM based upon line reclose relay parameter settings and/or the calculated wide-area angle separation limits.

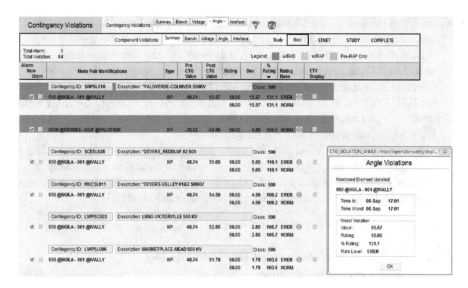

Fig. 2.20 CA simulation results with phase-angle pair limit exceedances. (Source: Peak Reliability)

Otherwise, the RCSOs might obtain adequate situational awareness on the phase-angle pair limit exceedances from RTCA ahead of time. Figure 2.20 is a screen-shot for RTCA solutions reproducing the disturbance on September 8, 2011.

Third, there are hundreds of RAS identified by WECC RASRS and required for modeling and monitoring them in EMS and/or other real-time tools, but only a few of those RAS were modeled active in RTCA for real-time assessment. Missing of bulk RAS models made RTCA violation results less credible and inaccurate and hence misled the RCSOs and engineers to spend more time vali-dating false contingency violations and distracted them from dealing with true cascading failure risk effectively. In addition, to enable the RCSOs obtaining RAS real-time operation awareness from EMS vendor displays is not possible because the RTCA RAS displays do not show relevant topology and RAS logic diagrams and separate a RAS from a group of the associated contingencies. One can't quickly verify RTCA RAS solution results to be correct or not by reviewing and checking the vendor-designed EMS displays.

Fourth, there are thousands of network elements, i.e., transmission lines, trans-formers, and buses that do not have appropriate ratings/limits modeled in the WSM before 2014, because those are either the ones with dynamic rating in use for operation or the limit setting is different between Peak RC and the owner entities. None of relay tripping limits were received from entities and modeled in the WSM for limit exceedance validation. There was no term in the SOL meth-odologies to recommend a threshold, e.g., 125% emergency thermal limit for potential IROL screening test guidance.

Fifth, the RTCA lacks the capability to automatically evaluate [N-1-1] and [N-x] cascading scenarios and report the cascading failure cases. Also there is no predictive or look ahead RTCA to identify system security risk a few hours ahead, so that RCSO has much more time to validate the risky contingencies and develop effective mitigation plans as needed. Due to nature of steady-state power flow, the RTCA is unable to model transient and dynamic or nomogram-driven RAS accurately; it's necessary to leverage other real-time tools, i.e., online TSAT and RT-VSA to complement RTCA's limitation.

Finally, loss of massive wind and PV solar generations could happen on either a local power plant or a large area/region, i.e., weather condition diminishing wind or solar generators of many plants. How to model and monitor those wide-area contingencies in RTCA remains an open question in industry.

2.5.5 Software Enhancement Implementation and Recommendations

2.5.5.1 Overview of the Current Approaches to Identify Cascading Risks

In the RC control centers, cascading risk assessment process is implemented per guidance of the NERC requirements:

- The NERC Operating Committee (OC) highlighted three major gaps/risks in terms of grid reliability operations: outage coordination, governor frequency response, and situational awareness.
- NERC Project 2009–02: Real-Time Monitoring and Analyses Capabilities defines the requirements for RTCA models/performance expectations.
- NERC TOP-001-3, R13: Each transmission operator shall perform a R-T assessment at least once every 30 min.

    Best practice in Peak RC and industry includes:

- Perform more operations planning studies.
- Implement near real-time assessments using voltage and transient stability tools.
- Manually run cascading outage testing using a post-contingency threshold such as 125%.

    - Assume equipment trips at this level.
    - Perform additional studies to determine where tripping stops.

- Use EMS RTCA application with RAS models as major real-time assessment tool for situational awareness.
- Conduct an evaluation of system conditions using real-time data to assess existing (pre-contingency) and potential (post-contingency) operating conditions.

    - The assessment shall reflect inputs including, but not limited to, load, generation output levels, known Protection System and Special Protection System status or degradation, Transmission NERC Operating Committee Response to NERC Standards Committee/RISC Triage of IEPR Gaps 2 outages, generator outages, interchange, Facility Ratings, and identified phase angle and equipment limitations.

- Real-time assessment may be provided through internal systems or through contracted services, such as Peak HAA model.
- Focus on essential capabilities: Monitoring and analysis. Effective R-T SA is supported by monitoring and analysis that:
  - Is performed with sufficient frequency.
  - Provides awareness of information quality.
  - Provides indication when processes are not operating normally.
- Take corrective actions in case of bad data quality or abnormal process status.
- Support inclusive RAS modeling capabilities:
  - Prevent RAS backfire undervoltage collapses. Be able to model path limit nomograms.
  - Support user-defined calculation for evaluation of arming/triggering points as needed.
- Calculate reactive reserve group impact by the worst post-contingency condition.

To conclude, as of today, significant manual efforts are needed to perform cascading risk assessment in daily operation at RC control centers.

2.5.5.2   Future RTCA Enhancements for Consideration

The existing RTCA need be improved in software capabilities and RAS model quality. The software enhancement recommendations are documented in Table 2.7.

In particular, there are specific requirements from user usability perspective:

- The user shall have the capability to enable a certain subset of contingencies for cascading. There shall also be a global flag to enable/disable this functionality. The number of CA reruns resulting from cascading shall be a configurable parameter. Simulation of cascading will stop when there are no more branch overloads exceeding the trip limit or when the maximum rerun threshold is reached. Displays shall also be updated to show which contingencies caused cascading.
- Support needs to be added to contingency analysis to support simulation of cascading contingencies. If a contingency violation exceeds the trip limit for a branch, CA should automatically remove the overloaded component and rerun. The trip limit shall be modeled as a new limit value on the branch records (the existing trip limit setting from dynamic ratings will be used to populate this during runtime).

Table 2.7 CA cascading contingency enhancements

EMS/RTCA enhancements	Descriptions
Add tripping setting limit in RTDYN (dynamic ratings application) transferrable to NETMOM (network model) only to facilitate RTCA [N-1-1] (cascading outage) analysis	**Peak reliability** to add the trip setting limits (determined by protective relay tripping parameters) in dynamic rating and transferable to NETMOM. When the trip rating is met, the line or transformer equipment will be forced outage
Enhance RTCA RAS logic to allow modeling of all required RAS and non-RAS protection schemes	**Enhancement requirements** Add tripping limit data field in NETMOM and allow user to enter the limit in RTNET limit displays on the fly Tripping limit violation will be reported in RTCA solution as the same as other limits RTCA will perform [N-1-1] post-contingency cascading analysis after a tripping limit violation is detected RAS logic need be modified to support RTCA [N-1-1] analysis
Add logic in RTCA to perform N-1-1 cascading evaluation with RAS models	**Use case and alternative solution** The new tripping limit parameters will be retrieved into RC workbook/RTCA summary view for each branch violation record for limit exceedance alarming. It's beneficial for both RTCA compliance and reliability monitoring Alternative solution is to modify RC Workbook for adding the capability to enter tripping limit
RAS display enhancements, i.e., output cascading contingency ID, cascading failure level: controlled or non-controlled, violation severity level per NERC standard, system impact defined by total gen/load loss amounts, a list of substations /lines lost from cascading outages, and RAS/SPS actions triggered by the cascading contingencies	**Benefit** The full CA [N-1-1] logic implemented with tripping limits will enable RC to evaluate potential cascading scenarios in RTCA and STCA applications Reduce human workloads, minimize human errors, and increase responsive and corrective actions against a system event

- Logic will have to be added to remove the elements with security violations caused by the contingency and to then rerun power flow for that contingency. These removed elements will be added to the outage list for the contingency and marked as a cascading element. Cascading contingencies will be noted on the contingency summary display as well. Violations are cascaded when they are in excess of the 15-min or load shedding limit. Cascaded elements become part of the contingency definition, and the contingency topology is reevaluated using the decoupled power flow. Further violations are cascaded if they violate the 15-min limit. This process iterates until no new contingencies are created, or the limit on the number of cascades has been reached.

- Display changes required for cascade are similar to that of RAS processing. There is a flag to enable CASCADE processing in a control field to enter the maximum number of cascading iterations (MITERCASC_ITEMS). Cascading flag can also be enabled/disabled at contingency level using "Enable Cascading" option in the CTG definition display. The contingency elements added during cascade processing while running CA are added to the definition of the particular contingency with the CASCADE flag set to true.

2.5.5.3   Considerations for [N-1-1] RTCA Implementation

Correctness of predictive cascading analysis depends on completeness and accuracy of RAS/SPS and impactful non-RAS protection schemes modeled in real-time assessment and monitoring (primary) and planning study (secondary) tools. The credibility of RAS models is affected by

- Vendor RAS modeling and screening software capabilities.
- Understanding on what other "non-RAS" automatic actions exist on the system.
- Modeling coordination to ensure appropriate levels of modeling exist.
- Real-time RTCA tool to eventually tell us when the system is at risk of cascading automatically and timely.
- A full network representation enabling RAS/SPS and relays to be modeled correctly without compromising due to network model reduction.
- Peer review on RAS modeling changes is needed to ensure the model work quality.
- Validating RTCA cascading analysis results through collaborative efforts around industry, including but not limited with EMS vendors, operating entities, RCs, and WECC RASRS, whoever are needed to address the risk of cascading outages.

2.6   Predictive RTCA Implementation

Peak RC implemented the real-time version of the Study Multi-Time Network Analysis (SMTNET) application called RMTNET. It performs a look-ahead security analysis for multiple future time points. RMTNET uses a load profile, interchange schedules, generation schedule, and outages for each of the future time points of study. RMTNET is configured to run every 15 min and performs studies for the current hour and 2 h in future. RMTNET copies NETMOM from last valid RTNET solution and CTGS, RAS, and DPF databases from RTCA. Any changes to RTCA are clone.

The following figure illustrates the architecture and data flow diagram (Fig. 2.21).

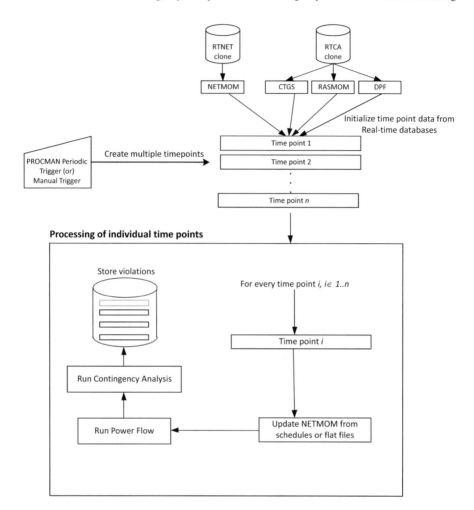

Fig. 2.21 RMTNET system architecture and data flow diagram. (Source: Peak RC)

2.6.1   RMTNET External Data Files

From the control options in Fig. 2.22, the user can enable/disable the option to read unit commitment (UCFILE), load forecast (LFFILE), net interchange (NIFILE), equipment in service (ISFILE), out of service (OSFILE), and path (interface) de-rate (IFFILE) information from text files, depending on data availability. To enable any particular file interface, check corresponding box, then the file name text field gets enabled automatically. Enter the input file name which contains the related data. Once the checkbox is enabled, RMTNET automatically reads the data from text file during next execution. Click on "View" button to review the input data on iGrid displays.

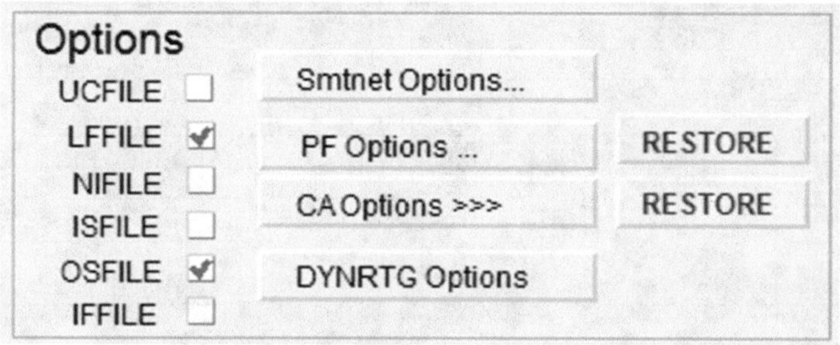

Fig. 2.22 RMTNET input file options display

2.6.2   Viewing RMTNET Results

RMTNET runs every 15 min and performs studies for next 2 h and stores the base-case and contingency violations. To view the violations results, one can navigate the displays shown on Fig. 2.23

The violations directory contains the total number of basecase and contingency violations for each time point and total number of violations in each category (branch, voltage, angle, interface, unsolved, etc.). There is a color coding of the violations.

- Click on the button under "Total" column, to see all basecase violations for any particular time point. A particular category of violations like branch, voltage, angle, or interface violations can be viewed by clicking on button under the category for any given time point.

Additional information about a particular violation can be accessed by clicking on information icon (🛈) next to violation to call up the display shown on Fig. 2.24.

- A particular category of violations can be viewed under different tabs of summary display.
- Click on directory (🗂) button on top right corner, to go back to violations directory display.

2.6.3   Input File Selection Options

User can enable/disable the option to utilize load forecast, generation profile, outages, and net interchange schedules data from text files. Below is the description of each available flag:

- UCFILE – Unit commitment data
- LFFILE – Load forecast data

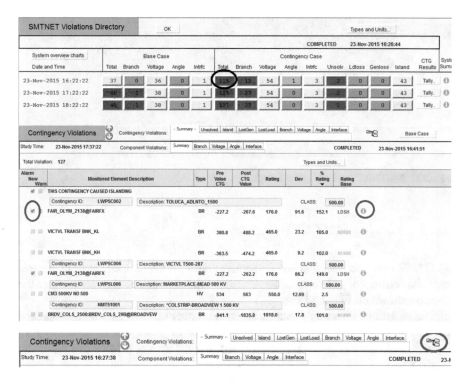

Fig. 2.23 SMTNET/RMTNET violation results displays

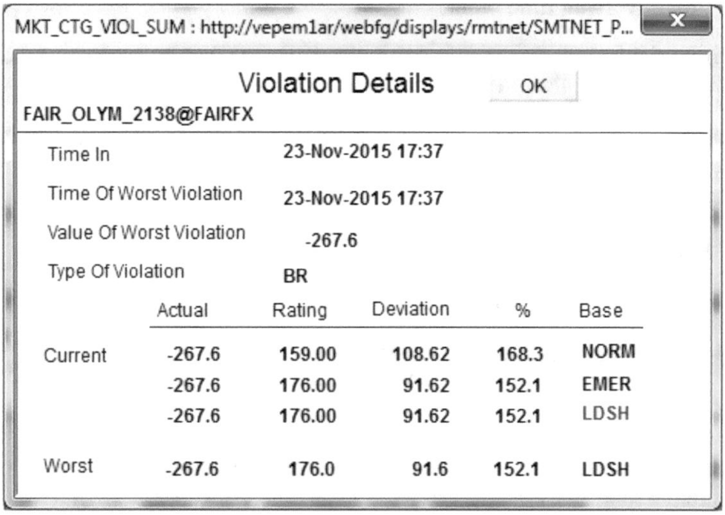

Fig. 2.24 RMTNET/SMTNET Violation Details display

- NIFILE – Area net interchange data
- OSFILE – Outage schedule data
- ISFILE – In-service data

If any of this data is available in text file format and like to use it in multi-time-point study, then user must check the corresponding flag.

2.6.4   Workflow for a Multi-Time-Point Study

1. If SMTNET task is not running, start SMTNET process.
2. Restore PF options by clicking on "RESTORE" button next to PF Options.
3. Restore CA options by clicking on "RESTORE" button next to CA Options.
4. Check "Enable to Run" box only for PWRFLOW task – this is to ensure that base case power flow is valid for all time points.
5. Check ISFILE, OSFILE, and/or any other input file flags depending on available time dependent data.
6. Retrieve NETMOM from autosaved RTNET savecase. In Windows Explorer, navigate to RTNET savecase directory on current production server (e.g., \\vepem1ar\savecases). Note the title of the RTNET auto savecase that is to be used as basecase for multi-time-point studies. (e.g., if the savecase name is "case_rtnet_netmom.autosave_08142014085129,"   then   auto-save_08142014085129 is the title of the case).

   - Click on "Data Retrieval" -> "Network Copy".
   - Enter "RTNET" in Retrieve Case from Application filed.
   - Enter "NETMOM" in "Casetype" field.
   - Enter the title (autosave_08142014085129) of the auto savecase in "Title" field.
   - Click on "File Open" icon to retrieve the case.
   - Upon successful retrieval, "CASE RETRIEVED" message will be displayed.

7. Retrieve Master Contingency List: Click on "CTGS" button to retrieve contingencies from Peak RC master savecase.
8. Upon successful completion, the following message will be displayed. Click OK on message.
9. Retrieve Master RAS List: Click on "RAS" button to retrieve RAS from Peak RC master savecase.
10. Upon successful completion, the following message will be displayed. Click OK on message.
11. Enter the first hour of the multi-time-point study in "From" filed.
12. Enter the end time of the multi-time-point study in "To" field.
13. Enter the time interval (in hours) between each time point in "Step" field.
14. Click on "Create Time Points." Upon successful creation of time points, Process Status will be changed to "INITIALIZED TPs." If there is any error, Process

Status will be changed to "TP INIT FAILED." Review SMTNET message log to see more details about the error. Sample error message looks as below:

15. Click on "Create and Link Default Cases" – It creates NETMOM, CTGS and RASMOM savecases (EMS_DEFAULT) with the data in the clone and link all the time points with EMS_DEFAULT cases.
16. SMTNET status changes to "Loading Cases," and after completion it is changed to "Listing Done."
17. Click on "Initialize Time Points."

2.7 Dynamic Rating Application Assures Correct Network Limits

2.7.1 Introduction

The Real-Time Dynamic Ratings (RTDYN) application provides the ability to manage facility limits of transmission lines, transformers, and zero-impedance elements. All limits for branches are centrally processed by the application including:

- Seasonal-based ratings
- Temperature-based ratings
- Telemetry ratings

The limit set includes a warning (WARN), normal (NORM), emergency (EMER), and load shed (LDSH). The warning is simply 95% of normal. The RTDYN application is a stand-alone application within EMS. It computes the appropriate limit set to be used for each branch defined and writes those limits to SCADA. The appropriate limit set is determined every 5 min.

Path and voltage limits are not addressed by RTDYN and will remain in SCADA without change.

The dynamic rating application reads data from the NETMOM and SCADA. It then writes the applicable limit back to SCADA. The data flow is shown on the right figure.

Once SCADA receives the limit from RTDYN, it provides the advanced applications RTNET/RTCA, QKNET, and STNET (Fig. 2.25).

2.7.2 Limit Source and Prioritization

The dynamic rating set can be populated by a number of data sources. RTDYN analyzes the different sources to determine which one is applicable at the time. The limit source options are:

– *Manual* – The limits are manually entered by an operator.

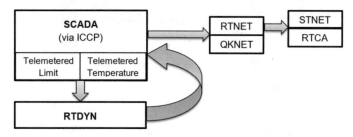

Fig. 2.25 Dynamic rating application data flow diagram

- *Telemetry* – The limits in MVA or Amps are received directly from the entity via ICCP. The source ICCP value can be sent in MVA or Amps.
- *Temperature* – The limits are determined by a weather station (WST) measuring the current ambient temperature for the element and then choosing the respective limit for the given temperature by using lookup table.
- *Season* – Seasonal-based limits are applied per NETMOM division.
- *Nominal* – The nominal limit is the active seasonal limit at the time the RTDYN DB was built, during the model build process.

RTDYN utilizes a prioritization system to determine the applicable limit set to be used. The same prioritization hierarchy is applied regardless of how many rating sources are available for the element. The types of limits are described in the picture below. The type of limit applied is color-coded as seen.

Manual limits will always be used if they are available. If a telemetered limit is available, without a manual entered limit for that element, it will be used and so on. For this reason if a telemetered limit via ICCP is lost, the next in priority will be utilized, which is temperature. If a temperature is not being received via ICCP, the season limit will be used. A seasonal limit is modeled for every element in the model, so this should be the default if manual, telemetry, and temperature are not available (Fig. 2.26).

2.7.3 Peak RC Custom Enhancement

Peak RC deployed the following software enhancements to the dynamic ratings application

- Expiry time on manual override limits – If the manual limit is to be overridden only temporarily and needs to be reverted back to its original limit after certain time, then the user can enter the time at which manual override limit to be expired. If user doesn't enter anything in expiry time, the manual override limit stays permanent until it is put back to normal.

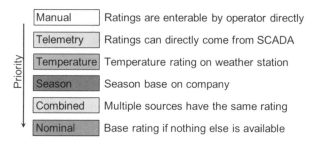

Fig. 2.26 Color code for RTDYN limit sources

- Maximum time duration fields for emergency and load shed limit – User can enter the times corresponding to emergency and load shed limits for each dynamic element and for each company. Note that these values are only for system operator awareness. Dynamic ratings application does not use in any processing.
- Tripping limits in dynamic ratings – User can enter the relay trip limits corresponding to each dynamic element. Note that these values are only for system operator awareness. Dynamic ratings application does not use in any processing.

2.7.4   RTDYN Alarms

Three RTDYN alarms can be received in alarm bucket 4 (medium level).

- Delta threshold
- Nominal threshold
- Limit overrides expired

The delta threshold alarm indicates that the device's rating has changed more than the defined threshold. The nominal threshold alarm indicates the rating computed by RTDYN is below the defined delta nominal rating threshold. The limit overrides expired alarm indicates that the manual limit override has been cleared for that device.

The RTDYN application automatically runs every 5 min and transfers the latest equipment limits to RTNET (State Estimator) and RTCA. The Last Periodic Processing Time is given on the top right-hand corner of the screen. If a new RTDYN limit set is needed quicker than the next run, it can be manually run using the process real-time button.

2.8 RAS Visualization for Real-Time Operation Awareness

To manage real-time monitoring specific impact of hundreds of RAS and similar protective schemes on thousands of contingencies in RTCA, the existing vendor EMS displays are insufficient for RSCOs to receive clear and correct awareness about the RAS operations on post-contingent conditions in a timely manner. A delay on real-time assessment of RAS operation impact to RTCA violations by RCSOs and ROEs could affect control room personnel from making timely corrections on stressed system operations. Since early 2017, Peak RC started to develop custom visualization displays in the SCADA and PI platform to enhance WI RAS operation awareness. This section presents an overview of the RAS visualization work and a few examples of RAS displays. Most figures are purposely made blurred to protect sensitive RAS information.

2.8.1 Overview

RAS displays in Peak RC EMS are categorized by BA and are currently under the process of information updates, including:

- Scheme names
- Scheme arming statuses (sent via ICCP)
- Document links
- PI RAS display availabilities

On EMS RAS displays, RCSO can obtain real-time RAS scheme arming statuses and related RAS documents. An example of EMS TOP RAS overview displays is presented on Fig. 2.27.

RAS PI displays are built exclusive for control room use. The primary objective of these displays is to help RCSOs and ROEs to better understand the RAS logic and to have a situational awareness of the RAS arming status. The displays are published in RAS tab of the WECC ProcessBook on EMS Workstation in reference to Fig. 2.28. Peak RC implemented the RAS displays for BC Hydro, BPA, PG&E, NWMT, PAC, PGAE, WASN, and WECC-1 et al.

Main informative items for operators on RAS displays:

- RAS document links
- RAS IDs in Peak RC RTCA and related contingency IDs
- Substations with modeled generation/load shedding actions and total armed amount
- Detailed arming information on which generator/load is armed for shedding
- RAS triggering logic diagrams

Benefits it brought to RC real-time operation:

- Becoming a communication platform between operators and engineers
- Acting as an alternative resource for validating RTCA results

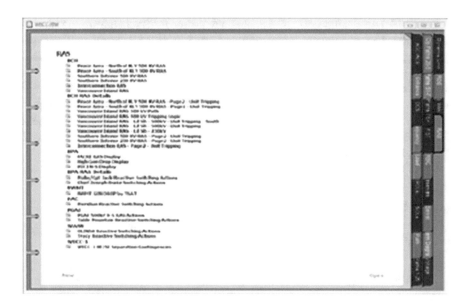

Fig. 2.27 EMS TOP RAS overview display sample

Fig. 2.28 RAS display tab in Peak RC PI ProcessBook

- Handling arming amount calculations cost-effectively
- Presenting RAS logics modeled in RTCA straightforwardly
- Updating display promptly
- Maintaining easily by a handy PI in the background

2.8.2   Conceptual Design of PI RAS Displays

A PI RAS display consists of a few sections: contingency IDs on the left, contingent element status (red/green – in service/out of service and RAS action devices (unit gen drop arming status, line switching status and load shedding, etc.) on the right. If analog measurements are input to RAS triggers or RAS protection actions, a PI trend could be added on the display for tracking the changes over the last 24 h or 2 h. The following two figures demonstrate the RAS PI display layout and settings (Fig. 2.29).

2.8.3   WECC-1 RAS Visualization Examples

2.8.3.1   Overview

The NE/SE separation scheme is one part of a larger group of RAS' collectively known as WECC-1 RAS and is designed to split the WECC system into two islands in a controlled process. Scheme monitors the line status of various 500 kV lines that comprise the Pacific AC Intertie and can be initiated by loss of four parallel lines north of Malin, loss of the COI, or loss of three parallel lines between COB and Tesla/Tracy.

The RAS uses multiple controllers at various utilities and has different levels of trip signal security based on where the initiating outage occurs. The controllers are listed below:

1. BPA – DM A & MR
2. PGAE – SF A and B & VV C and D
3. APS – FC A and B

The RAS requires two of three RAS signals to be received by APS to initiate – referred to as "2 of 3 voting," (RAS can also use "1 of 2 voting" if one of the three circuits is out of service). BPA sends one signal each from DM and MR directly to APS. PGAE sends four signals (one from each controller) to MD – at MD those signals are supervised by power rate relays – two signals leave MD on one communication path (thus considered one signal) to APS. APS then sends trip signals to all remaining separation points not tripped by BPA or PGAE directly.

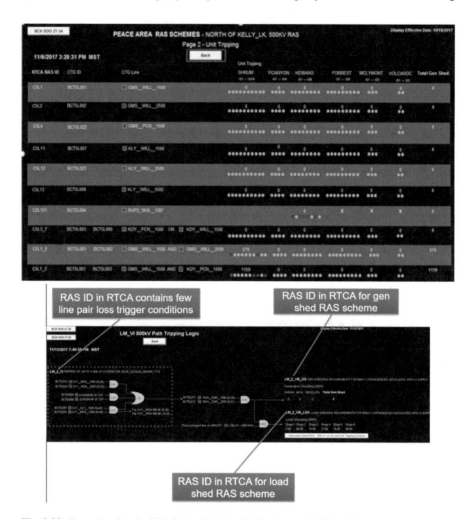

Fig. 2.29 Example of typical RAS visualization display layouts built in PI

2.8.3.2   Displays

The RAS display shows the possible combinations of line outages that can trigger the NE/SE separation scheme. Some 500 kV line sections (e.g., MR to CJ) have been combined into one logic point to reduce repetition.

Logic is built into triggering contingencies section and will indicate which contingency would cause the separation scheme to fire. The same logic is tied to display points that show when the system is an N-1 or N-2 contingency away from NE/SE separation.

Remedial actions are the same for any separation contingency and are shown at the top of the display. Real-time arming points and logic are used to highlight those actions that are active with corresponding MW amounts.

The underlying WECC-1 RAS displays are extensively used by RCSOs to monitor the RAS operation and system islanding conditions (Fig. 2.30).

2.9   RTCA Monitoring for [N-2] and MUC Contingencies

2.9.1   Introduction

Peak RC modeled 9000+ contingencies and 515 RAS/SPS/non-RAS schemes in RTCA, including 500+ N-2 or N-X/multiple outage contingencies (some are always credible, but most are conditionally credible). Currently, only a small portion of those N-2/MUC contingencies are enabled in RTCA for real-time screening at the present time.

Some of N-2/MUC CTGs involve transient RAS and UFLS models, which are unable to model in RTCA correctly. It takes more effort to validate impact of MUCs. In reality, N-2 or event MUC outages occurred sometimes. For instance,

- On December 2018 during N-2 (GS-BV), ATR tripped only 1 unit and on March 2019 tripped 2 units.
- On June 2018, two 230 kV lines relayed and the RAS tripped 1100 MW in BC Hydro area.
- On October 2017, N-1-1 scenario happened on the double lines connecting BC Hydro and US systems, resulting in an islanding condition.

Historically, major blackouts were triggered by N-2 or MUC or cascading outages, i.e., N-1-1 and so on. To evaluate impact of N-2/MUC outages to system operations, Peak RC conducted a multi-month testing in Test RTCA. The results are summarized in Table 2.8. It highlights importance of monitoring credible N-2 and MUC in RTCA.

From the testing results in Table 2.8, it's noticed major RTCA violation occurrences were repeated for a handful of MUCs, particularly,

- A few CTGs that separate BC Hydro and/or AESO from main system following RAS actions.
- 10 CTGs triggering Colstrip RAS operation
- Most of N-2/MUCs solved fine with proper RAS except for some MUCs under further validation.

2.9.2   Integrating RTCA and TSAT for [N-2]/MUC Monitoring

To be able to model and monitor those N-2 or MUCs in control room real-time assessment tools, Peak RC took actions as follows:

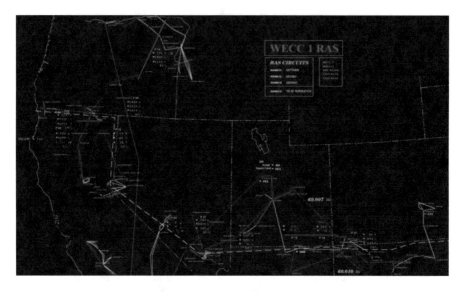

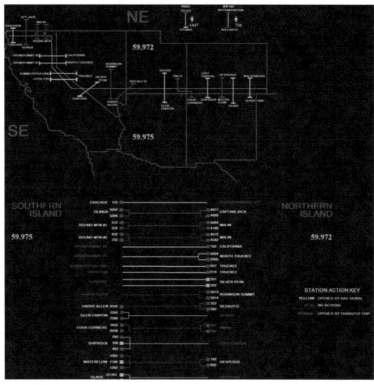

Fig. 2.30 Example of WECC-1 RAS displays in SCADA and PI

Table 2.8 N-2/3 CTG and MUC impact testing at Peak RC

	Prod RTCA w 10% MUCs enabled			TEST RTCA w all MUCs enabled		
	Total MUC records	Unique MUC CTG/ ME pairs	Unique MUC CTG	Total MUC records	Unique MUC CTG/ ME pairs	Unique MUC CTG
3/21/2019–4/16/2019						
Solved	15,317	354	22	92,027	624	71
Solved >125% exceedance	319	17	4	4,463	29	8
Unsolved	114	n/a	11	6,383	n/a	19
Island	543	n/a	4	10,249	n/a	16
2/8/2019–3/20/2019						
Solved	280,917	437	34	821,240	1,643	138
Solved > 125% exceedance	7	5	1	9,936	57	18
Unsolved	447	n/a	18	40,381	n/a	119
Island	227	n/a	3	48,325	n/a	16
12/19/2018–1/2/2019						
Solved	10,312	344	17	231,401	1,098	98
Solved > 125% exceedance	15	5	3	5,340	37	17
Unsolved	23	n/a	9	17,641	n/a	54
Island	44	n/a	2	13,561	n/a	9

- Implemented TSAT software enhancements to enable TSAT being a backup to RTCA:

  - Split basecase screening and transfer analysis.
  - Write out RAS gen drop unit IDs to a *.csv file.

- Created internal Filelink process to transfer TSAT results in *.csv file to the EMS/SCADA.
- Built SCADA TSAT output visualization displays and alarms.
- Modeled ATR transient RAS in the RTCA by using the TSAT solved RAS arming data as input.

  - Integration of the TSAT and the EMS/RTCA

The following diagam describes a framework to build a bridge between the TSAT and RTCA (Fig. 2.31).

The framework includes

- Transfer the TSAT solution results to the EMS via Filelink.

  - Insecure CTG IDs if any
  - CTG Insecure Reasons
  - Colstrip RAS gen drop arming data for input to RTCA

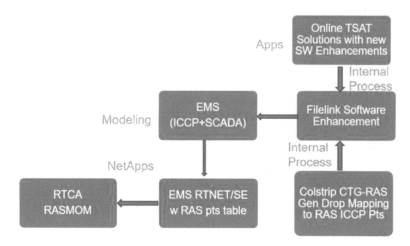

Fig. 2.31 Framework of integrating TSAT with EMS for ATR RAS monitoring

- Alarm TSAT Insecure CTG in EMS and use TSAT results in RTCA RAS modeling.

Peak RC implemented the custom process on integrating TSAT and EMS/ RTCA. It resolved many known issues with ATR operation monitoring:

- Colstrip RAS (ATR) is a transient RAS operated by NWMT to prevent instability and reduce unit tripping.
- The RTCA monitors/reports exceedance of limit, so RAS objective is not aligned 100% with MUC objective.
- The RTCA reported false violations frequently for ten associated CTGs as no RAS gen drop arming measurements is available from the entities.
- RCSOs had to validate those false RTCA violations manually and repetitively.
- Need to fix false violations for [N-1] and credible [N-2].

In the past a RCSO called a NWMT operators for the RTCA violations of concern, and they said ATR will resolve problem (but it is not always the case). We ran into a few occasions when ATR acted, violation happened, and then output of remaining units was runback by operators of plant. NWMT operators do not know if and how many units will be tripped. If ATR does not prevent it, plant output needs to be curtailed (before violation to prevent it); Peak RC cut off the TSAT use case for ATR RAS monitoring in October 2018 and shared online TSAT solved ATR gen drop info with entities via a web portal shown on Fig. 2.32.

The work brought up significant benefits:

- Using TSAT solved real-time RAS gen drop data, Colstrip RAS can be modeled in RTCA correctly per the RAS-operating procedure.
- ATR RAS model relies on arming data from TSAT and topology-based trigger for firing to avoid false RTCA violations:

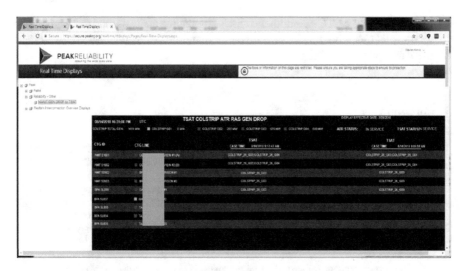

Fig. 2.32 ATR RAS operation monitoring display shared via a secured website

- Apply ATR gen drop automatically and correctly.
- No manual validation is needed typically.
- Account for both [N-1] and [N-2] CTGs.

- Several N-2 contingencies lead to islanding of certain areas with massive UFLS deployed.

2.10 Summary

Real-time monitoring on credible contingency exceedances or limit violations with inclusive RAS/SPS/non-RAS protection interactions is one of the most critical control room functions in accordance with the NERC standards/requirements. Accurate RAS/SPS models, dynamic rating limits, and cascading outage scenario assessment in real time are the weakest areas in the existing RTCA application at control rooms.

Peak RC has implemented an advanced RTCA tool to monitor the whole WECC system footprint by 8000+ contingencies and 515+ RAS/SPS/non-RAS protection schemes modeled active in the WSM. RTCA implementation framework and existing operation monitoring experience are introduced with specific examples. Lessons learned from the September 8, 2011 Pacific Southwest blackout event are provided from the perspective of the RTCA tool at Peak RC control rooms.

New challenges for RTCA to monitor N-2/MUC credible contingencies and loss of massive renewable generations due to weather conditions are analyzed by leveraging ERCOT and BPA wind generation cases that adversely impact the system operation reliability. Software and visualization enhancements for complex RAS modeling, predictive RTCA, automatic cascading outage analysis, and the framework integrating RTCA with TSAT for transient/dynamic RAS evaluation are presented by use cases proved by Peak RC.

References

1. NERC Report. (2002, August). *Review of Selected 1996 Electric System Disturbances in North America*. [Online]. Available: https://www.nerc.com/pa/rrm/ea/System%20Disturbance%20Reports%20DL/1996SystemDisturbance.pdf
2. Federal Energy Regulatory Commission. (2012, April). *Arizona-Southern California Outages on September 8, 2011: Causes and Recommendations*. FERC and NERC Staff. [Online]. Available: https://www.ferc.gov/legal/staff-reports/04-27-2012-ferc-nerc-report.pdf
3. Wangen, B., & Zhang, H. (2012, July 22–26). *Monitoring for post-contingency system operating limit exceedance in the western interconnection*. Proceedings of IEEE PES general meeting.
4. Initial review of methods for cascading failure analysis in electric power transmission systems IEEE PES CAMS Task Force on Understanding, Prediction, Mitigation and Restoration of Cascading Failures. Proceedings of IEEE PES general meeting, Pittsburgh, July 2008.
5. Zhang, H., & Wangen, B. (2012, July 22–26). *Implementation of a full Western bulk system operational model for reliability monitoring*. Proceedings of IEEE PES general meeting.
6. Vaiman, M., Hines, P., Jiang, J., Norris, S., Papic, M., Pitto, A., Wang, Y., & Zweigle, G. (2013, July 21–25). *Mitigation and prevention of cascading outages: Methodologies and practical applications*. Proceedings of IEEE PES general meeting.

Chapter 3
Real-Time Voltage Stability Analysis for IROL Assessment

Hongming Zhang

Organization Acronyms

BCTC	British Columbia Hydro Electric Service
BPA	Bonneville Power Administration
CAISO	California Independent System Operator
CENCE	Centro Nacional de Control de Energia
CFE	Comisión Federal de Electricidad
CHPD	PUD No. 1 of Chelan County
FERC	Federal Energy Regulatory Commission
GE	General Electric
IPCO	Idaho Power Company
NERC	North American Electric Reliability Corporation
NWCC	Northwest Power Pool
PAC	PacifiCorp
PEAK	Peak Reliability or Peak RC
PGE	Portland General Electric
PG&E	Pacific Gas and Electric
PowerTech	PowerTech Research Labs Inc.
PSE	Puget Sound Energy
SCE	Southern California Edison
SCL	Seattle City Light
SDG&E	San Diego Gas and Electric
V&R	V&R Energy System Research Laboratory
WECC	Western Electricity Coordinating Council

NERC/WECC Acronyms

ACE	Area control error
AGC	Automatic generation control
ATC	Available transfer capability

© Springer Nature Switzerland AG 2021
H. Zhang et al., *Advanced Power Applications for System Reliability Monitoring*, Power Systems, https://doi.org/10.1007/978-3-030-44544-7_3

AVR   Automated voltage regulator
BA   Balancing authority
BES   Bulk Electric System
COI   California-Oregon Intertie
DCS   Disturbance Control Standard
DER   Distributed energy resource
DG   Distributed generator
DTS   Dispatcher Training System
EIM   Energy imbalance market
EMS   Energy management system
FACRI   Fast AC Reactive Insertion
FTP   File Transfer Protocol
ICCP   Inter-Control Center Communications Protocol
IROL   Interconnection Reliability Operating Limit
LTC   Load tap-changing (transformer)
PDCI   Pacific DC Intertie (PDCI)
PMU   Phasor measurement unit
RAS   Remedial Action Scheme
RC   Reliability coordinator
RCSO   Reliability coordinator system operator
RTCA   Real-Time Contingency Analysis
SCADA   Supervisory Control and Data Acquisition
SCIT   Southern California Intertie
SE   State Estimator
SFTP   Secured File Transfer Protocol
SPS   Special Protection Scheme
SVC   Static VAR compensator
TTC   Total transfer capability
ULTC   Under loading tap changer
WAPS   Wide Area Protection Scheme
WI   Western Interconnection
WSM   West-wide System Model

Other Acronyms

DPF   Decoupled power flow
DSA   Dynamic stability assessment
EOC   Edge of congestion
LDC   Line-drop compensation
MUC   Multiple outage contingency
NPF   Newton power flow
NWNE   North West Net Export
NW-WA   North West Washington Area Import Limit
OPA   Operation analysis
OREX   Oregon Net Export Limit

PI	(OSIsoft's) Plant Information
POM	(V&R's) Physical and Operational Margin
RCC	Reactive current compensation
RFF	Regional flow forecast
ROE	Real-time operation engineer
ROSE	(V&R's) Region of Stability Existence application
RTDYN	(GE's EMS) Real-Time Dynamic Ratings application
RTNET	(GE's EMS) Real-Time Network Analysis (i.e., SE)
RT-VSA	(Peak RC's ROSE) Real-Time Voltage Stability Analysis
STNET	(GE's EMS) Study Network Analysis application
TSAT	(PowerTech's) Transient Stability Analysis Tool
VSA	Voltage Stability Analysis
VSAT	Voltage Stability Analysis Tool

3.1 Introduction

One of the main objectives in operating an electric power system is to maintain a proper voltage level throughout the system. Failure to do so can lead to equipment damage and blackouts.

3.1.1 Problem Definition

Voltage instability is generally characterized by loss of a stable operating point as well as by the deterioration of voltage levels in and around the electrical center of the region undergoing voltage collapse. Voltage collapse, an extreme form of voltage instability, commonly occurs as a result of reactive power deficiency. Unmitigated rotor angle instability can also result in fast voltage instability. In terms of time frame involved for an event, voltage instability is typically classified into three categories: short-term, mid-term, and long-term. Toward IROL assessment in near real time, long-term voltage instability analysis is primary focus and interest to be covered in this chapter.

3.1.2 History

Numerous voltage stability planning studies were performed by the Western Electricity Coordinating Council (WECC) and associated member entities using many offline simulation and analytical tools [7]. Since 2008 there were a few attempts in the industry toward online voltage stability analysis implementation for engineer use [8–14]. However, there was no online voltage stability analysis tool running in control rooms to perform real-time assessment of the IROLs during the September 8, 2011, Pacific Southwest blackout. This infamous blackout interrupted 2.7 million customers across Arizona, California, and Baja California for a period of time that ranged from 2 to 12 h. The event attested to the importance of identifying the IROLs in the WECC system and performing real-time assessment on the IROLs [6]. After the blackout, multiple IROLs were identified and established by Peak RC and its member entities. Since those IROLs are limited to a variety of voltage stability problems, it is imperative for Peak RC to implement an accurate and reliable RT-VSA tool for monitoring the following approved IROL Cutplanes for every 5 min (near real time).

- Northwest Washington Area (NW-WA) Import Limit
- SDG&E Import Limit (active for summer only)
- SDG&E/CFE (also called SDG&E/CENACE) Import Limit (active for winter and non-summer)
- Oregon Net Export (OREX) Limit, also called North West Net Export (NWNE)

There are a few other inter-area Cutplanes, such as El Paso/PNM Import, North Nevada IROL, etc., that Peak RC have also modeled in the RT-VSA tool to calculate the SOLs for real-time monitoring and day-ahead operation analysis (OPA) study.

Following FERC and NERC joint investigation report on the September 8, 2011, Southwest blackout, WECC RC (the forebear of Peak Reliability) started a project to implement a RT-VSA tool for operation use in 2012. V&R Energy System Research ("V&R") was selected to develop RC's VSA tool in both real-time and offline modes based on V&R's Physical and Operational Margin (POM) engine/Region of Stability Existence (ROSE) application (POM/ROSE).

3.1.3   POM/ROSE Overview

POM/ROSE defines the values of phasor measurements for which the system may securely operate. ROSE uses PMU, SCADA, and SE data for real-time calculation and visualization of the current operating point and its proximity to the stability boundary (see Fig. 3.1). The following constraints may be simultaneously included and solved for robust limits:

- Steady-state stability
- Voltage constraint (voltage range and pre-to-post contingency voltage drop)
- Thermal overloads

Each point on the boundary corresponds to a "nose" point on the PV curve or a thermal or voltage constraint being violated. Operating within this region is secure; operating outside this region corresponds to unreliable operation of a power system.

Use of PMUs allows users to incorporate the exact values of bus voltage magnitudes and angles. For the buses where PMUs are installed, the actual phasor measurements are used instead of power flow solution. Thus, state measurements, not

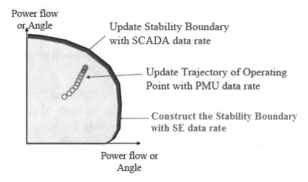

Fig. 3.1 V&R POM/ROSE function visualization (Source: V&R Energy)

state estimation, may be used in the ROSE. The ROSE automatically identifies the limit values based on real-time phasor measurements. The ROSE also computes and displays the current operating point. The location (e.g., coordinates) of the operating point is computed based on phasor measurements. The relationship between the current operating point and the boundary defines the "health" of the power system network and its proximity to steady-state collapse [8].

3.1.4 Challenges

V&R's POM-ROSE software was an offline study tool developed for planning engineers. To migrate offline POM-ROSE application to the RC's RT-VSA tool, there were a number of technical challenges that need be resolved since the project kicked off in late 2012 [15–16], such as:

- Integrate V&R software with GE (formerly Alstom) EMS platform, i.e., enable retrieving SE real-time snapshot files and sending VSA solution results to EMS/SCADA flawlessly.
- Import a full node-breaker basecase export file from real-time SE in a 5 min interval. Be able to interpret the system topology and network components and network schedules correctly as the EMS power flow engine does.
- Support parallel computing and fast power flow solution to ensure solving multiple transfer analysis scenarios in 5 min or less.
- Be capable of modeling complex Remedial Action Scheme (RAS) that may depend on a nomogram table accurately.
- Enhance visualization and alarming features to be acceptable for Reliability Coordinator System Operators (RCSOs).
- Archive and manage VSA output files to meet the NERC critical infrastructure protection (CIP) cyber security requirements.

3.1.5 Implementation

Since 2013, the RC's RT-VSA implementation project had successfully completed the following key deliverable milestones:

1. **Q2 2013–Q3 2014**, developed the framework integrating POM-ROSE with GE (formerly Alstom, Areva) EMS system via internal FileLink application. Implemented software enhancements for RAS modeling and automatic shunt switching features. The custom real-time version of V&R POM/ROSE was named to Peak RC ROSE-VSA tool.

2. **Q4 2014–Q1 2015**, completed extensive and stringent benchmarking test to validate correctness of ROSE-VSA solution results with PowerTech VSAT and PowerWorld's using the VSA limits calculated for NW-WA IROL. On February 22, 2015, Peak hosted a workshop with the entities from WECC North West region, including BPA, PSE, SEL, PacifiCorp, BC Hydro, etc., to update the RT-VSA tool validation results and demonstrate credibility of the WSM, State Estimator, and RT-VSA tools for IROL assessment in near real time. The workshop was well received.

3. **Q1–Q2 2015**, completed the RT-VSA scenario development and solution results validation testing against two new IROLs: SDG&E Summer Import and SDG&E/CFE Winter Import. On May 24, 2015, Peak RC hosted a conference call with the entities from WECC South West region, including CAISO, SDG&E, CFE, SCE, etc., to update the IROL calculation results by Peak RC ROSE-VSA vs. CAISO Bigwood-VSA tool. From the meeting, Peak RC received the agreement from CAISO and other impacted entities to roll out its RT-VSA tool for real-time assessment on two IROLs by mid-July 2015. Six weeks later, Peak RC successfully cut off the ROSE-VSA tool to calculate three IROLs for every 5 min for operation decision by Reliability Coordinator and System Operator of the impacted entities.

4. **Q3 2015–Q2 2016**, focused on investigating the root cause for SDG&E/CFE Import IROL miscalculation on September 20, 2015. The incidence of the RT-VSA tool failure resulted in mistaken load shedding in SDG&E service area. Peak RC worked with the entities to check and add missing RAS in the RT-VSA tool and supported V&R team to fix the software limitation for collective voltage reactive regulation, i.e., multiple units, SVC, and shunt devices regulate the voltage on the same bus. To validate the RT-VSA solution results, Peak developed an offline StudyVSA tool using EMS power flow engine. It was made available to Real-Time Operation Engineer (ROE) in March 2016 after extensive verification testing. After the RAS models and core engine of the RT-VSA were corrected, enhanced, and re-validated, Peak RC put back the RT-VSA tool in the control room for operation use again in May 2016.

5. **Q3 2016–Q1 2017**, tested and cut off V&R POM software upgrade from 2014 to 2016 release. Integrated the RT-VSA tool in Peak RC DTS environment. Supported V&R for successful deployment of the RT-VSA tool in SDG&E,

SCE, and IPC to enable them running separate voltage stability analysis using the real-time WSM SE savecases transferred over a secured FTP.

6. **Q1–Q2 2017**, prepared and delivered Peak RC and CAISO Joint RT-VSA Tool Training classes under DTS simulation mode. In the training sessions, both RT-VSA tools were running in parallel to facilitate communication and IROL drop validation. Cut off V&R POM software upgrade to 2017 release.

7. **Q3 2017–Q2 2018**, developed and validated the new VSA scenario for OREX IROL. During the validation testing, it's noted the RT-VSA tool failed to fire critical PDCI RAS to mitigate the system voltage instability issues due to post-contingency power flow diverging. This caused incorrect drop and oscillation of OREX IROLs calculated by the RT-VSA tool. Peak RC worked with V&R closely to enhance the software by adding two-stage RAS logic and test it out before RT-VSA calculation on OREX IROL was cut over in production for operation use by end of 2018.

8. **Q3 2018-Q3 2019**, keep improving the RAS models and validating the solution gaps between Peak RC's and CAISO's tools.

By December 3, 2019, Peak RC ROSE-VSA tool had been running for 4 years in the control rooms to perform near real-time assessment on the western IROLs effectively. Now it becomes a critical real-time application for RCSOs to obtain situational awareness on the security margin each of the IROLs and voltage stability constrained SOLs [17].

3.2 Objectives, Assumptions, and Model Representations

3.2.1 Objectives

Main goal for the RT-VSA implementation project is to develop a reliable and accurate real-time tool for online voltage stability assessments in consideration of credible [N-1] and [N-2] contingency constraints and effect of RAS interactions. The tool needs to handle the full AC nonlinear modeling of basecase power flow under a list of user-specified contingencies. It takes the general characteristics of power system operating environments into account and includes modeling of specific RAS affecting the monitored transmission interface. Many practical aspects of power system operations and characteristics are modeled in the tool, such as limits of real power generation, generating unit reactive capability curve so-called D-curve, governor and AGC power flow simulation with respect to AGC flags and unit participation factors, and automatic shunt switching with the AVR flag on. In addition to the main advantage of fast solution, the tool has shown the following capabilities:

- Generate PV and VQ curves with respect to a variety of scenarios of demand disturbances.

- Compute the exact loadability margin of the basecase or selected contingency to the voltage collapsing point, which can be either the saddle-node bifurcation point or the limited-induced bifurcation point.
- Identify the critical or limiting contingency, which can first lead to voltage collapse, from a credible contingency.
- Perform RAS arming evaluation and RAS screening on trigger conditions. Apply RAS protective actions if a RAS is armed and its trigger conditions are met at the current stressing level.
- Be able to run reverse transfer analysis to determine a minimal load shedding amount when the basecase is insecure against one or more specific contingencies.
- Respect economic Pmax for the concern of unit dynamic reactive reserve constraints.
- Be able to compute dV/dQ sensitivities.
- Support automatic shunt switching on selected areas.
- Node-breaker model representation at a local network or full system level.
- Custom visualization and alarming features.

3.2.2 Assumptions

3.2.2.1 General Assumptions

- All RAS impacting an IROL need be modeled based upon the operating procedures.
- Instability indicated by power flow divergence could come from numerical problems in an actual voltage collapse.
- The comparison between programs should provide sufficient evidence toward the cause of instability (it is unlikely that any issues with any particular program's numerical problems would exist in all VSA tools).
- Generator automatic voltage regulators (AVR) (remote regulating), dynamically controlled variable susceptance devices, reactive shunts, phase shifters, and LTCs shall be allowed to move in the basecase (pre-contingency).
- For post-contingency power flow solution, only units with AVR flag on and auto-switchable shunts are allowed for system voltage control.
- Generator AVRs (terminal regulating) and dynamically controlled variable susceptance devices will be able to move post-contingency (all others will be locked).
- SVC devices are equivalently modeled as synchronous condenser.

3.2.2.2 Generators (Source)

- Online generator units in synchronous condense mode that will maintain that mode will not be used in the "Source."
- The only units that are allowed to move are units which are currently online, generating a positive power output, and would be considered for frequency response, i.e., AGC control flag enabled.
- To pick up loss of generation against specific contingency or due to RAS gen drop, governor response or AGC response will be expected of all non-base-loaded units (i.e., AGC regulation flag on) within the entire system or island with their response proportional to their P_{MAX}.
- At each step or level of stressing transfer, Source generation is changed as follows:

  – Generation of online units on AGC with current real power output $P_{current}$ greater than 0 MW is increased in proportion to $P_{current}$:

  All units are scaled up.
  Pmax or real power limits of generators are honored.
  EPF (Economic Participation Factor) is considered if applicable.

  – When generation in item 1) above is exhausted, generation of online units on AGC with $P_{current} = 0$ MW is increased in proportion to maximum real power output P_{max}:

  All units are scaled up.
  Real power limits of generators are honored.
  EPF is considered.
  Units with $P_{max} = 9999$ MW are excluded.

  – A single slack generator determined in SE basecase solution picks up incremental system loss resulting from increased stressing.

- Line Drop Compensation (LDC)/Reactive Current Compensation (RCC) will not be modeled in V&R POM/ROSE due to program limitations.
- The use of LDC/RCC on the calculated output will be tested using PowerWorld to determine the impact.

3.2.2.3 Load (Sink)

- All load will be modeled as constant MVA through the voltage band.
- Effect of load voltage dependency is not considered.
- The limits of individual loads in "Sink" if applicable are respected during stressing.
- All loads in the "Sink" area will be scaled except the following:

  – Non-conforming loads and dynamic loads are identified by owner entities.

– Individual identified non-conforming loads and dynamic loads are excluded from the Sink area defined in scenario files.

This will help test and verify the impact that the use of non-conforming loads has on the results between the three programs.

3.2.2.4   Interchange Schedules and DC Schedules

3.2.2.4.1   Problem Definitions

- The network model typically consists of a single topological island and multiple operating areas/balancing authorities within that island.
- For operating areas that transact a pre-defined interchange (i.e., the "area interchange" flag is checked, thereby enforcing interchange for that balancing authority (BA)), generating units within such areas contribute toward a power balance within those areas, i.e., they regulate to meet load, losses, and interchange requirements for their area. A unit regulating MW will move up or down to lower mismatch as long as its MW output is within its Pmax/Pmin limits.
- Generating units within operating areas that do not enforce area interchange transactions contribute toward the power balance for the rest of the topological island, i.e., they regulate to meet island load and losses within their MW limits.
- DC flow schedule is a set point to determine how much MW flows on a DC pole during power flow study. It's controllable manually or by a control action such as RAS.

3.2.2.4.2   Peak RC Settings

Enforcing area interchanges is not enabled in the RT-VSA calculation process. All DC flow schedules keep the same values as in the SE basecase solution.

3.2.3   System Stressing Criteria

The system stressing in RT-VSA calculation is stopped for the following conditions:

1. **Maximum Source generation is exhausted** – When stressing a transmission interface, in some cases it is possible to maximize the Source generation in the simulation before approaching the nose point of a PV curve. If there is no more Source generation that can be used to serve as a source for simulation, the stressing process will be finished, and the maximum transfer capability is calculated by the RT-VSA tool. In this case, it can be concluded that no stability risks practically exist for the interface, and there is no reliability need to establish a stability limit for the interface or load area.

2. **Maximum transfer level is reached** – When stressing an interface between Source and Sink, it is possible to reach maximum transfer level (user configurable) before the RT-VSA tool is able to solve a voltage stability limit close enough to the nose point of a PV curve. In this case, the VSA limit is capped by the maximum transfer level.
3. **The nose point of a PV or VQ curve is reached** either under pre-contingency conditions or upon occurrence of a credible contingency. This condition indicates the presence of a voltage instability risk and thus the need to establish a voltage stability limit in real-time assessment or on an operating plan study.
4. It's an option for having the RT-VSA tool stop the stressing process when a lower bus voltage limit violation or a thermal limit violation occurs. In this setting, the transfer limits solved by the RT-VSA are much lower than the VSA limit.
5. If maximum transfer simulations are not governed by items 1 through 4, in order to identify potential cascading risks, the transmission interface or load area should be stressed to the point where post-contingency flow on a facility reaches the lesser of the following:

   - 125% of its highest applicable Facility Rating (this is typically equal to 125% of highest emergency rating)
   - The facility's loadability as determined by NERC Standard PRC-023, Transmission Relay tripping limit.
   - This analysis assumes that pre- and post-contingency flows are below applicable Facility Ratings prior to transfer analysis.

The first three stressing criteria are applied to the RT-VSA scenario definitions by Peak RC, though the POM/ROSE has the capability for adoption of other criteria such as voltage limit violation, thermal limit exceedance, etc.

3.2.4 Problem Formulation

Due to the nonlinear nature of interconnected electric systems, the load margin to voltage collapse of either the basecase or post-contingency condition depends on a number of factors such as the system topology/network configuration, system load profiles, and generation dispatch guidelines/assumptions. Hence, analysis of voltage collapse/stability must have adequately accurate power system modeling and scope to ensure that all equipment as well as system limits of the entire interconnected system network are properly taken into account. The modeling capability in V&R POM/ROSE tool includes:

3.2.4.1   Generators

Generators are modeled as active and reactive power sources which also provide voltage control. The MVAR output of each online generator is adjusted during power flow solutions in order to control the voltage of the local bus (the bus where the generator is connected to) or a remote bus. The generator's MW output has fixed limit. And the generator's MVAR output is limited by the so-called reactive capability curves "D-curves."

3.2.4.2   Loads

Loads can be modeled as constant power (PQ), constant current (I), constant impedance (Z), or any linear combination of them. The tool can also accept non-conforming or dynamic load models as long as they are defined in a VSA scenario.

3.2.4.3   Individual Control Devices

The following control devices are modeled:

- Switchable shunts and static VAR compensators
- ULTC transformers
- ULTC phase shifters
- ULTC phase shifting transformers (PST)
- DC poles/converters
- Interchange schedules

3.2.4.4   Wide Area Control Scheme-RAS/SPS

Remedial Action Scheme (RAS) is an automatic protection system designed to detect abnormal or predetermined system conditions and take corrective actions other than and/or in addition to the isolation of faulted components to maintain system reliability. Such action may include changes in demand, generation (MW and MVAR), or system configuration to maintain system stability, acceptable voltage, or power flows.

A SPS is similar to a RAS. However, SPS does not include (a) under-frequency or under-voltage load shedding; (b) fault conditions that must be isolated; and (c) out-of-step relaying (not designed as an integral part of an SPS).

As an integrated and automatic protection system, a typical RAS model includes arming point measurements received via ICCP, a number of triggering conditions and mitigation actions. RAS is usually fired by one or more selected contingencies. Hence, RAS models are typically incorporated into post-contingency power flow analysis for VSA limit calculation.

In Peak RC V&R POM/ROSE-VSA tool, RAS are modeled as external modules using VB scripts. The tool evaluates RAS arming data and triggering conditions during post-contingency power flow calculation. Once all required conditions are met, RAS protective actions will be applied on top of post-contingency operating point and re-run post-RAS power flow to verify if a VSA limit is reached. However, diverging of post-contingency power flow could prevent a RAS from firing properly. As a result, the VSA tool will probably solve for a lower VSA limit.

Voltage collapse/instability is generally caused by either of two types of system disturbances: demand or transfer increase and contingency disturbances.

Figure 3.2 is an example of RAS logic diagram. As one can see, it is challenging, if not impossible, to formulate the RAS models into basecase or post-contingency power flow questions directly.

3.2.4.5   VSA Study Modeling Requirements

1. A basecase power system network model with transmission facility, control devices, reactive power generation limits, real power generations due to participation factor, regulation schedules, etc.
2. The current operating condition (obtained from the state estimator and the topology processor, SCADA measurements for RAS arming points).
3. A set of proposed power transfer scenarios, which define the following:

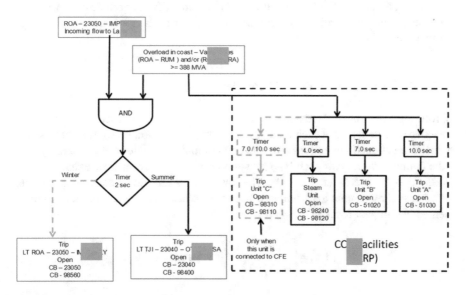

Fig. 3.2 Example of RAS logic diagram

- Source – it usually consists of generation areas and/or a group of individual units. When a generation area is selected, User has an option to exclude a subset of the units that do not participate in stressing analysis. Note scalable loads can be assigned to the Source though it's not common.
- Sink – it usually consists of load areas and/or a group of individual loads. When a load area is selected, User has an option to exclude a subset of the non-forming loads that do not participate in stressing analysis. Note units can be assigned to the Sink by reducing its MW output if applicable.
- Transfer interface – it's defined by a group of power lines and transformers with pre-specified flow direction/sign.
- Contingencies and relevant RAS mitigation control – a set of transmission or generation outages and RAS applied for transfer security analysis.
- Monitored buses – a set of buses sensitive to transfer increase. PV and Q-V analysis will be applied to evaluate and visualize voltage stability margin against the monitored buses.
- Control settings, such as maximal transfer/stress level, voltage constraints monitoring option, step size, shunt switching options, negative reactive load stressing options, VSA Import Limit alarm thresholds, etc.

Under the specified VSA scenario and the SE solved basecase solution, V&R POM power flow engine can reliably compute the load margin to the voltage collapse point of the basecase power system or the power system under a set of selected contingencies and accurately compute the voltage collapse point. If a voltage violation point or a thermal violation point occurs before the voltage collapse point, the corresponding points define the voltage isolation load margin or thermal limit load margin. Calculation of IROL in real-time aims to identify the voltage collapse point and determine the margin to the current operating condition, regardless of voltage or thermal violations.

3.2.4.6 Formulation of the VSA Problem

Given a power transfer scenario with a system load demand vector (i.e., real and reactive load demands at each load bus) and a real generation vector (i.e., real power generation at each generator bus), one can compute the state of the power system (the complex voltage at each bus) by solving the set of power flow equations. Let $P_i \equiv P_{gi} - P_{di}$ and $Q_i \equiv Q_{gi} - Q_{di}$. The lowercase g represents generation and the lowercase d represents load demand. The set of power flow equations can be represented in compact form as

$$f(x) = \begin{bmatrix} P(x) - P \\ Q(x) - Q \end{bmatrix} = 0, where\ x = (V, \theta) \tag{3.1}$$

where the vectors P and Q represent the real and reactive power injection at each bus, respectively.

Next, one can examine the power system steady-state behaviors under slowly varying loading and real power re-dispatch conditions. For example, if one needs to trace the power system state from the basecase load-generation condition specified by the following vector to a new load-generation condition specified by the following vector $\left[P_d^0, Q_d^0, P_g^0 \right]$ to a new load-generation condition specified by the following vector P_d^1, Q_d^1, P_g^1, then one can parameterize the set of power flow equations as the following equations

$$F(x,u,\lambda) = f(x,u) - \lambda b = 0 \qquad (3.2)$$

where u is the control vector and the load-generation vector b is

$$b = \begin{bmatrix} P^1 - P^0 \\ Q^1 - Q^0 \end{bmatrix} \qquad (3.3)$$

It is clear that the set of the parameterized power flow Eq. (3.2) becomes the basecase power flow equations when $\lambda = 0$,

$$F(x,0) = \begin{bmatrix} P(x) - P^0 \\ Q(x) - Q^0 \end{bmatrix} = 0 \qquad (3.4)$$

And when $\lambda = 1$, the set of parameterized power flow equations describes the power system steady-state behavior at the new load-generation P_d^1, Q_d^1, P_g^1 and is described by

$$F(x,k) = f(x) - b = \begin{bmatrix} P(x) - P^1 \\ Q(x) - Q^1 \end{bmatrix} = 0 \qquad (3.5)$$

Hence, one can investigate the effects of varying real power generations as well as varying load demands on power system steady-state behaviors by solving the set of parameterized power flow Eq. (3.2).

Definition The nose point VSA margin of a power system (3.1) under a contingency with respect to a load-generation vector (3.3) is the distance (in terms of MW generally) from the current operating conditions to the nose point of the parameterized power system (3.2) subject to the contingency. The definition of nose point can be characterized in Fig. 3.3.

The nose point corresponds to the voltage collapse point. Depending on several factors, the nose point can be a saddle-node bifurcation point or a limit-induced bifurcation point. The distance from the current operating point to the voltage collapse point is the VSA margin. If a voltage violation or thermal limit violation is reached before the voltage collapse point, then the corresponding distances are the voltage violation load margin and thermal limit load margin, respectively. The

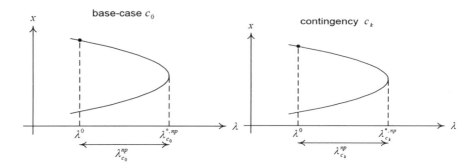

Fig. 3.3 The nose point VSA margin for the basecase power system $\lambda_{c_0}^{*,np}$ versus the nose point VSA margin $\lambda_{c_k}^{*,np}$ for the power system under contingency

voltage violation load margin and thermal limit load margin are not our concern in this chapter. Instead, the VSA margin is calculated by the first hit voltage collapse point or the last power flow solution point where the maximum power transfer level or maximum Source generation is reached.

There are generally two types of voltage collapse point. One type of voltage collapse point is the saddle-node bifurcation point at which there is no structure change in the set of power flow Eq. (3.2). Another type of voltage collapse point is the limit-induced bifurcation point which is characterized by a structure change in the power flow equations due to the reactive power limit of certain generator at the limit-induced bifurcation point. Under this situation, a generator has changed from a P-V-type bus to a P-Q-type bus due to its reach at the reactive power limit. The upper branch and lower branch of the power flow solution curve meet abruptly at the voltage collapse point. The solution curve is composed of two different solution curves.

Definition The VSA margin of a power system (3.1) under a contingency with respect to a load-generation vector (3.3) is the (minimum) distance in terms of MW from the current operating point to the state vector, of the parameterized power system (3.2) subject to the contingency, at which the voltage collapse at some bus occurs, the maximum transfer level is reached, or the Source generation resources in MW or the Sink load is max out.

A generalized flowchart of Peak RC POM/ROSE-VSA solution algorithm is given below (Fig. 3.4).

It must be pointed out that continuous stressing of basecase by a fixed or varying step size and/or evaluation of a contingency on the basecase can trigger relevant RAS actions, resulting in changes in load-generation pattern and network topology. At the level of original basecase and each newly stressed basecase, the ROSE-VSA software runs multiple power flow solutions: (1) basecase power flow, (2) post-contingency power flow, and (3) post-contingency and post-RAS action power flow. If any of those power flows is diverged, the ROSE-VSA will assume a secured operating point does not exist at the current stress level.

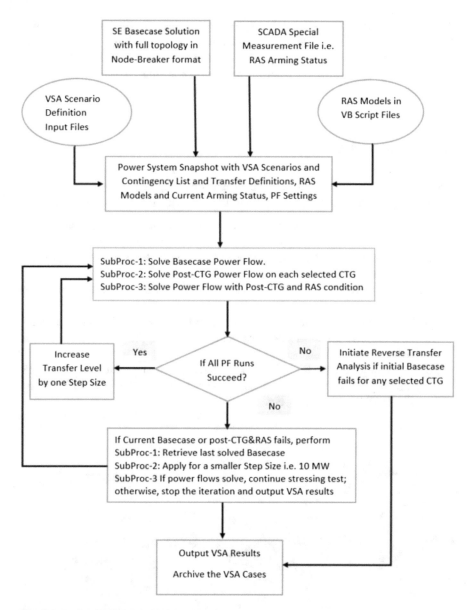

Fig. 3.4 Peak RC V&R POM/ROSE VSA flowchart

3.3   System Architecture and File Transfer Management

Peak RC uses GE EMS software suite to perform SCADA monitoring and advanced network applications analysis, including Real-Time Network Analysis (RTNET) or State Estimator (SE) and Real-Time Contingency Analysis (RTCA) [4, 5]. While the RT-VSA project was started, V&R Energy original Physical and Operational Margin (POM) Application Suite had no interface with any external EMS system. To integrate two separate vendor products under a real-time environment setting, Peak RC developed a custom framework to support real-time data streaming and file exchange between GE EMS and V&R POM servers.

3.3.1   System Architecture

The framework includes a few components shown in Fig. 3.5.
   The integration process is comprised of the following components [17]:

- Run custom scripts installed onto EMS servers to automatically retrieve the last solved SE solution snapshot into a WSM export file and special measurement output file in *.csv format. The SE files from primary EMS are sent over to all four VSA servers.
- The scripting process is scheduled to send the new SE files to V&R servers for every 4.5 min.
- The RT-VSA software automatically checks if there are new SE data files coming in and then kick off the next analysis run if the last VSA solution completes.
- After each new run, the RT-VSA tool generates solution output files, e.g., EMS alarms, VSA solution summary files, PV curves, VQ curves, and different log files like RAS, iteration logs, etc.

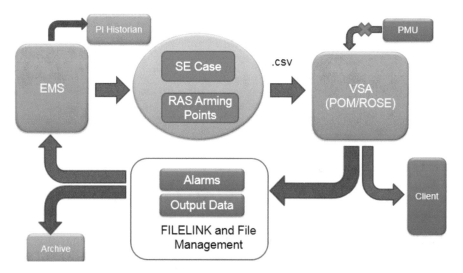

Fig. 3.5 EMS and RT-VSA system integration architecture and data exchange structure

- EMSAlarm file includes all critical VSA solution results, e.g., limits, critical contingencies, weakest buses, and solution failure code. The file is read into the GE's FileLink application for data interpretation per a used defined mapping table and data transfer to ICCP servers for real-time update on SCADA VSA points.
- SCADA has custom operator alarms and IROL monitoring displays for RC situational awareness.
- The RT-VSA results in SCADA are historized in PI.

A full dataset of POM/ROSE input and output files are auto-archived by a custom process into a zip file upon completion of each run, regardless of the RT-VSA study case being solved or not. The custom scripts are scheduled to automatically move 3 days older RT-VSA archived files from the POM servers to data servers for permanent storage.

3.3.2 FileLink Application

3.3.2.1 Overview

As a custom application, FileLink reads the POM/ROSE solution output files: EMSAlarm.csv and the PMUAlarm.csv. Then the application creates respective output EMSAlarm and PMUAlarm files in a format that ICCP or Open Access Gateway (OAG) server can read and send the data to EMS.

See Table 3.1 for a sample POM/ROSE EMSAlarm.csv and Table 3.2 for the associated sample output file created by FileLink. Since ICCP does not support Text format data such as contingency ID and alarm messages, FileLink is implemented to maintain a master mapping table for translation of Text format information into integer values, in order to be transferred from POM to EMS via ICCP.

Table 3.3 shows an example of the translation table between string key and integer number, where S001 represents VSA Scenario 1, followed by ".Base Case or Contingency IDs" and so on.

With new software enhancements, POM/ROSE need to store additional columns in EMS_Alarm.csv file. In order to transfer additional data to EMS, it is required to update FileLink mapping table and enhance existing FileLink software sometimes.

The FileLink software monitors a directory waiting for a new EMSAlarm.csv file or PMUAlarm.csv file from the POM/ROSE application. Once a file is done being written, then FileLink processes it.

Each input file is processed independently, and a new EMSAlarm or PMUAlarm csv file generated for each in a different directory. The output file has three output columns (ICCP Object ID, Value, and Quality code), but no header.

PMUAlarm files were integrated into the FileLink process and transferred to EMS likewise. However, presently the PMU data is not streaming into the RT-VSA for the IROL calculation in the control room.

Table 3.1 Peak RC-ROSE EMSAlarm file

	1	2	3	4	5	6	7	8	9	10	11	12	13	14	15	16	17	18
1	TimeStamp	10/18/2016 8:04																
2	Scenario #	Scenario Name	Level 1	Level 2	Invalid Input in Scenario	POM Sequence Failed in Scenario	Current MW	Limit	Import % of Limit	Margin	Load Shed	Load Shed with Margin	Pre-CTG Weakest Bus	Pre-CTG Weakest Bus Volt	LimitCtg1	Post-CTG Weakest Bus1	Post-CTG Weakest Bus1 Volt	LimitCtg2
3	S001	Summer SDGE	0	0	0	0	1002	4378	23	3376			13013 CRSTWD 69	0.95	*ECO_MIGUEL_1500	13238 SUNCREST 230	0.9	
4	S002	Winter SDGE	0	0	0	0	1051	4255	25	3204			4796 SMN115 115	0.77	*ECO_MIGUEL_1500	4796 SMN115 115	0.68	
5	S003	SCIT	0	0	0	0												
6	S004	WOCS	0	0	0	0												
7	S005	WOCN	0	0	0	0												
8	S006	BLEE	0	0	0	0												
9	S007	BORW	0	0	0	0												
10	S008	COI	0	0	0	0												
11	S009	Midpoint	0	0	0	0												
12	S010	Path 54	0	0	0	0												
13	S011	Path 47	0	0	0	0												
14	S012	Path 48	0	0	0	0												
15	S013	BCHYD	0	0	0	0												
16	S014	AESO	0	0	0	0												
17	S015	HW-NG	0	0	0	0												
18	S016	PNM	0	0	0	0	712	1102	65	391			10565 GUADLUPE 345	0.97	*FOURCORNERS-WES	10597 LA_MPF 115	0.78	
19	S017	NW-WASH	0	0	0	0	4371	8130	54	3760								
20																		
21	Code	Alarm	State															
22	G001	Unsolved Basecas	0															
23	G002	POM Sequence F	0															
24	G003	New SE Case Not	0															
25	G004	Reverse Transfer	0															

Table 3.2 Peak RC RT-VSA FileLink output file (Sample) sent to EMS ICCP server

	1	2	3
1	ICCP Object ID	Value	Quality
2	W004$VSA_G001_UNSOLVED_BASECASE	0	0
3	W004$VSA_G002_POM_SEQ_WRITE_FAIL	0	0
4	W004$VSA_G003_MISSING_NEW_SECASE	0	0
5	W004$VSA_G004_REVERSE_TRANSFER	0	0
6	W004$VSA_SCN_S001_LEVEL1_ALARM	0	0
7	W004$VSA_SCN_S002_LEVEL1_ALARM	0	0
8	W004$VSA_SCN_S003_LEVEL1_ALARM	0	0
23	W004$VSA_SCN_S001_LEVEL2_ALARM	0	0
24	W004$VSA_SCN_S002_LEVEL2_ALARM	0	0
25	W004$VSA_SCN_S003_LEVEL2_ALARM	0	0
40	W004$VSA_SCN_S001_INVALID_INPUT	0	0
41	W004$VSA_SCN_S002_INVALID_INPUT	0	0
42	W004$VSA_SCN_S003_INVALID_INPUT	0	0
57	W004$VSA_SCN_S001_POM_SEQ_FAILED	0	0
58	W004$VSA_SCN_S002_POM_SEQ_FAILED	0	0
59	W004$VSA_SCN_S003_POM_SEQ_FAILED	0	0
74	W004$VSA_SCN_S001_CURRENT_MW	1002	0
75	W004$VSA_SCN_S002_CURRENT_MW	1051	0
76	W004$VSA_SCN_S003_CURRENT_MW		32
91	W004$VSA_SCN_S001_IMPORT_LIM_MW	4378	0
92	W004$VSA_SCN_S002_IMPORT_LIM_MW	4255	0
93	W004$VSA_SCN_S003_IMPORT_LIM_MW		32
108	W004$VSA_SCN_S001_PERCENTLOADING	23	0
109	W004$VSA_SCN_S002_PERCENTLOADING	25	0
110	W004$VSA_SCN_S003_PERCENTLOADING		32
125	W004$VSA_SCN_S001_MARGIN_MW	3376	0
126	W004$VSA_SCN_S002_MARGIN_MW	3204	0
127	W004$VSA_SCN_S003_MARGIN_MW		32

Table 3.3 FileLink string key number mapping table (sample)

String key	Number
S001.	0
S001.Base case	1
S001.∗ IMVALLY_ECO_1500	2
S001.∗IMVALLY_N.GILA_1500	3
S001.∗N.GILA_HOODOOW_1500	4

After FileLink has created the transformed output file, being either an EMSAlarm file or PMUAlarm file, it is copied to the associated ICCP/OAG server. This file transfer crosses domain boundaries.

Table 3.4 The mapping table for mapping alarm names

Alarm	GCodeSuffix
Unsolved basecase	_UNSOLVED_BASECASE
POM sequence failed to write output	_POM_SEQ_WRITE_FAIL
New SE case not received	_MISSING_NEW_SECASE
Reverse transfer analysis performed	_REVERSE_TRANSFER

3.3.2.2 EMSAlarm File

The EMSAlarm.csv input file has two sets of codes S and G that are in the same file with different header, and both need to be processed into the same output file. The S codes are first; then there is a blank line and then the G codes.

3.3.2.2.1 G Code

The following columns for the G codes are being processed to the output EMSAlarm csv file:

Column 3 – State

The FileLink output EMSAlarm.csv file has three fields: ICCP Object ID, Value, and Quality code. Below are the details of each field corresponding to the column being processed:

Column 3 (State) The naming convention of **ICCP object ID** of this column should be "W004$VSA_" + ScenarioID + GCodeSuffix, where ScenarioID is first column value of the same row and GCodeSuffix is derived from the alarm name in the second column. The **value** should be the value of Column 3. **Quality code** should always be set to zero.

Table 3.4 is the mapping table for mapping alarm names of Column 2 to a GCodeSuffix.

3.3.2.2.2 S Code

The following columns for the S codes are being processed to the output EMSAlarm csv file:

Column 3 – Level 1
Column 4 – Level 2
Column 5 – Invalid input in scenario
Column 6 – POM sequence failed in scenario
Column 7 – Current MW
Column 8 – Import Limit
Column 9 – % of Limit

Column 10 – Margin

The FileLink output EMSAlarm.csv file has three fields: ICCP Object ID, Value, and Quality code. Below are the details of each field corresponding to the column being processed:

Column 3 (Level 1) The naming convention of **ICCP object ID** of this column should be "W004$VSA_SCN_" + ScenarioID + "_LEVEL1_ALARM," where ScenarioID is 1st column value of the same row. If Column 3 is empty, the **value** should be empty, and the **Quality code** should always be set to 32. If it is not empty, the **value** should be the value of Column 3, and the **Quality code** should be set to zero.

Column   4   (Level   2) The   naming   convention of **ICCP object ID** of this column is similar to Column 3 (Level 2), i.e., "W004$VSA_SCN_" + ScenarioID + "_LEVEL2_ALARM."

Column 5 (Invalid Input in Scenario) The naming convention of **ICCP object ID** of this column should be "W004$VSA_SCN_" + ScenarioID + "_INVALID_ INPUT," where ScenarioID is 1st column value of the same row. If Column 5 is empty, the **value** should be empty, and the **Quality code** should always be set to 32. If it is not empty, the **value** should be the value of Column 5, and the **Quality code** should be set to zero.

Column 6 (POM Sequence Failed in Scenario) The naming convention of **ICCP object ID** of this column should be "W004$VSA_SCN_" + ScenarioID + "_POM_ SEQ_FAILED," where ScenarioID is 1st column value of the same row. If Column 6 is empty, the **value** should be empty, and the **Quality code** should always be set to 32. If it is not empty, the **value** should be the value of Column 6, and the **Quality code** should be set to zero.

Column 7 (Current MW) The naming convention of **ICCP object ID** of this column should be "W004$VSA_SCN_" + ScenarioID + "_CURRENT_MW," where ScenarioID is 1st column value of the same row. If Column 7 is empty, the **value** should be empty, and the **Quality code** should always be set to 32. If it is not empty, the **value** should be the value of Column 7, and the **Quality code** should be set to zero.

Column 8 (Import Limit) The naming convention of **ICCP object ID** of this column should be "W004$VSA_SCN_" + ScenarioID + "_IMPORT_LIM_MW," where ScenarioID is 1st column value of the same row. If Column 8 is empty, the **value** should be empty, and the **Quality code** should always be set to 32. If it is not empty, the **value** should be the value of Column 8, and the **Quality code** should be set to zero.

Column 9 (% of Limit) The naming convention of **ICCP object ID** of this column should be "W004$VSA_SCN_" + ScenarioID + "_PERCENTLOADING," where ScenarioID is 1st column value of the same row. If Column 9 is empty, the **value** should be empty, and the **Quality code** should always be set to 32. If it is not empty, the **value** should be the value of Column 9, and the **Quality code** should be set to zero.

Column 10 (Margin) The naming convention of **ICCP object ID** of this column should be "W004$VSA_SCN_" + ScenarioID + "_MARGIN_MW," where ScenarioID is 1st column value of the same row. If Column 10 is empty, the **value** should be empty, and the **Quality code** should always be set to 32. If it is not empty, the **value** should be the value of Column 10, and the **Quality code** should be set to zero.

Current FileLink software is processing the information of the first ten columns in the EMSAlarm.csv file for the S codes. The new EMSAlarm.csv file has 18 or more columns of data for each scenario. Of those additional eight columns, it is required to add information from the following four columns to FileLink output EMSAlarm.csv file:

Column 11 – Load shed MW
Column 12 – Load shed with margin
Column 15 – Limiting contingency 1
Column 18 – Limiting contingency 2

The FileLink output EMSAlarm.csv file has three fields: ICCP Object ID, Value, and Quality code. Below are the details of each field corresponding to the new columns:

Column 11 (Load shed MW) The naming convention of **ICCP object ID** of this column should be "W004$VSA_SCN_" + ScenarioID + "_LDSHED," where ScenarioID is 1st column value of the same row. If Column 11 is empty, the **value** should be zero. If it is not empty, the **value** should be the value of Column 11. **Quality code** should always be set to zero.

Column 12 (Load shed with Margin) The naming convention of ICCP object ID should be "W004$VSA_SCN_" + ScenarioID + "_LDSHMR," where ScenarioID is 1st column value of the same row. If Column 12 is empty, the value should be zero. If it is not empty, the value should be the value of Column 11. Quality code can always be set to zero.

Column 15 (Limiting Contingency 1) The naming convention of ICCP object ID should be "W004$VSA_SCN_" + ScenarioID + "_LIMCTG_1," where ScenarioID is 1st column value of the same row. The value of this column is a string and OAG cannot support the strings. So the string needs to be mapped to an integer value. The combination of ScenarioID (column 1) and limiting contingency 1 (column 15) should be used as the key (ScenarioID + "." + Column 15) for mapping this column string to an integer value. If this column is empty, the value should be set to zero. Quality code should always be set to zero.

Column 18 (Limiting Contingency 2)   The naming convention of ICCP object ID should be "W004$VSA_SCN_" + ScenarioID + "_LIMCTG_2," where ScenarioID is 1st column value of the same row. The value of this column is a string and OAG cannot support the strings. So the string needs to be mapped to an integer value. The combination of ScenarioID (column 1) and limiting contingency 2 (column 18) should be used as the key (ScenarioID + "." + Column 18) for mapping it to an integer value. If this column is empty, it should be set to zero. Quality code should always be set to zero.

The FileLink master mapping tables mentioned above and internal file management scripts need be updated if new contingency IDs, new alarm codes, and new VSA solution data need be interpreted and transferred to EMS/SCADA, though overall framework remains the same.

3.4   RT-VSA Tool Validation on the IROLs

Since late 2014, Peak RC had conducted extensive V&R RT-VSA tool validation and comparison with other vendor software and CAISO's Bigwood RT-VSA tool. The whole RT-VSA tool validation process started with Northwest Washington (NW-WA) Load Import IROL, then moved on to SDG&E Import and SDG&E/CFE Import IROLs, and lately completed validation for Oregon Net Export (OREX) IROL. Each IROL solution validation identified different modeling and software enhancement requirements. At the end, the validation process helped improve both RT-VSA modeling accuracy and the functionality of a newer version of V&R POM engine. Each RT-VSA solution validation project against a specific IROL had to consolidate general assumptions from impacted operating entities to ensure all VSA tools used for comparison will adopt the same or similar assumptions.

3.4.1   NW-WA IROL Validation

3.4.1.1   Introduction

High loads and heavy imports in Northwest Washington Area may result in widespread post-contingency steady-state voltage instability in the area. Peak RC has classified the NW-WA Import Limit as an IROL.

Parties affected by the IROL are BPA, BC Hydro, CHPD, PacifiCorp, PSE, SCL, etc.

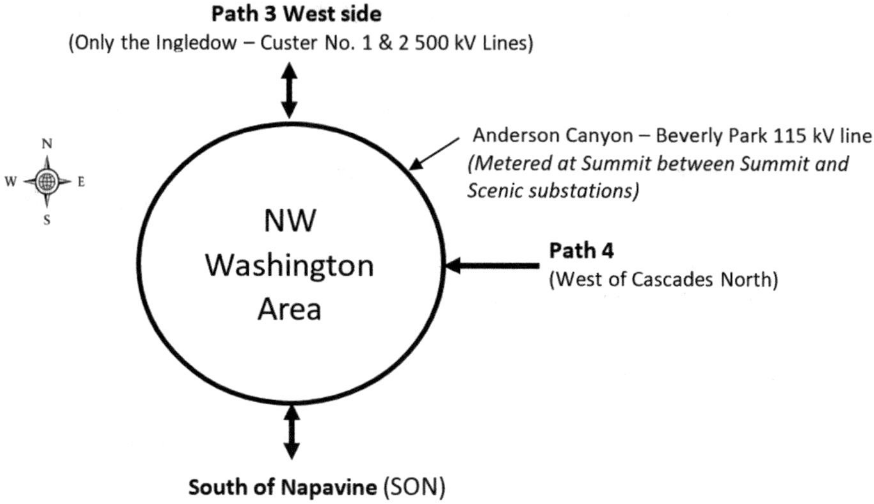

Path 3 West side
(Only the Ingledow – Custer No. 1 & 2 500 kV Lines)
N
W
E
S
NW
Washington
Area
Anderson Canyon – Beverly Park 115 kV line
(Metered at Summit between Summit and Scenic substations)
Path 4
(West of Cascades North)
South of Napavine (SON)

Fig. 3.6 Northwest Washington area bubble

CANADA
Path 3 (west-side)
WASHINGTON
Anderson Canyon – Beverly Park 115 kV Line
NW WA Load Area
Path 4: West of Cascades North (WOCN)
Spokane
Olympia
South of Napavine
Longview
Portland
The Dalles
Salem

Fig. 3.7 NW-WA geographical overview

Northwest Washington Area bubble and geographical overview are described in Figs. 3.6 and 3.7.

The IROL is defined by the Northwest Washington Area Import, which is the sum of flows into the Northwest Washington Area bubble, i.e.,

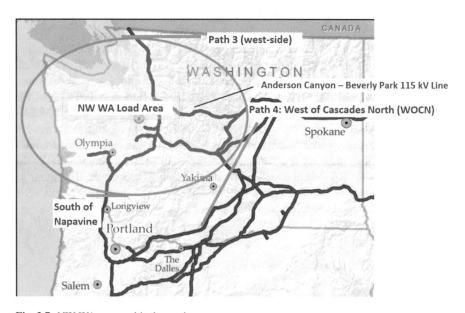

$$\text{NW WA Area Import} = \text{Path 3 Westside}\left(N > S\right) +$$

$$\text{WOCN}\left(E > W\right) + \text{SON}\left(S > N\right) + \text{AC} - \text{BEV}\left(E > W\right)^{i} \qquad (3.6)$$

The import interface consists of a bunch of Bulk Electric System (BES) elements. A number of [N-1] or [N-2] contingencies are identified by the entities to be impactful for the IROL calculation.

Peak RC need to determine the IROL in the Operational Planning Analysis (OPA) process ahead of real time and in real time based on the RT-VSA tool. Peak RC need to evaluate the IROL margins and update the IROL manually as necessary.

3.4.1.2   Validation and Benchmarking Test

To validate accuracy and robustness of the POM/ROSE core engine, we implemented the identical NW-WA IROL scenario among three VSA software tools (offline): (1) V&R POM/ROSE (VR), (2) PowerTech VSAT (VSAT), and (3) PowerWorld (PW). To improve test efficiency, we also set up an offline batch mode study process for each of the tools to run VSA study for a number of WSM SE snapshot cases automatically. The batch mode study has been extensively used for all IROL VSA calculation validation and troubleshooting for RT-VSA solution quality issues.

In order to test the voltage stability results of the ROSE software using Peak RC's State Estimator (SE) solution, it is first necessary to verify that the voltage stability tool is operating properly. Any discrepancies or differences found need to be explained. In order to have a reference point(s) for comparison, two well-known and vetted software packages were chosen to perform this verification. These are PowerWorld and PowerTech's Voltage Stability Assessment Tool (VSAT).

V&R ROSE reads the bus-branch model exported from Peak's Energy Management System through the topology processing performed by Alstom. However all zero-impedance branches (ZBRs) are maintained in the V&R model. V&R also maintains certain breakers for outages and pre-defined contingencies in order to properly model how the topology would change during contingencies.

PowerTech's VSAT reads a bus-branch model in the form of Siemens PTI v30 .raw file.

PowerWorld reads the full node-breaker model exported from Peak's Energy Management System (EMS) from Alstom and internally maintains a full node-breaker model.

3.4.1.3   VSA Limits vs. Margin

Peak's calculation is based upon the VSA results from V&R ROSE of the real-time output from the SE. The VSA tool's purpose is to provide not a limit but rather a "margin" (Margin), which is defined as how far the system is currently from the

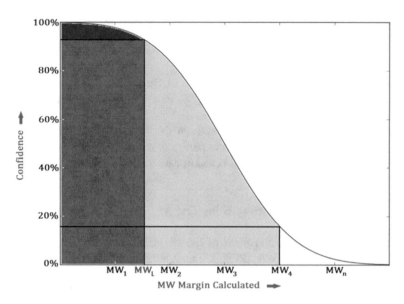

Fig. 3.8 Survival function example with a Gaussian (normal) distribution

limit. This is what the standing orders are based upon, so it is of particular concern to Peak.

A Margin is used because the approach of the V&R VSA tool does not necessarily produce an operating limit, but rather a "breaking point" where the powerflow fails to converge. In an offline voltage stability study, the system is able to be stressed as it would normally (i.e., maintaining generators within their normal operating band) resulting in the determination of an IROL. However, the approach taken by *any* online tool, which does not allow more units to come online as the system is ramped, results in a "breaking point" (the difference between a true stability limit and a stability limit that is due to a loss of reactive reserves). However, the closer the actual flow is to the IROL, the less difference exists between how the offline and online study is performed. This relationship can be illustrated by a simple survival function (complimentary cumulative distribution function) as seen in Fig. 3.8.

3.4.2   Testing Under Normal Conditions

These cases were taken from the actual SE solutions from various time periods. The Margins were calculated by each program.

3.4.2.1   Initial Flow Values

All initial flow values (interface flows) in the different programs have been verified. See the results below showing Monday morning load pickup data that continues through Tuesday evening (December 29 00:00–December 30 23:59). Forty-eight hours of data are shown below.

3.4.2.2   Margin Calculation Results – Test Case

During normal system conditions, the margins happened to be very high which did not necessarily provide the best comparison of results between the three different programs. For that reason, only one Margin calculation will be shown as it was the highest loading seen in the Pacific NW during the program testing. This occurred on a Monday morning load pickup and continued through Tuesday evening (December 29 00:00–December 30 23:59). Forty-eight hours of data are shown below.

3.4.2.3   Testing Result Under Outage Conditions

The system had to be heavily stressed in order to find a Margin, and as a consequence, the results begin to diverge from the other programs. This is evidenced by the V&R results recorded during these high periods of stressing. As discussed in the previous section "Peak Calculation Methodology," there is little confidence in the true calculation with a Margin which is high. To stress the system, Peak RC took several 500 kV outages on WECC Path 4 (West of Cascades North).

3.4.2.3.1   Margin Calculation Results – Test Case 1

Peak RC performed the same study under 3.4.1.5 but with the Shultz-Raver 3&4 500 kV lines out of service. This was performed by taking the WSMExport.csv and the .raw file and applying these outages before being imported into the program. These outages drop the limit considerably but also cause the previously seen divergences in V&R to become minimal.

Using this process, VSAT was producing unreliable values which skewed the results (discussed in the Conclusion under next section). This was witnessed in all outage cases. For this reason, PowerWorld will be the only program used for comparison for these artificial outage conditions.

3.4.2.3.2   Margin Calculation Results – Test Case 2

A similar test was performed previously with the Shultz-Raver 3&4 500 kV lines out of service but during a lower loading scenario (December 9 16:00–December 10 16:00).

There was also a correlation of the results between PowerWorld and V&R. These are shown in Fig. 3.13. A linear fit was used, and an almost direct correlation with a small offset shows that these programs are producing very similar results for the data set given. The slope and intercept are displayed as an equation on the plot.

3.4.2.4   Conclusions

Remarks Figure 3.10 gives 48 h of VSA solution data under normal system conditions; the margins happened to be very high which did not necessarily provide the best comparison of results between the three different programs. For that reason, only one low Margin calculation will be shown as it was the highest loading seen in the Pacific NW during the program testing. This occurred on a Monday morning load pickup and continued through Tuesday evening (December 29 00:00–December 30 23:59).

To stress the system, we took several 500 kV outages on WECC Path 4 (West of Cascades North) and re-ran the VSA study. As shown on Figs. 3.11 and 3.12, these outages drop the limit considerably but also cause the previously seen divergences in V&R tool to become minimal. Note the VSAT tool was excluded for outage case comparison. It has been shown that for the given data sets, V&R ROSE VSA tool has proven to be comparable with other software vendors for the outputted results. To mitigate VSA solution oscillations in high margin cases and ensure computation performance, Peak RC limited the max transfer level without overlooking actual operation risk. From the continuous validation test results, the ROSE RT-VSA tool has demonstrated adequacy in both its performance and reliability.

3.4.2.4.1   V&R Energy POM/ROSE

The studies provided similar results to the other programs as evidenced in Figs. 3.9, 3.10, 3.11, 3.12, and 3.13. This is the only program out of the three that Peak RC currently runs every 5 min from the SE solution. The availability of this process and results can be seen in Fig. 3.14. The top part of the graph is the import value as monitored, and the bottom (in green) is the calculated margin.

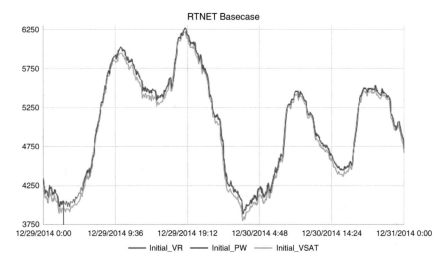

Fig. 3.9 Initial interface flows calculated by three tools

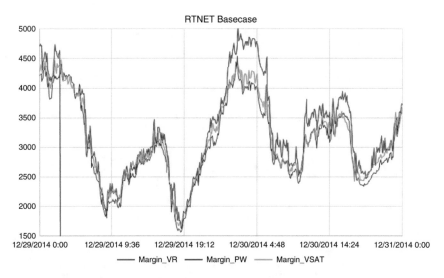

Fig. 3.10 Calculated margin December 29, 2014–December 30, 2014

3.4.2.4.2   PowerWorld

PowerWorld performed well and was numerically stable under all circumstances. It offered full visibility of the system and provided functionality that was needed to test the impact of given limitations for the other programs.

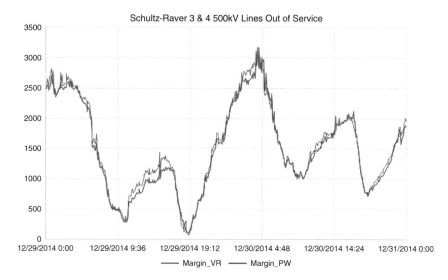

Fig. 3.11 Calculated margin December 29, 2014–December 30, 2014, with 500 kV lines outage

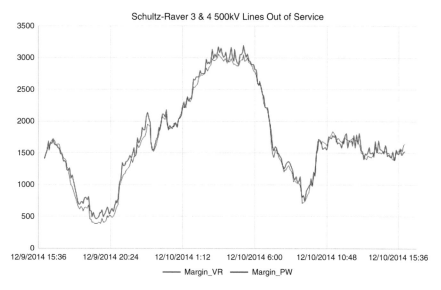

Fig. 3.12 Calculated margin December 9, 2014–December 10, 2014 with 3&4 500 kV line outages

All processing used was through the PowerWorld's own "Script" commands that were called in the program. Through the use of PowerWorld, Peak was able to perform the following:

• Verification that the inclusion/exclusion of pseudo-injections and non-conforming load appears to have negligible impact

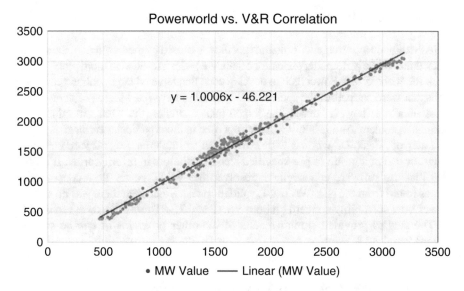

Fig. 3.13 Correlation of results for Test Case 2 (PW vs. VR)

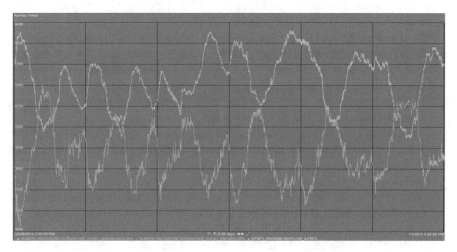

Fig. 3.14 Availability of the RTNET solution and subsequent V&R RT-VSA calculation

- Verification that the inclusion/exclusion LDC/RCC on these studies appears to have negligible impact

The studies provided similar results of the other programs as evidenced in Figs. 3.9, 3.10, 3.11, 3.12, and 3.13.

3.4.2.4.3 PowerTech VSAT

VSAT exhibited a convergence problem specifically caused by the selection of the system slack units. These had smaller generator capabilities which resulted in more problems, so Peak RC worked with PowerTech to develop code that targeted larger online generation units that were not part of the source. This seemed to have fixed these problems under most situations; however during the forced outage studies, the slack generator again became a problem. As a result, VSAT was not used in those particular verifications.

Later on, Peak RC worked with PowerTech to implement a software enhancement for use of a group of pre-specified slack generators in priority order. It's verified that the software enhancement resolved the VSAT power flow convergence issues found from the NW-WA IROL validation test. Since then there was no similar power flow solvability concern on both VSAT and TSAT tools of PowerTech.

The studies provided similar results of the other programs as can be seen in Figs. 3.9 and 3.10.

3.4.2.4.4 V&R POM/ROSE RT-VSA in the Control Room

It has been shown that for the given data sets, V&R ROSE RT-VSA tool has proven to be comparable with other software vendors for the output results.

The use of a VSA tool to monitor the system in real time is a significant improvement over the current model of using offline analysis which does not account for changes in the system as it occurs. From Fig. 3.10, where the blue/yellow trend on top are the actual flows and the green is the margin calculated from V&R ROSE RT-VSA, one can see the robustness of the tool. This figure is a representative sample from December 29, 2014, to January 4, 2015.

In conclusion, Peak RC engineering and management have determined that the V&R ROSE RT-VSA tool has demonstrated adequacy in both its performance and reliability and therefore ready for operational usage.

Offline version of the RT-VSA software was developed for ROE and operation planning engineer to validate the real-time tool results and calculate day-ahead IROL values manually. The offline POM/ROSE VSA software release was built to work in two options: a *.msi installer coded with user's work station IP address (soft key security) for installation in EMS environment and another installer works for installation in one's laptop that requires for a dongle drive.

Peak RC continually validated the RT-VSA-calculated NW-WA IROL value against BPA tool's results through the winter of 2016. BPA developed a three-dimensional nomogram for all lines in service (ALIS). It is dependent on NW-WA Gen, NW-WA Net Load, and Path 3 (Western 500 kV) MW values. However, they consider the Net Load vs. Peak's consideration of the import quantity. The difference being that the Net Load also includes any generation.

Since overall the RT-VSA tool performance and quality for the calculated NW-WA IROL were reasonable and trustworthy, Peak RC collaborated with impacted parties to develop an operating procedure for real-time assessment on the NW-WA IROL by using the RT-VSA tool.

Through the validation efforts, Peak RC rolled out *NW Washington Net Area Load IROL* in production on September 10, 2015, successfully.

After the cutover, there were a few NW-WA IROL VSA limit drops or oscillations reported by control room engineers. After investigation, it's noted that stressing step size has big impact on the VSA solution quality and performance, e.g., use of a smaller step size may produce a more accurate solution but yet takes longer solving time. For instance, 10 MW step size provides more accurate results, while 100 MW step size can solve all IROL scenarios in 5 min. As a tradeoff, a step size of 25 MW is chosen by default.

3.4.2.5 Lessons Learned

It was noticed the ROSE VSA tool sometimes compute a more conservative VSA limit for NW-WA IROL scenario when shunt capacitors are enabled for switch for basecase [N-0] and/or [N-1/N-2] contingency than the VSA limit solved with automatic shunt switching disabled.

The Northwest Washington region is a heavily compensated transmission network, where large shunt capacitors and reactors are installed at a number of 230 kV, 345 kV, and 500 Kv substations. In principle, enabling switchable shunt device will improve the long-term voltage stability limit margin to some extent. It's confirmed from other Peak RC IROL scenario VSA limit calculations. However, calculation of NW-WA IROL Cutplane VSA limit experienced some anomalous issues during initial validation. In comparison of Figs. 3.15 and 3.16, one can see a few solution issues:

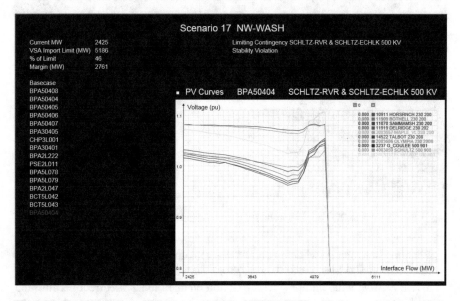

Fig. 3.15 The VSA limit and PV curves solved for NW-WA IROL with shunt switching enabled

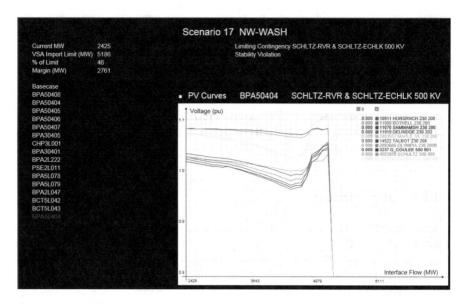

Fig. 3.16 The VSA limit and PV curves solved for NW-WA IROL without shunt switching enabled

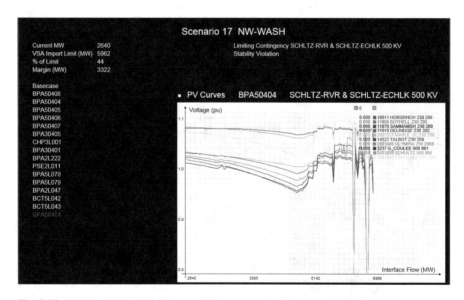

Fig. 3.17 NW-WA IROL VSA limits and PV curves solved by a step size of 50 MW

- Enabling auto shunt switching causes the VSA limit of NW-WA IROL Cutplane to be solved 500 MW lower than disabling auto shunt switching.
- The PV curve with shunt switching option enabled looks incorrect because the VAR compensation due to shunt switch successively boosts the critical bus volt-

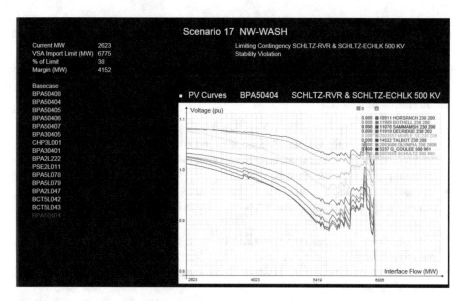

Fig. 3.18 NW-WA IROL VSA limits and PV curves solved by a step size of 25 MW

ages even above the basecase voltage profiles before having a sudden collapse. There is no oscillation on bus voltage change as anticipated to see in a real voltage collapse event.

- The PV curve with shunt switch option disabled looks more sensible, as a group of monitored buses continuously drop their voltage magnitudes to 0.80 p.u. while the system becomes collapsed.
- The VSA solution issues on the NW-WA IROL Cutplane scenario becomes worse when a smaller dead band (<0.02 p.u.) is adopted for auto-switchable shunt devices, because the shunt banks are forced to switch in and out more frequently to meet the regulation schedules.

Also, further simulation study shows the ROSE VSA solution robustness could be improved considerably while solving the VSA limit for the NW-WA IROL Cutplane at a smaller step size (say 25 MW). From the simulation results in Figs. 3.17 and 3.18, one can clearly observe a difference between two sets of VSA limits and PV curves solved by a step size equal to 50 MW and 25 MW, respectively. In particular, the VSA limits and margins are calculated higher by over 800 MW when a smaller step (25 MW vs. 50 MW) is applied.

The PV curves using a step size of 25 MW look most accurate and match with the text book, and show much less dropout, in comparison to the PV curves solved by a step size of 50 MW.

It's not a coincidence to note that the PV curves using a step size of 50 MW look more reasonable than what's solved at a step size of 100 MW (see Fig. 3.15).

3.4.2.6 ROSE Software Enhancements in Automatic Shunt Switching Logic

The shunt switching test on the NW-WA IROL VSA scenario promoted the POM/ROSE VSA software enhancement to implement automatic shunt switching logic requirements as follows:

- The user input flag must be set. This flag decides if individual shunts are eligible for switching. Note: Within EMS application software, there are global options that allow/disable shunt switching for all shunts in the model.
- A valid regulated node must be specified.
- A regulation voltage target in p.u. and an allowable deviation in p.u. must be specified for a switchable shunt. The target can be a scheduled value or manual entry.
- The regulated bus (bus corresponding to the regulated node) must be part of a solvable island.
- If switching by sensitivities is desired, the sensitivity field for each switchable shunt must be modeled.
- If the shunt has an associated switching circuit breaker, then the far side of the breaker (the terminal of the breaker that is not common with the terminal of the shunt) must be connected to a live bus, if the breaker is currently open, for the shunt to be a candidate for switching into service.

3.4.2.6.1   Shunt Model Processing Requirements

Before the network model is put online, running the model validation puts multiple switchable shunts at a bus into a bank so that the switching of those devices can be coordinated. When a single shunt regulates bus voltage, it is the lone member of the bank. When there are multiple capacitors or reactors or a combination of the two regulating the same bus voltage, they're banked together. If the shunt priorities are specified, then the priorities are used to order the shunts within the bank. If priorities are unspecified, the nominal MVAR is used to order the shunts. The shunts are ordered with the capacitors first (in descending order of priority or nominal MVAR), and then the reactors (if any) are ordered next in ascending order of nominal MVAR (i.e., a reactor having a nominal of −10 MVAR is considered as having higher precedence for switching than a −25 MVAR reactor, etc.). When ordered in this fashion, the shunt with the highest precedence is the bank master.

3.4.2.6.2   Two Switching Logic

The shunt switching logic can be optionally invoked in the following two ways:

- Switching by sensitivity

- – When this option is selected, the sensitivities must be modeled, and the rest of the criteria as specified in the section titled "Requirements for Switchable Shunts" are met.
- Switching based on priority
  - – When this option is selected, the individual shunt switching priorities for multiple switchable shunts banked together must be specified. If this option is selected but priorities are not specified, the nominal MVAR decide the priority with the shunts ordered in descending order of nominal MVAR. If two shunts have unspecified priorities and their nominal MVAR are the same, their priority order relative to one another is the same as the order in which they are modeled in the database.

Irrespective of the global option selected, there are some aspects of shunt switching that are common to both methods. They are as follows:

- The logic to perform shunt switching is performed in an "outer loop," i.e., outside of the iterative matrix update operations. If any shunts are switched by the logic, this is considered as a problem redefinition, and the iterative solution process is run again till the process converges or the maximum allowable iterations are reached.
- If there are multiple VAR sources (generators, under loading tap changing transformers (ULTC), switchable shunts, etc.) regulating the voltage at the same bus, they are switched according to a global priority. E.g., if the priority is set to generators, 1; shunts, 2; and LTC, 3, then at any bus that has two or more of these device types regulating the bus voltage, all devices of the type of the highest priority are moved till the regulated voltage is within an allowable deviation of the desired p.u. target or all devices are at their reactive limits. If it is the latter case, all devices of the type having the next lower priority are moved to bring the regulated voltage within allowable limits, etc. This means that shunts are eligible for switching only when devices of other types at the same regulated bus that have a higher specified priority are at their reactive limits, unless the shunts have the highest global priority among different device types.
- For every invocation of the shunt switching logic, only one shunt per station is allowed to switch in/out. This is to minimize switching operations.
- The preference is to allow switchable shunts to disconnect first, in an attempt to bring the voltage within the desired range. When all shunts at a bus have been disconnected and the regulated voltage is still not acceptable, then offline switchable shunts (if any) are switched in till the voltage meets the target requirements.
- Note: this indicates when the bus voltage violates the low kV limit, shunt reactors connected with the regulation bus disconnect first before switching in offline shunt capacitors to boost the voltage.

In addition to the above steps that are common to both types of shunt switching, the following switching actions are specific to each option.

• Functionality specific to switching by sensitivity

 – At the beginning of the solution, sensitivities are initialized to their modeled
   values (i.e., SENST_CP). As the solution progresses and a shunt is switched
   in/out, an acceleration factor is applied to the sensitivity values. Currently, the
   acceleration factor is a hardcoded value of 1.20. In general, it must be any
   value greater than 1.0, and a value between 1.0 and 2.0 is recommended.

• Functionality specific to switching by priority

 – At the beginning of the solution, if a bus has multiple switchable shunts in a
   bank, it is unacceptable for both capacitors and reactors to be connected at the
   bus. If there are connected capacitors, then all reactors are disconnected.
 – If the number of switching actions for a shunt over the course of the application solution process exceeds a user-specified globally defined threshold, then
   that shunt and all the other banked shunts are blocked from switching any
   further.

After the shunt switching logic enhancement was implemented in the ROSE
VSA tool, Peak RC can calculate the NW-WA IROL Cutplane VSA limits by the
default step size (i.e., 100 MW) correctly and within a cycle of 5 min.

3.4.3 Validating SDG&E Import and SDG&E/CFE Import IROLs

3.4.3.1 Introduction

High loads and heavy imports into the Southern California Edison, San Diego Gas
and Electric, and Centro Nacional de Control de Energia (CENACE) areas may
result in unacceptable pre- and post-contingency performance, including steady-
state voltage instability in the area. Peak RC has identified two IROL Cutplanes for
SDG&E voltage stability studies: the SDG&E (summer) Import Cutplane and
SDG&E/CFE (winter or non-summer) Import Cutplane (lately renamed to SDG&E/
CENACE Import Cutplane). Both IROL Cutplanes and adjacent areas are illustrated
in Fig. 3.19.

The SDG&E/CFE Import Cutplane is utilized when CFE RAS is set to cross trip
TL23050 or when TL23050 is out of service. The total import of SDG&E/CFE
Cutplane is calculated by adding the active power flows on the following lines:

• South of San Onofre Cutplane flow
• TL50001 flow metered
• TL50003 flow metered
• TL23050 flow metered

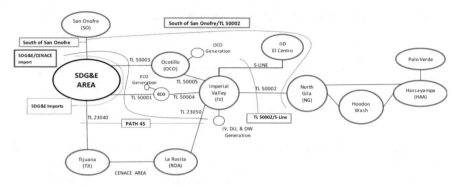

Fig. 3.19   SDG&E area overview (reference: GIP 2005)

The SDG&E Import Cutplane is utilized when CFE RAS is set to cross trip TL23040 or when the TL23040 is out of service. The total import of SDG&E Cutplane is calculated by adding the active power flows on the following lines:

- South of San Onofre Cutplane flow
- TL50001 flow
- TL50003 flow
- TL23040 flow

The information provided below are study assumptions agreed among all impacted parties to validate SDG&E and SDG&E/CFE Import VSA limits calculated by Peak RC and CAISO for every 5 min, respectively.

3.4.3.2   Study Assumptions

3.4.3.2.1   Scenario Definitions

Peak RC V&R POM/ROSE VSA tool models two scenarios to calculate the voltage stability limits at all times.

- SDG&E import scenario
- SDG&E/CFE import scenario

Apart from these two scenarios, CAISO Bigwood VSA tool further differentiates the scenarios by hardcoding the CFE cross trip RAS actions. This results in four scenarios:

- SDG&E import scenario with and without cross trip action from the "Overload of CFE's Valle – Costa Path SPS," respectively.
- SDG&E/CFE import scenario with and without cross trip action from the "Overload of CFE's Valle – Costa Path SPS," respectively.

Table 3.5 A summary of the IROL Cutplane scenario definitions

SDG&E import scenario		SDG&E/CFE import scenario	
Peak RC	CAISO	Peak RC	CAISO
Scenario-1SDG&ESummer.csv	SDG&E import with no cross trip	Scenario-2SDG&EWinter.csv	SDG&E/CFE import with no cross trip
	SDG&E import with cross trip		SDG&E/CFE import with cross trip

For the CAISO cross trip scenarios, the Bigwood tool will operate the "Overload of CFE's Valle – Costa Path SPS" to trip TL23040 or TL23050 as applicable.

While CAISO calculates the Import Limits with respect to cross trip and no cross trip scenario at the same time in the VSA tool, CAISO send out the higher Import Limit between no cross trip and cross trip scenario. Since cross trip actions occur only when thermal limit is exceeded, which is monitored by no cross trip scenario, cross trip scenario VSA limit calculated prior to thermal overload violation is not a valid Import Limit (Table 3.5).

3.4.3.2.2   Contingency List and Monitored Bus List

Five contingencies and a dozen of monitored buses being selected for VSA studies are identical for both Peak RC's V&R and CAISO's Bigwood VSA tools.

3.4.3.2.3   Source and Sink Definition

The Source is defined by generations in SCE and APS for both IROL Cutplanes. In both VSA tools, all Source generations do not include pump, renewable (wind, solar), and nuclear generations. Synchronous condensers are excluded from the source group as well if they are modeled as a generator. Online generator units in synchronous condenser mode are maintained in that mode and are not used in the "Source."

The Sink of the SDG&E Summer Import Cutplane is the loads in SDG&E area, excluding all non-conforming loads identified by SDG&E.

3.4.3.2.4   Stressing Parameters

Transfer parameters and options for the tools that are active are as follows (Table 3.6):

Table 3.6 A summary of stressing parameters adopted in Peak RC and CAISO VSA tools

Peak RC V&R POM-ROSE VSA tool	CAISO Bigwood VSA tool
Cutplane is stressed at maximum of 5000 MW more	No limit is set for the maximum MW the Cutplane can be stressed to
Transfer is increased by 100 MW each iteration	
Transfer is refined by 10 MW iterations around the nose point	Transfer is increased/refined by steps intelligently chosen by tool for each iteration; typically the tool chooses small step sizes as it nears the PV nose
Stop and issue a message when Source generation is exhausted message	
Reverse transfer feature, if basecase limit is reached; reduce transfer until the IROL margin meets the threshold.	

3.4.3.2.5   Automatic Shunt Devices

The Peak-ROSE and Bigwood tool is set to allow the switching of SDG&E and auto-switchable shunts both pre- and post-contingency. SDG&E's procedures TMC1005, TMC1005a, TMC1005b, and TMC1005c contain the automatic shunt devices information needed to model the shunt capacitors and reactors along with their automatic voltage control information that needs to be included in the model. These procedures are updated frequently to reflect new additions or other changes; many of these changes are not captured by the other processes. SDG&E distributes the procedures to Peak RC and CAISO every time there is a change in these procedures. Peak RC and CAISO should follow their internal process to accurately and expeditiously reflect such changes in the tools.

The Peak RC-ROSE tool and CASIO Bigwood tool also allow shunt devices with AVR enabled in SCE to switch during stressing.

In both VSA tools, LTC transformers in SDG&E, SCE, and CENACE areas are disabled for case stressing and post-contingency voltage regulating. This has been confirmed with subject matter experts at SDG&E, SCE, and CENACE. A full list of automatic shunt devices and their control parameters are accessible in the applicable entities operating document.

With the addition of the Imperial Valley phase shifter, both VSA tools should be configured such that there is no automatic action. Tap settings are locked. This is the preferred tool setting since market operations control the Imperial Valley phase shifter tap setting and runs every 15 min. No other phase shifters are used in operations for this interface. All other phase shifters will be locked in place.

3.4.3.2.6   Monitored Elements

In Peak RC ROSE VSA tool, the monitored buses for PV analysis are pre-defined by the user and are shown below under Peak-ROSE columns. In CAISO Bigwood tool, the PV curve of each monitored bus will be reported. It should be notice that the V&R ROSE tool will calculate the weakest bus of the system regardless of the monitored buses. The weakest bus is where the VSA tool indicates collapse, either by lowest voltage magnitude solution or by lowest reactive margin at a bus. Therefore, the list of monitored buses will not affect the final result. All scenarios share the same monitored buses. The list shows the monitored buses for SDG&E and SDG&E/CFE IROL scenarios.

3.4.3.2.7 RAS Modeling

Currently there are a dozen of impacted RAS modeled in Peak RC ROSE VSA tool using Visual Basic (VB) scripts, but two Safety Net schemes – "South of San Onofre Safety Net" and "North Gila Safety Net" – are not modeled. Those RAS were identified by CAISO, SDG&E, CFE, etc. impacted parties.

As noted in Sect. 3.2, the VSA tool at the CAISO has the CFE Costa Valley SPS modeled and hardcoded in the contingency definition, which means that in the two scenarios with cross tripping, the TL23040/TL23050 will always be cross tripped followed by the contingency of TL50001, TL50003, TL50003, or TL50005, regardless of the post-contingency flow on the CFE internal 230 kV lines.

The CAISO do not model the San Onofre Safety Net actions in the VSA because it is not the normal practice for CAISO to allow the Safety Net to operate following a single contingency.

In addition to the hardcoded CFE Costa Valley SPS, several other Remedial Action Scheme (RAS) logics that may affect SDG&E voltage stability are modeled explicitly in the VSA tool at the CAISO. The RASs are programmed using the script language provided by the tool.

3.4.3.3 Peak RC vs. CAISO VSA Calculation Validation

Starting around April 2015, Peak RC collaborated with CAISO to compare and validate both SDG&E (summer) Import and SDG&E/CFE (winter) Import IROLs calculated by Peak-ROSE vs. CAISO Bigwood tools in real time.

3.4.3.3.1 Initial Setup in Peak-ROSE-VSA Tool

- Run x critical N-1 500 kV contingencies + y N-1 500/230 kV transformer contingencies.
- Only model the identified powerful RAS in VB scripts.
- Control parameters were set upon CAISO & SDG&E procedures.
- Read Real-Time RAS Arming Status from EMS, and keep the Arming status unchanged during stressing.
- Unit governor response against specific contingency is proportional to Pmax.

- Unit AGC response to stressing or reverse transfer is proportional to the current unit output with respect to Pmax or Pmin.

  – Certain hydro units due to economic operation and/or maintain regulation reserves will not operate to Pmax.

3.4.3.3.2   Determination of the IROLs and Sharing of VSA Limits

- Peak VSA results – primary.
- CAISO VSA results – primary backup.
- In case both VSA tools fail, Peak real-time engineers will perform offline study using real-time cases and Peak-ROSE.
- Peak and CAISO results were shared with each other and impacted entities via ICCP.

Peak RC and CAISO compared both VSA results solved in real time and investigated significant factors resulting in the differences between two tools. Through close validation collaboration, we're able to identify a number of the modeling gaps and tool setting issues. Particularly,

- SE model scales (WSM full model vs. CAISO regional model) and SE solution differences (CAISO SE could solve a native active power load to minimize bus mismatch).
- RAS model representation – Peak model the RAS by VB scripts (more flexible and accurate), and use RAS Arming status from EMS for both basecase and stressing iterations. CAISO model RAS actions approximately in contingency definitions and enable re-evaluating RAS Arming status on the course of stressing.
- Shunt switch setting for reactive power regulation.
- Handling of loads with negative reactive power – due to reduction of a local distribution network with DERs, some equivalent loads are modeled at a lower kV bus of high-voltage substation in the WSM. The MW and MVAR load values in the base case are estimated from relevant line flow measurements. Under certain circumstance, MVAR line flows could become negative when non-modeled local shunt devices are switched in to boost local bus voltages.

3.4.3.3.3   RAS Modeling and Auto Shunt Switching Impact Validation

The VSA scenarios for both IROLs were initially validated between January 2015 and July 2015 and re-validated between October 2015 and April 2016 after the failure in Peak RC ROSE-VSA tool caused inappropriate load shedding in SDG&E service area.

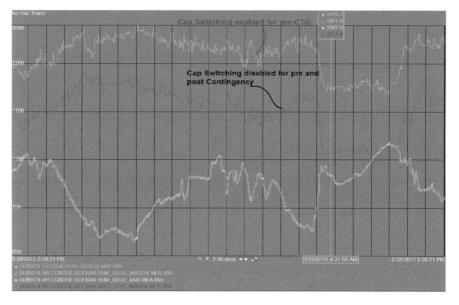

Fig. 3.20 Two-day SDG&E import VSA results with and without shunt switching

In February 2015, Peak RC conducted a series of testing to verify impact of shunt capacitor switching to SDG&E Summer Import VSA limit results. Figure 3.20 shows the plots of 2-day real-time VSA calculation results. Particularly,

- The SCADA flow measurement (green line) matched well with the ROSE-VSA solved the ROSE-VSA calculated interface flow (brown line).
- The VSA margin difference with shunt capacitor switching enabled (blue line) and disabled (pink line) in basecase (pre-contingency) varies from 250 MW to 700 MW. In a word, without enabling shunt capacitor switching, the SDG&E Summer Import Limit could be reduced by 10% at least.

Further validation test was performed for the SDG&E Import IROL scenario using two different control settings: conservative one with RAS and auto shunt switching models disabled vs. the default one with CGCC auto shunt switching and RAS modeled and enabled. Figure 3.21 shows 24 h real-time testing results on June 5, 2015. The results clearly indicate the VSA limits won't be valid and credible if the required RAS and auto shunt switching schemes are not modeled and enabled for VSA limit calculation.

The validation test and comparison were conducted for the VSA margins solved by Peak RC V&R and CAISO Bigwood tools, respectively. The testing results are summarized in Fig. 3.22, where CGCC is a centralized auto shunt switching scheme and can be enabled for basecase and/or post-contingency conditions. Conservative case has no RAS and auto shunt switching enabled. One can see from the plots that CAISO solved VSA margin is between Peak's conservative and default setting solution, and the margins by both tools and under different control settings vary with

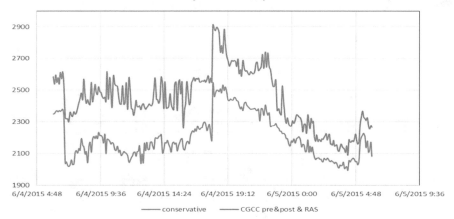

Fig. 3.21 Auto shunt switching and RAS impact study by Peak RC ROSE-VSA

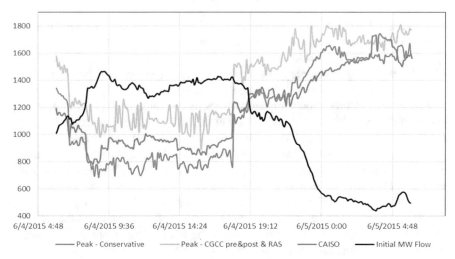

Fig. 3.22 Comparisons of Peak RC vs. CAISO VSA margins

changing basecase flow level. Overall Peak RC tool with preferable settings solves the highest VSA limits/margin.

In Fig. 3.22, note that Initial MW Flow is solved basecase flow on the IROL Cutplane; CAISO represents the Bigwood tool solved VSA margin; Peak conservative means the solved VSA margin without RAS and CGCC enabled; Peak-CGCC pre and post and RAS is the VSA margin solved with preferred RAS and CGCC models enabled.

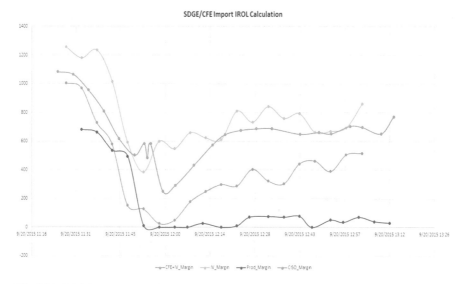

Fig. 3.23 RAS impact to the VSA results on September xx, 2015, load shedding incidence

There are 7 500 KV N-1 contingencies and 13 RAS associated with the South West IROLs identified by the entities till date. Before September xx, 2015, load shedding event, only seven RAS are modeled in Peak-ROSE-VSA tool. There were a few RAS missing or inappropriate modeling for some reason. The gap in the RAS modeling impacted the IROL calculation on this event case poorly.

Figure 3.23 clearly shows impact of RAS model settings to the SDG&E/CFE Import IROL calculations on a system event that led to unnecessary load shedding: (1) Prod_Margin is Peak RC production RTVSA results without CFE and IV RAS modeled; (2) IV_Margin is Peak RC offline VSA study results with IV RAS modeled; (3) CFE_IV_Margin is Peak RC offline VSA study with CFE and IV RAS modeled; (4) CISO_Margin is CAISO production Bigwood VSA tool results (no CFE RAS modeled). Per coordinated ad hoc event analysis among Peak RC, CAISO, SDG&E and CFE, both CFE and IV RAS shall be modeled in the RTVSA tool, but CFE RAS won't be fired to reduce the IROL. By validation, IV_Margin produces the best estimate of the SDG&E/CFE IROL on the event.

Following this incidence, Peak RC temporally pulled down the real-time VSA tool and left the IROL calculation to offline manual study for a while. The EMS team worked with entities and V&R to make a few improvements:

- Identified missing RAS models and made them up in the VSA tool
- Re-validated RAS models, contingency definitions, and control parameter settings for all activated VSA scenarios
- Improved V&R POM/ROSE engine robustness in multiple areas: fixed limit drop to zero when there is no VSA margin, collectively reactive power regulation logic (Var control coordination among units and shunt devices regulating the

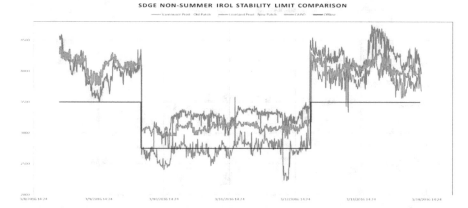

Fig. 3.24 SDG&E non-summer IROLs: real-time calc. vs. day-ahead study on March 8–March 14, 2016

    same bus); scaling generation by the unit current MW during stressing to avoid large change of individual output

- Added logic to enable reverse transfer study when the basecase is not solved against a contingency
- Implemented an EMS power flow-based StudyVSA tool for engineers to validate the ROSE-VSA results as needed

    After extensive re-validation efforts, Peak RC put back the ROSE-VSA tool for real-time assessment on the IROLs in May 2016.

3.4.3.3.4   Real-Time VSA Limits Under Outages

It's important to validate the credibility of real-time VSA results under major system outages that are sensitive to the IROL Cutplanes. It's anticipated to see VSA limit drop when major outages relevant to the IROL Cutplane occur. On the other hand, stressed system conditions could make the VSA tools run into power flow diverging issues earlier before approaching the nose point, i.e., a false voltage collapse point.

    Figure 3.24 shows the real-time VSA results on the week of March 8, 2016, including a 2-day outage period between August 10 afternoon and August 12 morning. From the plots, the highest VSA limit curve during the outage period was solved by the new V&R POM/ROSE software installed at Peak RC Loveland office, while the lowest VSA limit (blue curve) during the outage was solved by the old V&R POM/ROSE software installed at Peak Vancouver office; the lowest VSA limit curve without outage (flat line) was calculated by CAISO offline study tool; the CAISO real-time Bigwood tool solved the VSA limits in a green curve between two VSA curves solved by Peak real-time tools.

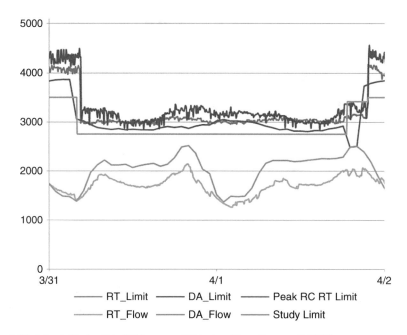

Fig. 3.25 March 31–April 1, 2016, outage VSA plots [Data Source: CAISO]

In March 31 through April 2, another major outage occurred and lowered the SDG&E Non-Summer IROL calculation results significantly. Multiple VSA solution results are compared in Fig. 3.25, including CAISO real-time interface flows and VSA limits (denoted by RT-Flow and RT-Limit), CASIO day-ahead study interface flows and VSA limits (denoted by DA-Flow and DA-Limit), and Peak RC calculated real-time VSA limits (denoted by Peak RT VSA limit). The outcome shows both real-time VSA tools line up well for the calculated VSA limits and day-ahead study tools give the most conservative VSA limits, regardless of normal or outage conditions.

Overall both real-time and CAISO day-ahead VSA study tools provided similar results as shown in Fig. 3.25.

3.4.3.3.5 Negative Reactive Load Stressing Test

Negative reactive loads (Q load) in Sink was estimated by SE, mostly resulted from WSM model reduction on sub 69 kV networks, which contain distributed generating resources or DERs and shunt. When stressing the SDG&E Import Cutplane, raising negative Q loads could lead to solve VSA limits too higher or too optimistic.

The ROSE VSA tool increases P/Q load in proportion to solved base values, regardless of negative Q load estimated in basecase. There are many distributed generators (DG) installed with capacitor banks. As the WSM does not model sub-100 kV typically, those DG resources are simply represented by equivalent

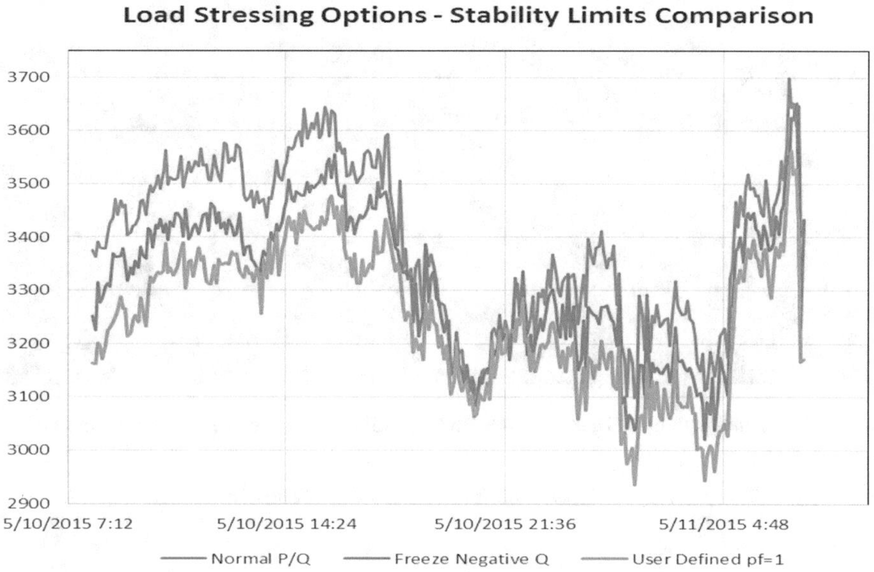

Load Stressing Options - Stability Limits Comparison

Fig. 3.26 SDG&E Import VSA limits under three negative Q load stressing options

loads. As a result, those reactive loads could be estimated either positively or negatively. Stressing reactive loads makes the IROL solve noticeable higher mistakenly.

To evaluate impact of negative Q loads, Peak RC worked with V&R to implement a software enhancement to allow for three options for stressing negative
Q loads:

- Normal P/Q ratio – scale up P and Q loads by basecase load values. This option
  is pro-optimistic.
- User-defined PF – scale up P and Q loads based on user-defined power factor.
  This option is pro-conservative.
- Freeze negative Q – keep negative Q loads constant during stressing. This option
  is relatively moderate.

From Fig. 3.26, one can see how much gaps are there among three options for
stressing negative loads. Based on the validation results, Peak RC selected option
"froze negative Q load" while calculating two IROLs. The negative reactive load
estimation issues become more obvious for the CAISO footprint, where more DG
or DER resource have been deployed thus far.

3.4.3.3.6   Continuous VSA Validation Through Collaboration with CAISO

After September xx, 2015, load shedding incidence, Peak RC started to host
biweekly conference call with external stakeholders including CAISO, SDG&E,
CFE, and SCE to review problematic VSA results between Peak RC and CAISO

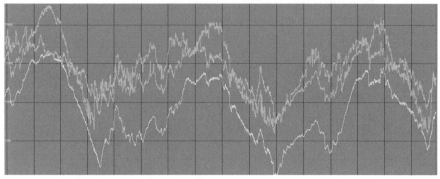

Fig. 3.27 Benchmarking of Peak RC V&R VSA, CAISO Bigwood VSA, and day-ahead study

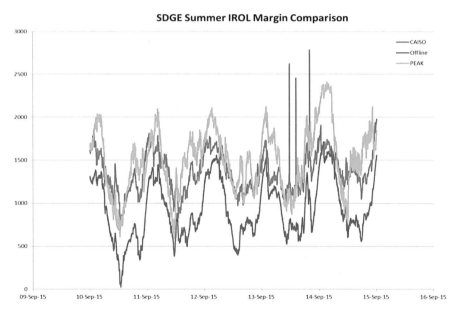

Fig. 3.28 Weekly SDG&E Import VSA margins. CAISO means CAISO RT-VSA results, offline means CAISO day-ahead VSA study results, Peak-means Peak RC RT-VSA results

tools and identify action items and deliver the resolution plan timely. Thanks to those collaborative efforts, Peak RC V&R and CAISO Bigwood VSA results were getting closer and closer.

Figures 3.27 and 3.28 show PI trend on SDG&E/CFE Import VSA limits in mid-2016, indicating both real-time VSA tools provide similar results (green and blue lines) higher than the day-ahead VSA study limits (yellow line). Without loss of generality, the day-ahead VSA study results are most conservative.

At certain time points, the VSA margin shown in Figs. 3.27 and 3.28 is less than 200 MW or even close to zero. In the meanwhile, both real-time VSA tools show more than 600 MW margins. The validation results confirm the advantages of real-time VSA over offline VSA study for accurate assessment of the IROLs.

3.4.3.4   Lessons Learned

From VSA results validation on SDG&E Import and SDG&E/CFE Import IROLs, it's understood that neither the models nor the software is flawless and correctly solvable for all system conditions. Therefore, a thorough and rigorous test must be performed for each EMS/VSA model update and V&R ROSE software patch. In the meanwhile, operation procedures and communication mechanism must be clearly defined on how to handle extreme situations such as VSA tool failure. Principally,

- It's agreed among the affected parties that Peak RC real-time VSA results are primary and CAISO real-time VSA results are backup for reliability monitoring. Both share real-time VSA results via ICCP. When both tools fail or do not match each other for a lower VSA margin case, offline VSA study needs to be performed timely by on-shift real-time operation engineer (ROE) for proper validation and analysis study for decision-making.
- Check if there is any SE solution quality issue causing the VSA tool failure or solving a very low VSA limit and margin. In case of extremely low margin or no margin solved by the RT-VSA tool, validate RT VSA results with RTCA before actions are taken.
- Review the QV curve to verify whether the weakest bus runs out of reactive power margin at the last step of stressing test. Otherwise, the VSA limit is likely a false or local maximum solution.
- If offline POM/ROSE VSA tool also obtains a similar solution, use Peak RC in-house developed StudyVSA tool to perform additional validation study. The in-house tool is built on top of GE EMS power flow engine and the RAS models in RTCA and is able to import and adopt the same VSA scenario definition files to run stressing analysis. The StudyVSA tool will identify the VSA limit upon diverging of power flow. The UI display of the tool is illustrated in Fig. 3.29 for reference.
- Under validation of Peak POM/ROSE RT-VSA solution results, we also captured a few incidences showing sudden VSA limit drop. Peak RC developed separate mitigation plans to address those issues. For example,
  - July 14, 2015, RTVSA limit dropped intermittently for a few times due to a three-winding modeling error causing negative equivalent impedance. As a mitigation action, a troubleshooting procedure was created to identify/remove similar issues quickly.
  - September 16, 2015 (10:37 AM–12:42 PM), topology error at one major station caused VSA unsolved for a critical contingency, while SE solves for a lower bus voltage mistakenly.

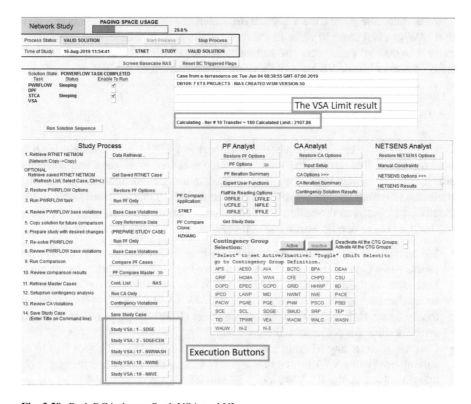

Fig. 3.29 Peak RC in-house StudyVSA tool UI

- – Sometimes RT-VSA solved lower limits because the unit D-curve was modeled incorrectly, and manual override values on Qmax and Qmin were not
  imported and interpreted properly by RT-VSA. We worked with vendor to
  deliver a software fixed to preserve and respect various manual overridden
  values in the RT-VSA tool correctly.
- Improved auto shunt switching logic for robust power flow solution. Add new
  logic to accommodate three options for reactive load stressing: (1) per basecase
  P/Q ratio; (2) per user-defined constant power factor (PF); and (3) freeze negative reactive load while increase active power load.
- Implemented multiple software enhancements to improve user's analytical capability and situational awareness on VSA solution quality.
- Those changes in software and IROL scenario definitions and RAS scripts significantly improved both SDG&E Import and SDG&E/CFE Import IROL solution quality.

3.4.3.5 ROSE-VSA Software Enhancements

In light of the findings from validating both IROLs, Peak RC worked with vendor to make corresponding software enhancements and reviewed/updated the RAS modeling scripts in collaboration with the entities.

3.4.3.5.1 Peak-ROSE Reverse Transfer PV Analysis

Reverse transfer analysis is performed when Peak-ROSE identifies a contingency or multiple contingencies that fail to solve at zero transfer level (base case condition) or if a margin is less than 5%. The new feature is developed to cover two typical cases.

Case 1: Initial Basecase "Insecure" on One or More Contingencies

In order to let the RC know how much load they may need to shed to bring the system back to within 95% of the actual stability limit, reverse transfer PV analysis is performed.

When some contingencies fail to solve at the zero transfer level, Peak-ROSE starts the reverse transfer PV analysis for all contingencies. Reverse transfer analysis starts if parameter **EnableReserseTransfer** is set to **1** in file WECC Scenario List.csv.

The analysis is performed as follows:

1. Reduce load-generation with a user-defined step. The step is defined by parameter ReverseStep in file *WECC Scenario List.csv*.
2. Apply all contingencies at that step.
3. Repeat items (1) and (2) above until we either reach a transfer level at which there are no post-contingency violations or (b) reach a user-defined threshold for maximum reverse transfer increase. A user-defined threshold for maximum reverse transfer increase is specified by parameter **ReverseIncrease** in file *WECC Scenario List.csv*.
4. After we determine transfer level at which there are no post-contingency violations, increase the transfer using a new step size 10 MW until the transfer level at which contingency(ies) fail to solve is reached.
5. From the new stability limit point, further perform reverse transfer analysis, while applying all contingencies, with **ReverseStep** until the interface flow is reduced by 5% of the limit (to meet the 5% margin requirement). The 5% is a user-defined value in the scenario file. If the 5% margin requirement can't be met, Peak-ROSE issues alarm **G004**.
6. Report how much load is reduced compared to the basecase which will be the amount of load that needs to be shed. The amount of load shed is listed in column **Load Shed** in file *EmsAlarm.csv*. The amount of load shed to meet the 5% mar-

gin requirements is listed in column **Load Shed 5%** in file *EmsAlarm.csv*. If the 5% margin requirement can't be met, column **Load Shed 5%** is left blank.
7. The values of **Load Shed** and **Load Shed 5%** are being output only when performing reverse transfer PV analysis.

Case 2: No Sufficient Margin

When Peak-ROSE still can calculate the stability limit, but the margin is less than the 5% margin requirement, reverse transfer PV analysis should be triggered to calculate the limit which meets the 5% margin requirement.

The analysis is performed by doing items 5 through 7 of Case 1. Please note that the analysis starts from **Current MW**.

Computing Import Limit, Margin, and % of Limit During Reverse Transfer PV Analysis

For reverse transfer analysis, the **Import Limit** should be the stability limit after the reverse transfer PV analysis. The **Import Limit** is the last "healthy step." Since **Import Limit** is less than **Current MW**, the **Margin** is negative. The value of **% of Limit** is computed using the following formula:

$$\% \text{of Limit} = \frac{\text{Current MW}}{\text{Import Limit}} * 100\% \tag{3.7}$$

Since **Current MW** is greater than **Import Limit** for reverse PV transfer analysis, **% of Limit** exceeds 100%.

Addition of New Alarm Code "G004. Unsolved Basecase – Reverse Transfer"

If pre-contingency case cannot be solved when performing the reverse transfer PV analysis, Peak-ROSE stops the analysis and issues alarm **G004. Unsolved Basecase – Reverse Transfer** (e.g., the state is equal to **1**) in the *EmsAlarm.csv* file (Table 3.7).

Table 3.7 VSA failure alarm code

Code	Alarm	State
G001	Unsolved Basecase	0
G002	POM Sequence Failed to Write Output	0
G003	New SE Case Not Received	0
G004	Unsolved Basecase - Reverse Transfer	1

Table 3.8 Sample of ROSE VSA output file

	A	B	C	D	E	F	G	H	I	J	K	L	M	N	O	P	Q	R
1	TimeStamp	12/30/2015 10:08																
2	Scenario #	Scenario Name	Level 1	Level 2	Invalid Input	POM Seq	Current MW	Import Limit	% of Limit	Margin	Load Shed	Load Shed 5%	Pre-CTG Weakest Bus	Pre-CTG Weakest Bus Volt	LimitCtg1	Post-CTG Weakest Bus1	Post-CTG Weakest Bus1 Volt	LimitCtg2
3	5001	Summer SDGE	1	0	0	0	3161	3431	92	270			12842 LARKSP 69		0.55 *IMVALLY_ECO_1500	12965 SLNCREST 230		0.93 *ECO_MIGUEL_1500
4	5002	Winter SDGE	1	1	0	0	3217	2914	110	-303	300		12842 LARKSP 69		0.96 *ECO_MIGUEL_1500	12731 CAMERN 69		0.73

Table 3.9 Sample of bus list input file

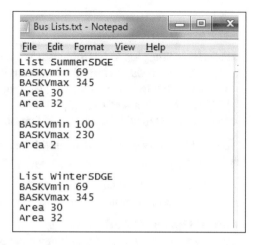

Limiting CTG(s) and the Weakest Buses

The limiting CTG(s) and weakest bus information is saved to the *EmsAlarm.csv* file as follows (Table 3.8):

This information is written at the last "healthy step." The weakest bus information is written both pre- and post-contingency for limiting CTG(s). The weakest bus naming convention includes bus number, name, and base kV.

The weakest buses are selected from the buses specified in *Bus Lists.txt* file. The file has the following block structure:

A sample *Bus Lists.txt* is shown below (Table 3.9).

POM Client During Reverse Analysis

If a contingency causes stability violation at **Current MW** and reverse analysis is performed, PV/QV curves for this scenario are not built.

To enable stopping at the first "unhealthy" step, the following changes have been made:

- Added a capability to stop stressing the system after the first "unhealthy" step is identified. All contingencies at the last "unhealthy" step are applied, and all contingencies causing a violation are identified. These are limiting CTG(s).
- On a per-scenario basis, added an option that will allow the user to perform stressing until:
  - Base case stability violation is identified.
  - Last "unhealthy" step is determined, and all contingencies are applied at that step (see Change 1 above).

Table 3.10 Sample of area
shunt switching options

I
CapSwitchingControlAreas
30#32

- Parameter **StoppingAt1stUnhealthyStep** is added to file *WECC Scenario List. csv*. When **StoppingAt1stUnhealthyStep** is set to 0, stressing is performed until basecase stability violation is identified (see Option 2.a above). When **StoppingAt1stUnhealthyStep** is set to 1, stressing is performed until the last "unhealthy" step is determined (see Option 2.b above).

To enable cap switching on a control area basis, we added flexibility to specify control area(s) where cap switching is enabled on a per-scenario basis as follows:

- Cap switching is enabled in specific control areas if **Shunts Switching** option is set to **1** in Section

11. POM Options of the Scenario file.
Control area numbers where cap switching is enabled are defined by parameter **CapSwitchingControlAreas** in file *WECC Scenario List.csv*.
If multiple control area numbers are listed, they should be separated by a **#** sign; please see example below (Table 3.10).

Format of File *WECC Scenario List.csv*
The following five columns were added in file *WECC Scenario List.csv*:

- StoppingAt1stUnhealthyStep
- CapSwitchingControlAreas
- EnableReserseTransfer
- ReverseIncrease
- ReverseStep

3.4.4   Validating Oregon Net Export (OREX) IROLs

3.4.4.1   Introduction

The Oregon Net Export, also called Northwest Net Export (NWNE), is monitored to stay under the OREX IROL limit. Historically Path 66, Path 76, Path 75, and Path 14 simultaneous exports are studied seasonally to establish coordinated operating boundaries to protect the North West grid from voltage collapse. The seasonal studies confirm simultaneous operation of the paths, up to accepted WECC ratings. They are also used to establish a nomogram if simultaneous operation is not achievable.

NW Net Export Diagram

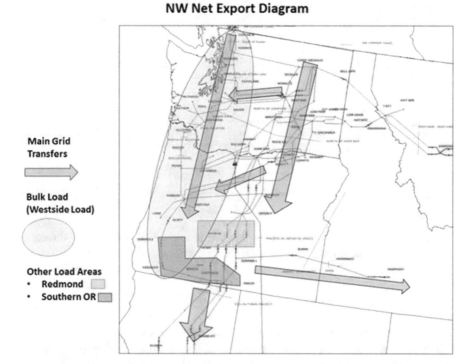

Fig. 3.30 OREX area overview

3.4.4.2   Overview of OREX Area Operations

The export Cutplane out of the BPA area is shown in the following figure (Fig. 3.30).

Peak RC has identified one IROL Cutplane for OREX voltage stability studies. Peak RC has implemented the VSA tool suite created by V&R Energy for real-time and offline modes. This tool interfaces with the Peak RC energy management system (EMS) to obtain network snapshot data from state estimator (SE) for conducting voltage stability analysis. The VSA tool runs on four nodes to provide system redundancy. The primary node data is archived on the server, and all important data transferred to EMS is archived in PI historian.

3.4.4.3   OREX IROL VSA Scenario Definition

3.4.4.3.1   Transfer Interface/Cutplane Definition

The total export of the OREX Cutplane is calculated by adding the active power flows on the following three established WECC paths according to the Cutplane definition as follows:

Table 3.11 OREX IROL contingency IDs

Peak RC contingency name	Peak RC contingency ID
PDCI BiPole	MUC2L000
Single Palo Verde unit trip	APS0U005, APS0U006, APS0U007
Single Diablo Canyon unit trip	PGA0U014, PGA0U015
PDCI MonoPole	LWPDCP3, LWPDCP4

Note: For multiple facility contingencies, follow the system performance requirements as per SOL methodology when requiring mitigation actions

For offline VSA analysis done for the OPA, there are additional breaker failure contingencies (more than ten in total) that should be run in the VSA tool

- Path 66
- Path 75
- Path 14

The power flow is assumed to be positive if the actual power flow direction matches with the positive direction specified in the above tables.

3.4.4.3.2 Contingency List

The contingencies that are selected for RT-VSA OREX VSA studies are listed in the following tables. Any contingency found to be the limiting contingency from an offline study not in this list would be run for that operation day in RT-VSA (Table 3.11).

3.4.4.3.3 Source Definition

The Source/Sink definition for each scenario is listed below. In Peak VSA all Source generations do not include pump, renewable (wind, solar), and nuclear generations. Synchronous condensers are excluded from the source group as well if they are modeled as a generator.

Gen areas defined in Source
BPA
CHPD
PSE
PGE
A subset of BC Hydro units participating North West transfer

For this scenario specific additional generator exclusions exist. For PSE and PGE, non-EIM (Energy Imbalance Market) generator resources are excluded. For BPA area Lower Snake river generators are excluded to best represent river operation restrictions. The following assumptions are made for the generation in the Source and Sink areas.

Generation

Source	Sink
1. In case stressing, the generators in the Source areas are scaled up based on their available capacities, i.e., $P^{max} - P^{online}$. Priority is given to online units on AGC with $P^{online} > 0$ MW and then to online units on AGC with $P^{online} = 0$ MW 2. All nuclear, pump, and renewable generators are excluded from scaling up in case stressing 3. Only the online generators (on AGC) will participate in the re-dispatch, according to the participation factors assigned in the model. No offline generators will be committed	1. The real outputs of the generation in the Sink areas are held constant during case stressing. The reactive outputs of the generation in the sink areas follow the reactive capability curve

3.4.4.3.4   Sink Definition

The sink definition for the scenario will be the WSM model areas for LADWP, SCE, SDGE, and PGAE. The following assumptions are made for the load in the Source and Sink areas.

Load

Source	Scenario sink
1. Load in the source areas is held constant during case stressing	1. All loads in the Sink areas will be scaled up in case stressing excluding the non-conforming loads provided by the entities 2. For loads that have negative Q, the Q component of the load will be locked during case stressing. Only P component will be scaled up

3.4.4.3.5   Case Stressing

Transfer parameters and options for the tools that are active are as follows:

V&R ROSE RT-VSA
Cutplane is stressed at maximum of 2000 MW or more
Transfer is increased by 100 MW each iteration
Transfer is refined by 10 MW iterations at the nose point
Source area is exhausted message
Reverse transfer feature, if basecase limit is exceeded, reduces transfer until expected limit is found

The SOL methodology for the operations horizon v8.1 describes stressing parameters when doing transfer analysis. Key concepts from SOL methodology include:

- Conditions when Source area is exhausted or sink is depleted, before the nose of a PV or VQ curve is reached
- Stressing significantly beyond current interface flows exceeds the 5% historical load for the real time horizon.

RT-VSA meets or exceeds these criteria, even though the methodology states "While the stressing methodology may optionally be applied to Operational Planning Analyses and Real-time Assessments, it is not required."

3.4.4.3.6   Operational Awareness of OREX IROL

In Peak EMS monitoring is achieved by using the SCADA and RTCA applications. In SCADA there is an interface display (PATHNWNE) showing the actual flow (as calculated by Peak EMS) and the IROL (as sent via ICCP from BPA). EMS alarms are issued when the actual flow is at 85%, 90%, 95% of the IROL and when over 100% of the IROL. The EMS IROL ICCP point is sent from BPA.

3.4.4.3.7   Related RAS

Three major RAS operated by BPA have been identified impactful to VSA limit calculation for the OREX IROL Cutplane:

- Fast AC Reactive Insertion (FACRI) – this BPA scheme was installed to improve transient and voltage stability performance in the Northwest for larger generation loss contingencies south of the Northwest. The scheme will switch reactive devices at Captain Jack, Malin, and Meridian and insert Fort Rock series compensation groups 1, 2, and 3.
- AC high generation drop – this BPA scheme performs generator dropping (at specific Northwest powerhouses) to prevent thermal overloads, voltage instability, and transient instability on the power system for loss of certain 500 kV lines during high north to south power transfer across the Northwest transmission system. This scheme will also suspend AGC with the generation dropping.
- Pacific DC Intertie (PDCI) generation drop – this BPA scheme performs generator dropping (at specific Northwest powerhouses), high-speed series capacitor insertion (at Fort Rock), and shunt capacitor and reactor switching to maintain stability on the WECC interconnected AC systems in the event of partial or total power loss on the 1000 kV HVDC line. This scheme can also block BPA AGC.

3.4.4.3.8 Automatic Devices

Peak RC ROSE tool is set to allow the switching of SCE and SDG&E auto-switchable shunts for pre-contingency but not for post-contingency.

In the Peak RC ROSE, LTC transformers in the model are disabled. A full list of automatic shunt devices and their control parameters are accessible in the applicable entities operating document.

Phase shifters in the VSA tool should be configured such that there is no automatic action. Tap settings are locked for all phase shifters.

Generator automatic voltage regulator (AVR) devices are modeled for all generator types in line with the parameters provided by the generator owner. This includes set points, voltage ranges, and D-curve data. For regulation targets the Peak model uses the node identified by the generation owner. If it is unclear or unknown WSM models to the closest node in the model, the node at the low side of the generator step up transformer is on regulation by default.

3.4.4.3.9 Monitored Elements

In Peak RC ROSE VSA tool, as all other PV analysis software, the monitored buses for PV analysis are pre-defined by the user and are shown below under Peak-ROSE columns. It should be noticed that the V&R tool will calculate the weakest bus of the system regardless of the monitored buses. The weakest bus is where the VSA tool indicates collapse, either by lowest voltage magnitude solution or by lowest reactive margin at a bus. Therefore, the list of monitored buses will not affect the final result. The list shows the monitored buses for OREX IROL scenario.

CO	Bus name	Bus kV
BCTC	Ingledow	500
BPA	Malin	500
BPA	Klamath Co-gen	500
BPA	Captain Jack	500
BPA	Summer Lake	500
BPA	Grizzly	500
BPA	Celilo	500
BPA	John Day	500
BPA	Alvey	500
BPA	Dixonville	500
IPCO	Midpoint	500
WASN	Olinda	500

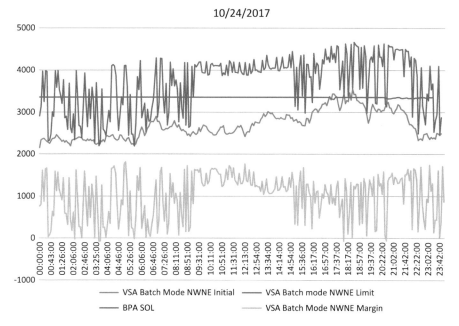

Fig. 3.31 Initial OREX IROL VSA validation results by batch mode study

3.4.4.4 OREX IROL Validation

3.4.4.4.1 Challenges

The OREX consists of three WECC paths that may compete with each other, e.g., Path 66 (COI), Path 14, and Path 75. This IROL has large impact on power transfer capability among multiple areas. Peak RC started to develop the IROL scenario including RAS models in the RT-VSA tool and validate it against BPA offline study tool results in October 2017 for the first time.

The validation study was performed between BPA offline study and Peak RC ROSE VSA results (run batch mode study using the real-time SE snapshot cases). From the solution comparisons given in Fig. 3.31, one can see Peak RC VSA limits experienced many drops. This is because two RAS modeled in the VSA tool PDCI and FACRI could not be fired properly against double or single DC pole outages. The RAS triggers require post-contingency power flow to be solved before the RAS actions are applied. However, the power flow can't be solved without RAS actions being applied. It's similar to the old dilemma – "Chicken or Egg, which one is the first to come?"

RAS not firing could cause low or even zero OREX IROL margin frequently.

In Fig. 3.31, RT_Initial and RT_Limit are Peak RTVSA tool calculated initial interface flow and IROL limit, where the RTVSA could not solve the IROL for hours, because RAS was not modeled. BPA SOL is BPA offline study tool solved

VSA limits using the nomogram, e.g., RAS was modeled. June Batch Initial and June Batch Limit are the results of the new RTVSA software, with new RAS added to RT-VSA.

3.4.4.4.2   Software Enhancements for RAS Modeling

The real-time voltage stability tool Peak-ROSE models Remedial Action Schemes (RAS) for the OREX Interconnection Reliability Operating Limit (IROL).

For OREX-specific RAS stage 2, processing has been created to mitigate the issue of RAS not fired properly. This allows for certain RAS actions to be assumed to calculate an accurate limit. RAS stage 2 is "topology-based" RAS. The two-stage RAS approach applies RAS stage 1 first. If contingency is unsolved, RAS stage 2 is applied. RAS stage 2 is also included in the reverse transfer. The logical flowchart is shown below:

Two-stage RAS logic is illustrated in Fig. 3.32. A key point is to ensure the required RAS protection actions can be applied correctly in three steps: (1) apply RAS stage 1 (full RAS model) if post-contingency power flow solves and RAS trigger conditions are met; otherwise go to (2) restore to the basecase at the last healthy stress level, and then apply contingency and RAS stage 2 (partial RAS model); if RAS stage 2 does not solve the power flow, then (3) stop and report the last solved IROL value. The new RAS logic resolves the most RAS backfire issues and produces more credible IROL assessment than BPA offline study SOL results.

A number of scenarios were used to test correctness and robustness of the new two-stage RAS logic, including validation by batch mode study and real-time VSA results, respectively. The test results are shown in Figs. 3.33, 3.34, 3.35, 3.36, and 3.37.

From Fig. 3.38, we can see the new software calculates more reliable IROLs, due to addition of two-stage RAS logic.

3.4.4.4.3   Validating VSA by TSAT Results

PowerTech DSA Manager/TSAT is a prevalent time domain simulation tool for transient stability analysis. Peak RC started to run online TSAT tool for operation decision on October 2018. All the IROL transfer scenarios have been fully modeled in the TSAT using real-time Arming measurements received from ICCP, including accurate and non-compromised representations of PDCI and FACRI RAS models. In the TSAT both RAS are able to fire correctly with the RAS Arming on and the triggering conditions met. This is because the TSAT runs dynamic simulation to solve governor power flows under a system fault. It does not require to solve conventional power flow against a post-contingency before evaluating the RAS conditions. Peak RC leveraged online TSAT solutions to verify if RT-VSA tool fails to solve post-contingency power flow from the daily real-time application performance metrics, as shown in Fig. 3.39.

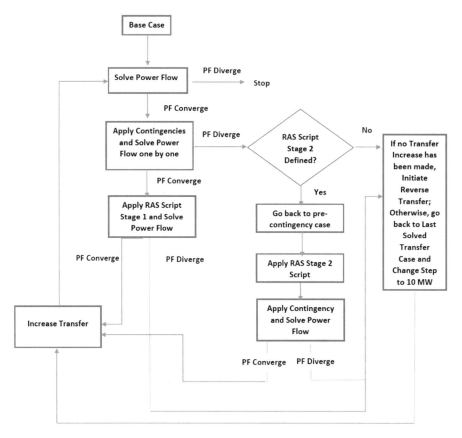

Fig. 3.32 Two-stage RAS modeling and screening algorithm

Fig. 3.33 Two-stage RAS algorithm testing for May 1, 2018, SE snapshot cases

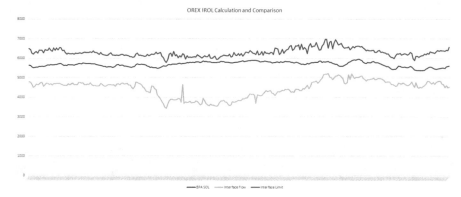

Fig. 3.34 Two-stage RAS algorithm testing for May xx, 2018, SE snapshot cases

Fig. 3.35 Two-stage RAS algorithm testing for June 4, 2018, SE autosave cases

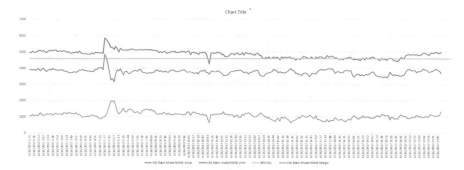

Fig. 3.36 Two-stage RAS algorithm testing for June 15, 2018, SE autosave cases

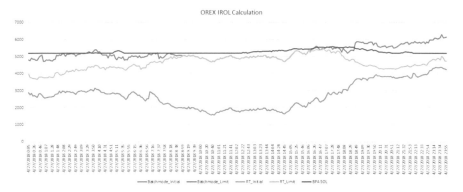

Fig. 3.37 Two-stage RAS algorithm testing for April 27, 2018, SE autosave cases

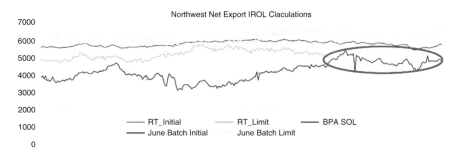

Fig. 3.38 Oregon net export IROL calculations

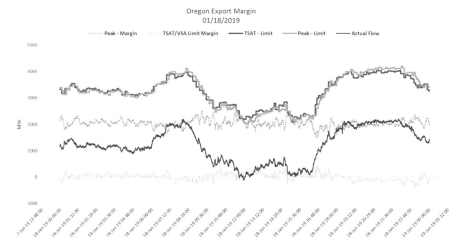

Fig. 3.39 Validation and comparison of ROSE RT-VSA and PowerTech TSAT IROL solutions

After extensive validation testing, the new two-stage RAS logic enables ROSE RT-VSA tool to solve the OREX IROL VSA limit correctly. The OREX IROL operating procedure was developed to guide RCSO and ROE to make operation decision upon the RT-VSA solution results. The online TSAT solved OREX IROL limit has been used for backup and additional validation sources in case the ROSE RT-VSA fails under certain circumstance.

3.5 Special Case Studies and Troubleshooting Practice

3.5.1 VSA Limit Drop Resulting from Local Voltage Instability

This case study identified the local voltage support issue in part of CEN load area and how it can affect the VSA calculated interface limit. A sequence of events on March 19, 2017, were studied to show how insufficient voltage support in a local area with insignificant amount of loads can cause the interface limit to drop significantly, due to power flow divergence and mask potential greater operational risks facing the system. Based on the outcomes of the study, mitigation solutions are suggested for the VSA tool to avoid this situation.

3.5.1.1 Event Description

The VSA SDG&E Winter IROL dropped around 1000 MW from 8:50 AM to 18:35 PM on March 19, 2017. ROE didn't notice any major operating or topology changes occurring around 8:50 AM. In initial studies, NetApps engineers also did not identify any major network changes or manual overrides, e.g., AVR flag and Qmax/Qmin associated with such big drop. At 12:15 PM, OCOTILLO-SUNCREST 500 kV line (OCOT_SUNC_1500) went out of service, but since the limit was already low (3600 MW), the limit only slightly decreased due to this outage (around 200 MW). Therefore, the real operational effect of the outage was masked by the existing low limit. When the OCOTILLO-SUNCREST 500 kV line restored back in service around 18:31 PM, the limit went back to normal (4900 MW). Figure 3.40 shows the stability limit, margin, and interface flow for the day of the event extracted from PI.

3.5.1.2 Study Results

The SDG&E limit drop investigations revealed that the KON115-CIP230 115 kV line (KON_CIP_1115) outage happened around 8:45 AM, and after this outage when the loads in that area were stressed over 80 MVA, the basecase powerflow could not be solved due to insufficient voltage support in the area. This caused the

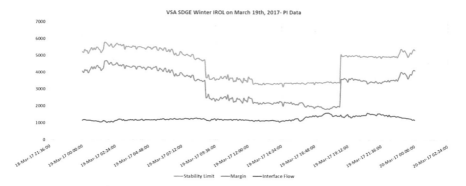

Fig. 3.40 VSA SDG&E Winter IROL on March 19, 2017 – PI data

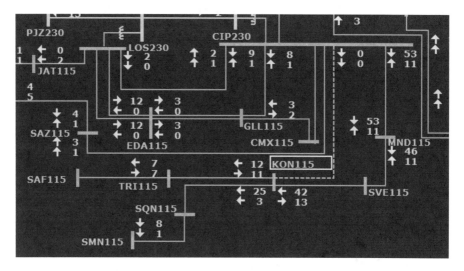

Fig. 3.41 Snapshot of the CEN load area close to the outage

voltage stability limit to drop significantly. Figure 3.41 provides the snapshot of the CEN load area close to the outage.

Figure 3.42 shows the VSA SDG&E winter limit changes and the flows on KON115-CIP230 and OCOTILLO-SUNCREST 500 kV line to clarify the event.

Figure 3.43 gives a summary of the network abnormal voltage at the last step of the VSA study. The critical abnormal voltages all belong to the mentioned CEN load bucket. Power flow solution iteration and power flow iteration mismatches for the last step of the **StudyVSA** also point to the same troubling buses.

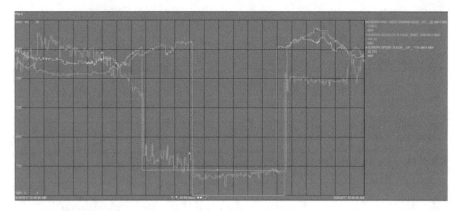

Fig. 3.42 VSA SDG&E winter limit changes and related line flows

Network Abnormal Voltage Summary			-- Bus --	Branch	Low Volt Threshold:	0.90	Update List
					High Volt Threshold:	1.20	

						Study	Run	STNET	STUDY	VALID SOLUTION		
			- Low Voltage -						- High Voltage -			
Island	Area	Station	KV	BS#	Voltage(pu)		Island	Area	Station	KV	BS#	Voltage(pu
1	CEN	SMN115	115	4822	0.636							
1	CEN	SQN115	115	4823	0.647							
1	CEN	SAF115	115	4818	0.746							
1	CEN	TRI115	115	4831	0.752							
1	CEN	KON115	115	4776	0.754							
1	CEN	SVE115	115	4826	0.782							
1	PSE	ELECTHTS	2.3	11771	0.887							
1	LADWP	CASTIC	11.5	5835	0.896							
1	SDGE	MIGUEL	12	13251	0.897							

Fig. 3.43 Network abnormal voltage summary at the last step of the VSA study

3.5.1.3 Limiting Contingency

As displayed in Fig. 3.44, the VSA solution results show that before KON115-CIP230 115 kV line (KON_CIP_1115) outage happened around 8:45 AM, the limiting contingency was "ECO_MIGUEL_1500." But after the outage due to basecase powerflow going unsolved, a credible limiting contingency cannot be calculated by VSA, which could potentially mask more serious operational risks due to voltage support issue in the 115 kV level. Figure 3.45 shows that after OCOTILLO-SUNCREST 500 kV line outage, the VSA is again reporting the credible limiting contingency. The VSA calculated weakest bus which is "4822 SMN115 115" also points to the insufficient voltage support in the mentioned CEN 115 kV load area.

Time	Initial	Stability L	Margin	Pre-CTG Weakest Bus	Pre-CTG Weakest Bus Volt	LimitCtg1	Post-CTG Weakest Bus1
3/19/2017 8:23	1265	4777	3512	4822 SMN115 115	0.74	*ECO_MIGUEL_1500	4822 SMN115 115
3/19/2017 8:28	1246	4757	3511	4822 SMN115 115	0.74	*ECO_MIGUEL_1500	4822 SMN115 115
3/19/2017 8:33	1297	4696	3399	4822 SMN115 115	0.74	*ECO_MIGUEL_1500	4822 SMN115 115
3/19/2017 8:38	1236	4738	3503	4822 SMN115 115	0.77	*ECO_MIGUEL_1500	4822 SMN115 115
3/19/2017 8:43	1239	4741	3502	4822 SMN115 115	0.76	*ECO_MIGUEL_1500	4822 SMN115 115
3/19/2017 8:48	1250	3695	2445				
3/19/2017 8:53	1235	3678	2443				
3/19/2017 8:58							
3/19/2017 9:03	1158	3598	2440				
3/19/2017 9:08	1151	3592	2441				
3/19/2017 9:12	1156	3596	2441				
3/19/2017 9:17	1214	3656	2441				

Fig. 3.44 VSA limiting contingency before and after KON115-CIP230 outage

Time	Initial	Stability L	Margin	Pre-CTG Weakest Bus	Pre-CTG Weakest Bus Volt	LimitCtg1	Post-CTG Weakest Bus1
3/19/2017 12:05	1156	3602	2446				
3/19/2017 12:10	1154	3599	2445				
3/19/2017 12:15	1170	3315	2145	4822 SMN115 115	0.71	*ECO_MIGUEL_1500	4822 SMN115 115
3/19/2017 12:19	1140	3324	2185	4822 SMN115 115	0.72	*ECO_MIGUEL_1500	4822 SMN115 115
3/19/2017 12:24	1107	3364	2257	4822 SMN115 115	0.68	*ECO_MIGUEL_1500	4822 SMN115 115
3/19/2017 12:29	1167	3332	2165	4822 SMN115 115	0.7	* IMVALLY_ECO_1500	4822 SMN115 115
3/19/2017 12:34	1148	3355	2207	4822 SMN115 115	0.66	*ECO_MIGUEL_1500	4822 SMN115 115
3/19/2017 12:39	1138	3283	2145	4822 SMN115 115	0.68	*ECO_MIGUEL_1500	4822 SMN115 115
3/19/2017 12:44	1149	3305	2156	4822 SMN115 115	0.67	*ECO_MIGUEL_1500	4822 SMN115 115
3/19/2017 12:49	1121	3265	2144	4822 SMN115 115	0.69	*ECO_MIGUEL_1500	4822 SMN115 115

Fig. 3.45 VSA limiting contingency before and after OCOTILLO-SUNCREST 500 kV line outage

3.5.1.4 QV Analysis

Figure 3.46 shows the VSA QV analysis results for the SDG&E Winter Scenario right before the outage. The reactive margin went to zero on bus TJUANA 230.

3.5.1.5 CEN Area Insufficient Voltage Support Issue

To confirm that the voltage support issue in CEN 115 kV load area is independent of the load profile and topology of the system on the event day, a real-time RTNET case was retrieved, and after applying the similar KON115-CIP230 115 kV line outage, the loads in the given area were stressed to the point that power flow goes unsolved.

The study using the real-time case also showed that stressing the loads in that specific load bucket over 90 MVA will result in the power flow to go unsolved due to voltage collapse in the study area. The Network Abnormal Voltage Summary and Powerflow Solution Iteration displays show the same low bus voltage buses and power flow-struggling buses when the specified loads were stressed over 90 MVA.

QV-Curve Analysis - Real Time (RAS)		
Last Run:03/19/2017 08:43:23		
WSM Time:03/19/2017 08:41:40		
Scenario Number: 2		
Scenario Name: Winter SDGE		
Transfer: 3430 MW		
Contingency: *ECO_MIGUEL_1500		
Interface Flow: 4741 MW		
Bus: 2013244 MIGUEL 500		
Calculation is not available		
Bus: 4829 TJUANA 230		
VM (p.u.)		Q (MVAR)
	0.9387	0
Reactive Margin: 0 MVAR		
Voltage Margin: 0.000 p.u.		

Fig. 3.46   QV analysis results

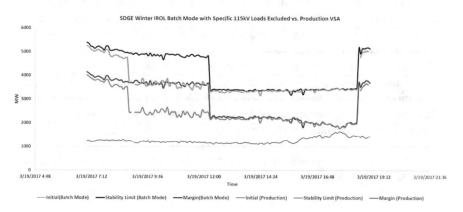

Fig. 3.47   SDG&E Winter IROL batch mode with specific 115 kV loads excluded vs. production VSA

3.5.1.6   VSA Batch Mode Study

To explore the effect of load stressing in the specified load bucket on VSA solution results, we ran a VSA batch mode study for March 19, excluding the loads in Table 1, in SDG&E Winter Scenario file. Figure 3.47 compares SDG&E Winter IROL for the VSA batch mode study and VSA results from production. The batch mode results show that when the loads in the specified bucket are not stressed in the scenario, the limit drops after the OCOTILLO-SUNCREST 500 kV line outage.

3.5.1.7   Discussion with CFE

Per Peak's discussion with CFE, they confirmed the insufficient voltage support issue in the given local area. However, the maximum load in that local CENACE pocket has been reported to be 85 MW in summer (specifically at Maneadero [MND115] pocket). For providing voltage support, Canyon 115 kV [KON115] shunts switch in at higher loading which have been currently modeled in Peak EMS, and Valle De La Trinidad [TRI115] has three new capacitors (normally in) that are going to be added to Peak's model.

3.5.1.8   Proposed Mitigation Plans for VSA Tool

As explained in this report, the insufficient local voltage support issue can cause the SDG&E Winter IROL to drop due to unsolved power flow and mask more serious system operational risk. Therefore, there is a need to develop a mitigation plan for VSA tool to avoid this situation. Since based on CFE historical data, the loads in the specified area have not been over 85 MW, we can limit the amount of stressing for these loads to avoid unrealistic high stressing. For this purpose, an enhancement to the software needs to be requested from V&R to add the option of selectively limiting load increase, like the generation.

In EMS individual load points can be capped by real power load limits, i.e., Pmax/Pmin, and reactive power load limits, i.e., Qmax/Qmin, if applicable. The load limits can be imported into POM/ROSE VSA tool for stressing analysis. In the meanwhile, new code can be added into the POM/ROSE VSA software to enable checking both Pmax and Qmax/Qmin while stressing individual loads defined in a Sink group, to avoid any exceedance of active and reactive load limits.

The load limits can be imported from the EMS load table or only defined in a VSA scenario definition file. For example

Loads		Pmax	Qmax	Qmin
Station	Load ID			
AGNEW	AGNEW	100	20	-15
ARTNDL_E	115	200	50	-20

Once the sensitive loads causing local voltage instability are limited by 120% of the historical high measurement values, or simply excluded from stressing calculation, the SDG&E/CFE IROL VSA limit calculation was not affected by local voltage stability.

3.5.1.9   Lessons Learned and Potential RC Training Case

Insufficient voltage support in local load areas can significantly affect the VSA calculated interface limit in higher levels of stressing and mask greater operational
risks facing the system.

After this incidence, Peak RC experienced similar VSA limit drop for a few more
times, due to insufficient voltage support on other local load areas. Thanks to internal training on the VSA tool and the troubleshooting procedure, Peak RC engineers
were able to identify the root cause and troublesome areas effectively by using the
StudyVSA validation tool and checking the weakest bus and VQ curves included in
the ROSE RT-VSA output results.

Voltage collapse in local load areas from VSA and power flow solutions might
correspond to real insufficient voltage support in the field, like the case presented
here. Therefore, it is always a good practice to study these cases thoroughly and
check the issue with the respective entity.

3.5.2   Load Modeling Causing SDG&E/CFE IROL VSA Limit Drop

On the week of September 22, 2017, Peak RC reported several SDG&E/CFE IROL
calculation issues:

- The ROSE VSA tool showed a 700 MW limit jump in 5 min. See PI trend in
  Fig. 3.48.
- The ROSE VSA tool solved low margins for SDG&E/CFE IROL Cutplane last-
  ing for a few days.
- No major outages and system changes were reported during the incidences.

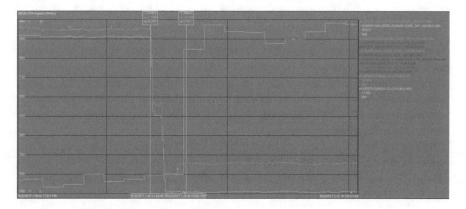

Fig. 3.48   SDG&E/CFE IROL VSA limit drop

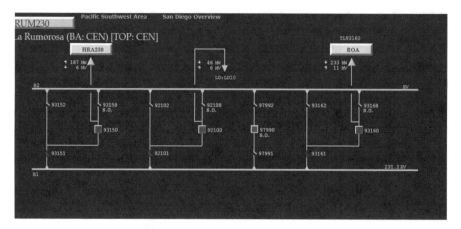

Fig. 3.49 The one-line display with the VSA limit = 3219 MW

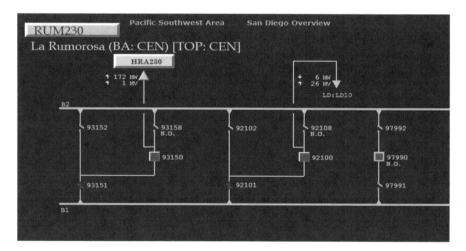

Fig. 3.50 The one-line display with the VSA limit = 3836 MW

Troubleshooting was initiated to identify the root cause for the VSA solution issues.

Two SE real-time snapshot cases (5 min before and after the VSA limit jump point) were retrieved into the StudyVSA tool for analysis. The problem was narrowed down to a 230 kV substation-RUM230, where a load changed significantly between two SE snapshot cases.

While the VSA limit was solved for 3219 MW from the first SE case, the real power load was 46 MW. While the VSA limit was solved for 3836 MW from the second SE case, the real power load was 6 MW. Substation one-line displays are shown in Figs. 3.49 and 3.50.

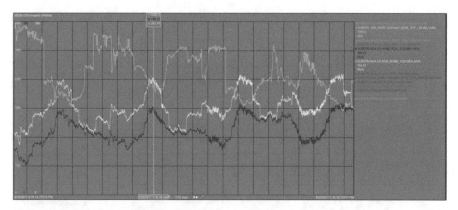

Fig. 3.51 VSA limit vs. CFE Gen pattern

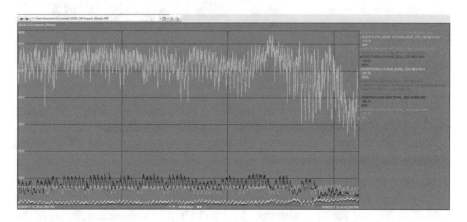

Fig. 3.52 Correlation between SDG&E/CFE IROL VSA limit and CFE plant total generations

By reviewing historical SCADA measurements on the load, we noticed the load is a non-conforming load varying from a near-zero value to 40 MW or higher value suddenly.

From network sensitive analysis (NETSENS) results, we also identified the strong correlation between the MW value of the load and the Valley-Coast 230 kV line flows.

From the plots in Fig. 3.51, one can see the pattern that the SDG&E/CFE IROL VSA limit is highly dependent on the Valley-Coast 230 kV line flow level. CFE cross trip RAS will be fired when the line flow exceeds its rated 388 MVA limit to lower the IROL significantly.

To conclude with the finding, the line flow's margin from 388 MVA limit causes the cross tripping RAS to fire. The closer to 388 MVA, the lower the VSA limit.

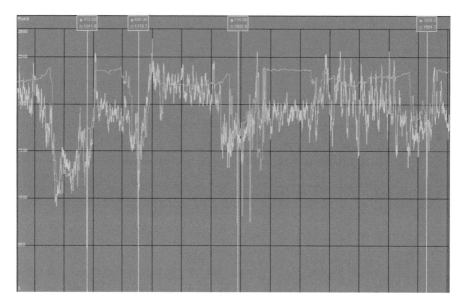

Fig. 3.53 Forty-eight hours OREX IROL margins solved by RT-VSA (green) vs. PowerTech TSAT (blue) (unit, MW)

Figure 3.52 provides a 3-month pattern. It shows lower gen at PJZ230 causes more flows on Valley-Coast lines, thereby decreasing the VSA limit.

To mitigate the issue, we took the following actions to improve the RT-VSA tool accuracy and situational awareness:

- Exclude the non-conforming loads in RUM230 substation from Sink to avoid triggering CFE RAS incorrectly by load increase.
- Add PI trend of sensitive elements, i.e., the RUM230 load measurement, ROA-RUM2 line flow in MW, PJZ230 plant total MW generation, etc. in the SCADA SDG&E/CFE IROL monitoring display to enhance monitoring of CFE RAS operation to the IROL VSA limit calculation.

3.5.3 OREX IROL VSA Limit Drop with Outages in External Areas

3.5.3.1 Background

On April xx, 2019, on-shift ROE reported the ROSE RT-VSA solved the limits for the OREX IROL with sudden and non-sustained drops (even close to a zero margin), while no major outages were identified in the North West grid during the incidence. Below is a summary of observations from PI trend in Fig. 3.53:

- It's confirmed the RT-VSA had momentarily limit drop recently while TSAT solved fine in the meantime. Both tools line up on the VSA limit solutions until the last 2 days.

- This indicates (1) RT-VSA might run into the numerical instability issue due to certain system conditions, which NetApps need work with vendor to debug into the cause. (2) PDCI N-2 CTG showed the most impact to the OREX VSA limit in the past. Specially, in 2018, RT-VSA ran into more solution issues starting April through June. We've improved the RT-VSA software and PDCI and FACRI RAS modeling since then. But it looks there is still something wrong with the RT-VSA tool. Peak RC team will look into it from perspective of software robustness and RAS modeling to find a resolution or mitigation plan.
- In the TSAT tool, we modeled PADCI and FACRI RAS in a more accurate way due to the nature of the transient simulation tool. It can be used for a reference for RT-VSA solution validation on the fly in this case, i.e., RTVSA limit has a momentary drop, while TSAT solves the OREX limit normal. The thing will be more critical for your validation if RT-VSA continuously shows limit drop.

To enable ROE validating such OREX IROL calculation errors, Peak RC engineering rolled out the new version of VSA study tool to improve accuracy of OREX VSA calculation.

3.5.3.2 Investigation Findings

The incidences occurred when Path 3 and Path 83 had major outages for maintenance, which resulted in weak electrical connection between Canada system and US main system. In such cases, the contingency of double pole outages in PDCI triggered a drop of 2500 MW generations in BPA; all AGC units in the entire WECC system were forced to adjust generations to account for loss of the PDCI RAS gen drop, in proportion to Pmax. Given the outages and missing of NATAL RAS (out of step relays for separating AESO system from the rest) in EMS/RTCA, AESO local 138 kV networks suffered low voltage issues, which could make power flow unsolved while calculating the OREX VSA limit without reaching the nose point at all. Note that NATAL RAS was modeled correctly in Peak RC online TSAT tool. That explains why TSAT did not suffer a OREX IROL VSA limit drop during the period of the outages.

It's confirmed by StudyVSA results post-contingency power flow diverges immediately when the same SE auto archive cases are imported for VSA simulation study. RT-VSA tool was less impacted than StudyVSA as it increases post-CTG generation in proportion to the current MW output. In those cases, BC Hydro was importing large generation from US systems. It's likely their gen output level was relatively low and had more spinning reserves as backup.

During the path outages, major western inter-area oscillation modes were affected as well. From Figs. 3.54 and 3.55, Peak RC's Mode Meters application shows behavior changes:

- North-Source Mode A dropped the dominant frequency to 0.2 Hz.
- The Mode Energy decreased significantly.

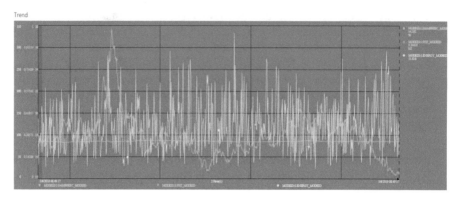

Fig. 3.54 N-S mode A normal oscillation behavior

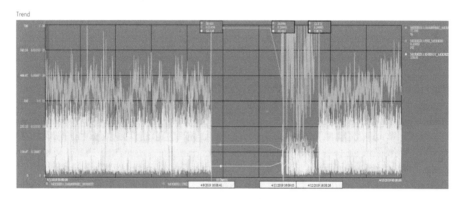

Fig. 3.55 N-S mode oscillation behavior under the outages

3.5.3.3 Mitigation Plan

Following the event investigation findings, we narrowed down the root cause for the OREX VSA limit drop under US-Canada intertie outages which is incorrect modeling of unit governor power flow and AGC regulation function in EMS SE basecase. Here are a few improvements made to mitigate the OREX VSA limit drop issues:

- Review all unit normal participation factors modeled in the WSM, and make necessary corrections upon unit types (i.e., DERs or DG should have AGC flag disabled), Pmax, and baseload flags in WECC planning basecase file.
- Set AGC flag exclusively for a subset of AESO and BC Hydro units that used to perform primary system frequency response under a disturbance. This will limit excessive power transfer from US system to Canada system.
- Use TSAT dynamic models and solution results to tune up unit participation factors and AGC flag setting in the EMS.

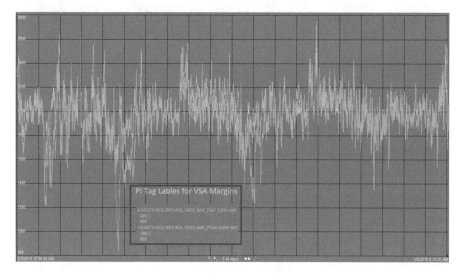

Fig. 3.56 Real-time benchmarking test results with the changes in AGC flag and unit participation factors

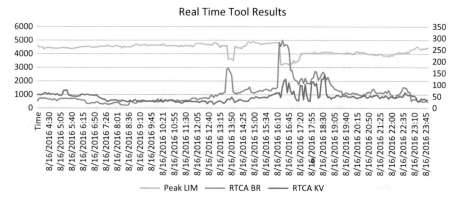

Fig. 3.57 RT-VSA tool results

- Enhance StudyVSA software to mitigate power flow diverging issues under major outages of US-Canada Interties, resulting in weak system connection in between.

After the mitigation measures, both ROSE RT-VSA and in-house StudyVSA tools become more robust against system disturbances. We successfully performed both real-time RT-VSA testing (see Fig. 3.56) and offline VSA study on the problematic cases (see Fig. 3.57) after implementation of the changes and deployed the new models in production in May 2019. Since then the RT-VSA has been solving well for the OREX and other IROL Cutplane limits.

Below are the offline VSA study results with correction in AGC flags and unit participation factors, which show significant improvements of the VSA limit calculation.

Timestamps	Original VSA limit	New VSA limit with AGC flag update
20190409_230000	2816	3763
20190410_055832	2995	3697
20190410_222735	2896	3749
20190411_210647	2883	3609

3.6 Proof of Concept for Future Use Cases

Peak RC have implemented V&R POM/ROSE RT-VSA tool for near real-time assessment of the IROLs successfully. In addition to calculation of VSA limits, we also proposed and evaluated for two new use cases: Edge of Congestion (EOC) and Reactive Power Sufficiency Monitoring.

3.6.1 EOC Calculation

Peak RC was working in conjunction with V&R Energy to develop a proof of concept to enhance the V&R ROSE tool. The proof of concept will deliver a number, to be called the Edge of Congestion (EOC) in this initial prototype, which can be used to represent the cap of flow on a given flowgate or interface where the flow if exceeded would cause post-contingency exceedances to occur. The proof of concept will calculate the number based on both thermal and voltage exceedances.

3.6.1.1 Background

Peak and some members of the Northwest Power Pool (NWPP) worked together to develop the Regional Flow Forecast (RFF) tool to provide situational awareness to reliability and merchant staff in the Western Interconnection. One of the primary purposes of the tool is to provide merchants with awareness of when curtailments are imminent as a result of flow on a flowgate so that merchant entities could take proactive steps to mitigate their exposure to curtailment. Prior to April 1, 2017, this was accomplished by assessing the difference between actual or forecasted flows and the System Operating Limit (SOL) defined for the flowgate. After the retirement of TOP-007-WECC-1a, most flowgates defined in the RFF tool no longer had an SOL value. SOLs are now only monitored as Facility Ratings, voltage limits, or stability limits on specific elements.

Entities not associated with the development of the RFF tool also expressed interest in the results of this proof of concept to use as an input to TTC calculations and to provide to reserve sharing groups in place of TTC so that reserve deliverability is not considered constrained if flow on an interface is above the currently calculated TTC but no post-contingent issues are seen.

3.6.1.2   Overview of Enhancements

The high-level milestones that are deliverable by vendor are described in this section. Supporting information for these are found as follows.

3.6.1.3   Adding Constraint Monitoring

V&R will produce a tool with constraint monitoring which includes:

- Incorporating additional inputs for voltage/thermal constraint monitoring
- The ability to monitor the new limits including thermal, pre-post contingent voltage limits, and %Delta voltage exceedance

3.6.1.4   Changing the Stressing Approach

V&R will produce a tool with additional stressing capabilities added beyond what is available in the Peak-ROSE tool:

- Gen/Gen stressing
- Incorporating different stopping criteria for stressing
- Adding the ability to continue stressing after one type of exceedance is identified (voltage limit, pre-to-post contingency voltage deviation, or thermal limit)
- Confirmation that 5 min calculations are appropriate or recommendation on what time interval is appropriate for performing the calculation

3.6.1.5   Outputs and Reporting

V&R will need to enhance reporting capabilities:

- Output the value of the Edge of Congestion (EOC) for three interfaces specified by Peak.
- During stressing, for each of the defined voltage criteria, record the most sensitive monitored bus(es).
- Record thermal limit exceedance.

- Record the ten transmission lines which are closest to exceeding their thermal limit (by percentage) for specific limiting contingencies.
- Alarming.
- Archiving.

3.6.1.6 Scope

The proof of concept shall perform calculations for the current operating hour. Calculations for future operating hours may be scoped in a future phase. The target is that the calculations will run every 5 min. The tool shall produce:

- Edge of Congestion (EOC): a number which represents the flow, measured in MW, where post-contingency exceedances may occur if flow on a flowgate exceeded this amount. This number shall represent the step where violations occur.
- The transmission lines which are closest to exceeding their thermal limit (by percentage) for specific limiting contingencies that are defined in the scenario. The number of transmission lines reported shall be configurable. The tool shall report on BES elements for the entire WSM. At the point of first thermal limit violation, list the top highest loaded elements (relative to limits) as a percentage margin. Continue stressing looking for Vlim exceedances and/or collapse, but thermal limits no longer need to be evaluated.
- Limiting element or condition associated with the limiting contingency (e.g., line that will exceed its thermal limit or bus where a pre- or post-contingency voltage exceedance would occur). All limiting elements/conditions shall be reported.

The tool shall accommodate configurable parameters on selected monitored buses:

- Pre-contingency Vmin
- Pre-contingency Vmax
- Pre-contingency allowable %Delta-V exceedance (expressed as a percentage)
- Post-contingency Vmin
- Post-contingency Vmax
- Post-contingency allowable %Delta-V exceedance (expressed as a percentage)

The tool shall check pre-contingency limits for the basecase and each transfer step prior to contingencies being applied. In order to determine pre-contingency allowable %Delta-V exceedance, the tool shall compare voltage at the current transfer step with basecase voltage at a bus. In order to determine post-contingency allowable %Delta-V exceedance, the tool shall compare post-contingency voltage at a transfer step with pre-contingency voltage at this transfer step.

Allowable %Delta-V limits represent a % voltage deviation allowed on a monitored bus from the base unstressed state (positive or negative). During stressing, for each of the defined voltage criteria, record the first monitored bus (most limiting) that exceeds the voltage criteria and the interface flow at the point of exceedance.

The first monitored bus is the bus with the lowest voltage as a % (likely the weakest or most sensitive bus). The tool shall report all voltage violations at the same stressing level listed by percentage.

Exceedances of any defined voltage limit criteria should only result in the tool recording the exceedance, the type of exceedance, and the limiting contingency, and the tool shall continue stressing the case toward point of collapse or a thermal limit exceedance.

The tool shall monitor voltage constraints after we encounter the first voltage exceedance and before the thermal limit exceedance. %Delta-V criteria (pre/post) and Vlimit (pre/post) criteria should all be treated separately. Once one of the criteria has been violated, it no longer needs to be monitored, but the others should continue to be monitored. For example, don't stop when a %Delta-V violation is encountered, but continue to stress and test Vlimits (pre/post). As an option:

- %Delta-V and Vlim (post-contingency) – Continue stressing and testing with other contingencies and recording violations for each contingency (don't necessarily have to reevaluate the same contingency in future stressing).

If a thermal limit exceedance occurs before voltage exceedance, keep stressing until voltage exceedance, but record thermal limit violation number.

Line limits (LNLIM) and transformer limits (XFLIM) are available in WSMExport file already provided to the V&R ROSE tool. Buses where voltage is monitored shall be defined in the scenario file. Vmin, Vmax, and allowable %Delta-V limits shall be defined in a configuration file.

Include the ability to have an inclusion or exclusion list of elements monitored against LNLIM and XFLIM.

The tool shall allow gen-gen stressors or gen-load stressors and shall stress the interface in both directions (incremental and decremental).

For Source generation:

- All the units are scaled up simultaneously.
- User-specified units may be excluded from stressing.
- Real power limits of generators are honored.
- EPF (Economic Power Factor) is considered.

To decrease generation in the Sink, the tool shall allow:

- Pro-rata decrease based on decremental capacity down to Pmin
- Treat the gens in the sink as loads with negative sign, ignoring their operational limits such as Pmin or even shutdown of the unit when $P = 0$

To set up the scenario, groups of gens will be specified, Group 1 and Group 2. For each flowgate two calculations/scenarios are executed:

- Scenario 1: Group 1 is Source. Group 2 is Sink. Transfer is increased from Source to Sink.
- Scenario 2: Group 2 is Source. Group 1 is Sink. Transfer is increased from Source to Sink.

For one flowgate, there will be two transfer scenarios.

The tool shall allow the scenario to run with AVR on or off for automatic shunt switching.

The tool producing these calculations will run in a separate environment from the ROSE RT-VSA tool. This will allow for thermal and voltage stressing on interfaces.

The data shall be provided from the ROSE tool to Peak via a .csv file. The .csv file shall be provided from Peak to the customer via SFTP transfer. The tool shall produce the EOC as well as a flow margin (difference between base flow and stressed flow at the point of exceedance) and the amount of incremental flow relative to the base flow at the point of exceedance.

3.6.2   Reactive Power Sufficiency Monitoring

3.6.2.1   Introduction

Peak RC ROSE RT-VSA tool includes VQ analysis for calculating reactive power margins at monitored buses. Peak RC attempted to evaluate the reactive sufficiency in major load areas in the Western Interconnection by the enhanced ROSE RT-VSA tool. This consists of results from the case study performed with the Blue Cut Fire event from August 2016.

A pilot project for proof of concept was performed to demonstrate the use case of monitoring reactive sufficiency on key load centers in near real time. A major component of the proof of concept report is to analyze the performance of the tool, and the results have been discussed later in this report.

3.6.2.2   Blue Cut Event

The Blue Cut Fire started on August 16, 2016, at 10:36 AM in the Cajon Pass along Old Cajon Blvd. north of Kenwood Avenue west of Interstate 15. The fire quickly spotted across Cajon Creek and grew into a large wildland fire. During the course of the firefight, railroad lines, local roads, highway 138, and Interstate 15 were closed along with a large evacuation area that included Lytle Creek, Wrightwood, Summit Valley, Baldy Mesa, Phelan, and Oak Hills.

As a result, Lugo-Mira Loma #3 500 kV line tripped around 14:10. All times are in Pacific Time. Later, around 15:15, Lugo-Mira Loma #2 and Lugo-Rancho Vista 500 kV lines tripped.

This stress became apparent in many toolsets. Peak RTCA showed many new thermal and voltage violations. Peak RT-VSA showed a dropped in transmission capacity for the Pacific Southwest Imports. Peak control room received calls from multiple entities affirming key RTCA thermal and voltage violations for specific contingencies.

Figure 3.57 shows the status of three 500 kV lines and the limit calculated by Peak RT-VSA tool for SDGE import scenario. The times shown in the figure are in Mountain Time. At 1610, all three lines are tripped. These outages coupled with imports into the Southern California transmission system caused system stressing.

Figure 3.57 shows the RT-VSA calculated limit for SDGE scenario along with the number of RTCA branch and voltage violations. Around 1610, the number of branch violations and voltage violations reported by RTCA are about 290 and 125, respectively.

One of the key lessons learned from this event is the necessity for reactive sufficiency awareness. It is imperative for the system operator to have situational awareness about the reactive margin at key buses in a major load center. In this case, the concerned load center is Southern California load center.

3.6.2.3   Concept of VQ Analysis

Voltage stability analysis determines the effect of change in reactive power (Q) on the system voltages (V). QV relationship shows the sensitivity and variation of bus voltages with respect to reactive power injections or absorptions. A system is voltage stable if VQ sensitivity is positive for every bus and voltage unstable if VQ sensitivity is negative for at least one bus.

In order to create a QV curve, a fictitious generator is placed at the bus which is being analyzed. The voltage set-point of this generator is then varied, and its VAR output is allowed to be any value needed to meet this voltage set-point. The vertical axis depicts the reactive output of the fictitious generator in MVAR. The horizontal axis depicts the corresponding voltage under this condition. We actually plot the "VQ" curve, but this is traditionally still called the QV curve, and hence the terms have been used interchangeably in this report.

3.6.2.4   Scenario File

In Peak RT-VSA tool, the scenario file is a csv file which has the Source and Sink definition (for transfer analysis), monitored bus information for PV and QV analysis, contingency definition, and other transfer analysis parameters.

Since, QV analysis is performed at basecase and without any additional load stressing, the Source and Sink parameters are not necessary.

Six contingencies and 18 buses (500 kV or 230 kV transmission facilities and buses) were identified essential for the QV analysis and reactive margin calculation, respectively.

3.6.2.5 ROSE VSA Tool Modifications

Traditionally, the POM/ROSE VSA tool is used to identify a transfer limit for a pre-defined interface by stressing the Sink load and increasing the Source generation, thereby increasing the interface flows against a set of pre-defined contingencies. Once limiting contingency(s) and a limit have been identified, QV analysis is performed for the pre-defined buses at the highest interface flow with the contingent element out of service.

On the other hand, for this reactive sufficiency project, the idea is to identify the reactive margin at the basecase flow without additional stressing. As a result, the scripts were modified to accommodate this requirement, and the initial studies were performed. It was identified that there was plenty of VAR margin at all the buses even after the line trips during the Blue Cut Fire event.

In the next phase, a set of contingencies were introduced, and the scripts were modified to determine the reactive margin at the monitored buses against this set of contingencies. As a result, some potential VAR margin issues were identified. It is important to understand that this set of contingencies was performed over the existing 500 kV line outages. So, a total of six contingencies were identified for the SCIT scenario. The calculation time for just the SCIT scenario for basecase and six contingencies was about 22 min. In contrast, Peak-ROSE/RT-VSA tool solves three IROLs for <5 min in real time.

It was identified from the previous phase that the highest VAR margin for the 500 kV buses was around 5000–6000 MVAR. Since the idea of this project is to determine the buses with least margin, it was decided to limit the highest MVAR margin to 1000 MVAR in order to cut down the calculation time, assuming 1000 MVAR QV margin is sufficient to prevent a bus from voltage instability. The QV analysis would stop for a particular bus when the reactive margin exceeds 1000 MVAR and would continue to the next bus. This change was accomplished by minor script changes, and as a result, the calculation time for each run was reduced to about 5 min.

Restrictions with Peak VSA Tool

As of now, VSA tool can perform the VQ calculation at basecase flow for pre-defined contingencies for only one load center at a time. But, a software enhancement has been made from V&R to modify the script to allow load center calculations to be performed in parallel. This may increase the calculation time and will need to be tested.

3.6.2.6 Blue Cut Fire QV Analysis Results

This section contains the reactive margin at each bus plotted for each timestamp during the Blue Cut Fire event with major transmission outages. At basecase flow with no contingent outage, it can be observed that all monitored buses have sufficient reactive margin of at least 1000 MVAR, except a dead 230 kV bus with zero VAR margin. It is evident that CR-PV 500 kV line was the most impacting contin-

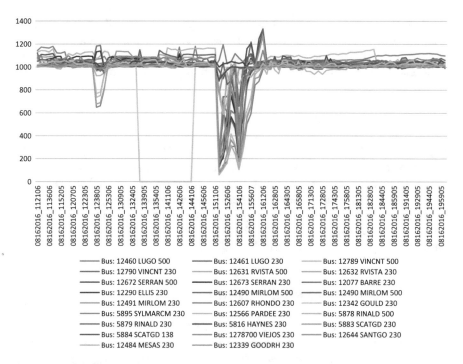

Fig. 3.58 Reactive margin for CR-PV 500 kV line outage

gency on that day after the three key 500 kV lines tripped. With the loss of this 500 kV line, the reactive reserves were depleted drastically, and the lowest margin of 59 MVAR at SCE's Mesa 230 bus was observed around 15:11 which is the event time of loss of Lugo-Rancho Vista 500 kV line. It can be observed that almost all buses show a steep drop in the reactive reserve at that time. This situation shows the real use case of this reactive sufficiency study where the operators can monitor the real-time VAR margins following major transmission or generator outage(s). Even though the margin recovered in about an hour, the system was exposed to instability, and there was a potential voltage collapse in the SCE area.

As shown in Fig. 3.58, reactive margin for the Devers-Valley contingency shows a reduction in reactive margin as well but not much as the Colorado River-Palo Verde 500 kV line contingency. The lowest margin in this CTG is about 500 MVAR at SCE's MESA 230 bus.

3.6.2.7   Real-Time Testing Results

SCIT scenario has been implemented in a test server to evaluate the performance of the tool and the scripts in real-time mode with real system data. The tool is set up to receive the input files from Test EMS server, and it runs about every 5 min.

Since the VSA tool has multiple thread functionality, typically it is possible to run up to x scenarios in a real-time environment. At this point, it is not exactly possible to say how many scenarios can be run in a single server as we can do some performance testing only after the enhancement is received from V&R.

3.6.2.7.1   Testing Results

In order to run the QV scenario and other existing VSA transfer scenarios in the same server, it was decided to test using second instance of POM in Loveland TEMS server. During the testing phase, two instances were running simultaneously, and the solve times were monitored and collected to determine the impact.

It can be seen that, during the testing period, the solve time for the existing transfer scenarios was increased by an average of 0.68 s. This information can be used during the implementation phase of the project.

3.6.2.8   Conclusions and Recommendations

It's essential to receive an enhancement from V&R to perform QV analysis on multiple load centers. It is a key step to proceed with this reactive sufficiency project.

Once the enhancement is received, it is important to determine the optimal number of load centers for which analyses can be performed.

It is crucial to determine whether the assumption of 1000 MVAR as a safe margin is realistic for QV analysis.

For the Southern California load center, contingencies and monitored bus list need to be re-examined to ensure the calculation time is optimal.

Once the major load centers are modeled for QV analysis, depending on the calculation time, it is important to determine the periodicity at which the analysis is performed.

The results from the TEMS server need to be archived in PI for future visualization steps.

Model necessary RAS actions to calculate accurate reactive margin for various load centers. RAS actions involving reactive device switching are extremely critical as they can directly alter the reactive margin at a bus.

Evaluate hardware specification requirements for implementing real-time analysis for multiple load centers.

3.7   Summary

Peak RC made a long haul to implement V&R ROSE RTVSA tool in RC control room through close collaboration with vendor and entities. After 4-year continuous validation and improvement, the tool is proven adequate for providing near real-time assessment on four IROLs effective as of today. Peak RC's RTVSA solution results are updated for every 5 min and are shared with CAISO, BPA, SDGE, and CFE via ICCP to improve operation situational awareness across the regions. Peak RC sends WSMExport cases to SDGE every 5 min, and SGDE also runs ROSE RTVSA every 5 min for SDGE Import and SDGE/CEN IROLs in their control center. There were lot of good practice and lessons learned from V&R Peak RC-ROSE RTVSA implementation projects, such as:

- Cross-validation between Peak RC's RTVSA tool and CAISO's is a critical and effective way to solve "real" SDGE Import and SDGE/CEN IROLs.
- Cross-validation of Peak RC's RTVSA and BPA's VSA study results is essential for calculating and monitoring NW Washington and NW Net Export IROLs.
- Biweekly conference calls on RTVSA tool setting and VSA results review are the key to build transparency and mutual trust among all stakeholders.
- Trustworthy collaboration between Peak RC and V&R Energy enables major RTVSA solution quality issues solved productively.
- Batch mode study process developed by Peak RC team is proven very useful in testing new software patches, VSA scenario, and RAS modeling script changes.
- It's important to develop practical RTVSA tool troubleshooting training modules from real system events. The training helps RCSO and control room engineers in building their skills and confidence in the RTVSA tool.
- Effective coordination and clear communication on RTVSA software and IROL scenario & RAS setting changes will minimize unnecessary human errors greatly.
- Presently the V&R Peak RT-ROSE RTVSA tool is the only one real-time tool solving four effective IROLs using a full western system operational model. The value of the tool has been widely recognized and applauded by internal and external customers/partners. However, no tool is perfect. There are a few areas we look forward to improving with V&R's support.
- It remains sensitive to massive unit Var regulation control and shunt switching operations. The VSA limit could drop incorrectly while power flow diverges due to numerical instability upon massive switch changes.
- The tool definitely provides better VSA results when a smaller step size (say 10 MW) is applied. But it causes the RT-VSA does not compute all IROLs in 5 min. Dynamic step size searching on the last healthy point needs be improved to balance the solution accuracy and solving time.
- It's not very convenient for users to debug problematic RT-VSA cases and invalid basecase solution issues. User-friendly error logs and debugging means are more than welcome to add on.
- The RTVSA tool might encounter an issue while multiple islands exist in the basecase.
- The IROL computation iteration will stop when a local voltage collapse issue occurs for some reason. There is no "SMART" logic to distinguish IROL-oriented voltage collapse or a local voltage instability issue.

The RT-VSA tool can be modified to be used for TTC and ATC calculation, as well as real system reactive sufficiency assessment in the future. Peak RC performed the pilot projects to expand the RTVSA use cases in this regard.

Acknowledgments The authors gratefully acknowledge the contributions of James O'Brien, Ran Xu, Madhukar Gaddam, Jiawei Ning, May Mahmoudi, and Hari Ramana who sequentially contributed to implementing and maturing Peak RC-ROSE RT-VSA tool. We would thank Saad Malik and Matthew Veghte for their operational input to development of various IROL scenarios and validation of the tool results. Peak IT team, specially Lon Kepler, Murat Uludogan, Peter Tang, and Steve Pharo, provided essential support for online and offline ROSE RTVSA software installation and integration. We also appreciate many utility partners who sponsored and participated West Interconnection Synchrophasor Program (WISP) and Peak Reliability Synchrophasor Program (PRSP) since 2011, respectively.

References

1. NERC Report. (2002, August). *Review of selected 1996 electric system disturbances in North America.* [Online]. Available: https://www.nerc.com/pa/rrm/ea/System%20Disturbance%20 Reports%20DL/1996SystemDisturbance.pdf
2. Taylor, C. W., & Erickson, D. C. (1997, January). Recording and analyzing the July 2 cascading outage. *IEEE Computer Applications in Power, 10*(1), 26–30.
3. Kosterev, D. N., Taylor, C. W., & Mittelstadt, W. A. (1999). Model validation for the August 10, 1996 WSCC system outage. *IEEE Transactions on Power Apparatus and Systems, 14*(3, August), 967–979.
4. Zhang, H., & Wangen, B. (2012, July 22–26). Implementation of a full western bulk system operational model for reliability monitoring. In: *Proceedings of IEEE PES General Meeting.*
5. Wangen, B., & Zhang, H. (2012, July 22–26). Monitoring for post-contingency system operating limit exceedance in the western interconnection. In: *Proceedings of IEEE PES General Meeting.*
6. Federal Energy Regulatory Commission. (2012). *Arizona-Southern California outages on September 8, 2011: Causes and recommendations.* FERC and NERC Staff, Apr 2012. [Online]. Available: https://www.ferc.gov/legal/staff-reports/04-27-2012-ferc-nerc-report.pdf
7. Taylor, C. W. (1994). Voltage stability. In *Power system voltage stability.* New York: McGraw-Hill.
8. Maslennikov, S., Litvinov, E., Vaiman, M., & Vaiman, M. (2014, July 27–31). Implementation of ROSE for on-line voltage stability analysis at ISO New England. In: *Proceedings of IEEE PES General Meeting.*
9. Corsi, S., & Taranto, G. N. (2008, August). A real-time voltage instability identification algorithm based on local phasor measurements. *IEEE Transactions on Power Apparatus and Systems, 23*(3), 1271–1279.
10. Abdelkader, S. M., & Morrow, D. J. (2012, May). Online tracking of Thévenin equivalent parameters suing PMU measurements. *IEEE Transactions on Power Apparatus and Systems, 27*(2), 975–983.
11. Burchett, S. M., Douglas, D., Ghiocel, S. G., Liehr, M. W. A., Chow, J. H., et al. (2017). An optimal Thévenin equivalent estimation method and its application to the voltage stability analysis of a wind hub. *IEEE Transactions on Power Apparatus and Systems, 33*(4, July), 3644–1804.
12. Vittal, E., O'Malley, M., & Keane, A. (2010, February). A stead-state voltage stability analysis of power systems with high penetrations of wind. *IEEE Transactions on Power Apparatus and Systems, 25*(1), 433–442.

13. Kawabe, K., & Tanaka, K. (2014, November). Analytical method for short-term voltage stability using the stability boundary in the P-V plane. *IEEE Transactions on Power Apparatus and Systems, 29*(6), 3041–3047.
14. Kawabe, K., Ota, Y., Yokoyama, A., & Tanaka, K. (2017). Novel dynamic voltage support capability of photovoltaic systems for improvement of short-term voltage stability in power systems. *IEEE Transactions on Power Apparatus and Systems, 32*(3, May), 1796–1804.
15. Malik, S., Vaiman, M. Y., & Vaiman, M. M. (2014). Implementation of ROSE for real-time voltage stability analysis at WECC RC. *2014 IEEE PES T&D Conference and Exposition*, 14TD0175. https://doi.org/10.1109/TDC.2014.6863542
16. Malik, S., Vaiman, M. Y., Vaiman, M. M. (2014). Implementation of RAS actions for real-time voltage stability analysis (VSA) at peak reliability. *2014 PAC World Americas Conference*, OP062
17. Zhang, H., Vaiman, M., Ning, J., Vaiman, M. (2019). Implementing the Real-Time Voltage Stability Ananlysis Tool on a Large Scale System Model for IROL Assessment in the Western Interconnection. *IEEE Power &Energy Society General Meeting (PES-GM)*, August 5–8, Atlanta.

Chapter 4
Real-Time Transient Stability Analysis Implementation

Slaven Kincic and Hongming Zhang

Organization Acronyms

BPA	Bonneville Power Administration
CAISO	California Independent System Operator
CENCE	Centro Nacional de Control de Energia
CFE	Comisión Federal de Electricidad
FERC	Federal Energy Regulatory Commission
GE	General Electric
IPCO	Idaho Power Company
NERC	North American Electric Reliability Corporation
PEAK	Peak Reliability or Peak RC
PG&E	Pacific Gas and Electric
Powertech	Powertech Research Labs Inc
SCE	Southern California Edison
SDG&E	San Diego Gas and Electric
V&R	V&R Energy System Research Laboratory
WECC	Western Electricity Coordinating Council

NERC/WECC Acronyms

ATR	Acceleration trend relay
ACE	Area control error
AGC	Automatic generation control
AVR	Automatic voltage regulator
BA	Balancing authority
BES	Bulk Electric System
COI	California-Oregon Intertie
DCS	Disturbance Control Standard
DER	Distributed energy resource
DFR	Data fault recorders
DTS	Dispatcher Training System

© Springer Nature Switzerland AG 2021
H. Zhang et al., *Advanced Power Applications for System Reliability Monitoring*, Power Systems, https://doi.org/10.1007/978-3-030-44544-7_4

EI   Eastern Interconnection
EMS   Energy Management System
FACRI   Fast AC reactive insertion
FRM   Frequency reserve measure
ICCP   Inter-Control Center Communications Protocol
IROL   Interconnection Reliability Operating Limit
LMTF   Load Modeling Task Force
MVWG   Modeling and Validation Work Group
NWMT   Northwestern Energy
PDCI   Pacific DC Intertie (PDCI)
PMU   Phasor Measurement Unit
PSS   Power system stabilizer
PV   Photovoltaics
RAS   Remedial Action Scheme
RC   Reliability Coordinator
RCSO   Reliability Coordinator System Operator
RTCA   Real-Time Contingency Analysis
SCADA   Supervisory control and data acquisition
SE   State Estimator
SPS   Special Protection Scheme
SVC   Static VAR compensator
TI   Texas Interconnection
UFLS   Under-frequency load shedding
UVLS   Under-voltage load shedding
WAPS   Wide-Area Protection Scheme
WI   Western Interconnection
WSM   West-wide System Model

Other Acronyms

ePMU   EPG's enhanced Linear State Estimator
HDR   GE's Historical Data Record application
MAS   Montana Tech's Modal Analysis Software
MUC   Multiple outage contingency
NWNE   Northwest Net Export IROL
NW-WA   North West Washington Area Import IROL
OREX   Oregon Net Export IROL
PI   (OSIsoft's) Plant Information
POM   (V&R's) Physical and Operational Margin application
ROC   Rate of change
ROE   Real-Time Operation Engineer
ROSE   (V&R's) Region of Stability Existence application
RTNET   (GE's EMS) Real-Time Network Analysis (i.e., State Estimator)
RT-VSA   (Peak RC's ROSE) Real-Time Voltage Stability Analysis

TSAT   (Powertech's) Transient Security Analysis Tool
VSA   Voltage Stability Analysis
VSAT   Voltage Stability Analysis Tool

4.1 Introduction

Power system stability has been recognized as an important problem for secure system operation since the 1920s [1, 2]. Many major blackouts are caused by power system instability issues. IEEE/CIGRE joint taskforce defined system stability as "the ability of an electric power system, for a given initial operating condition, to regain a state of operating equilibrium after being subjected to a physical disturbance, with most system variables bounded so that practically the entire system remains intact."

Historically, transient instability has been the dominant stability problem on most systems and has been one focus of the power industry over the last century. As power systems have evolved through continuing growth in interconnections, massive use of inverter-based generating resources with no or minimal inertia response, and the increased power energy transfer and path flow shifting in highly stressed conditions, different forms of system instability, i.e., voltage stability, frequency stability, and inter-area oscillations, have become greater concerns than in the past. For example, most of the existing IROLs identified in three major interconnections across North America are limited to voltage stability issues. In reflection of most recent blackout events reported in European Grids and South America countries, North America grids will probably experience more severe frequency stability and inter-area oscillation issues with retirement of traditional generating units and additions of massive DERs.

Traditionally, transient instability was mainly defined by rotor angle instability. Rotor angle stability refers to the ability of synchronous machines of an interconnected power system to remain in synchronism after being subjected to a disturbance. In history, transient instability could also lead to a large system disturbance, including cascading outages, fast voltage collapse, and non-damped inter-area oscillation/power swing.

Power system is constrained to operate within the system operating limits (SOLs) for N-1 and credible N-k or multiple outage (MUC) contingencies. By the approved definition from the NERC Glossary of Terms, system operating limit (SOL) is the value (such as MW, MVAR, amperes, frequency, or volts) that satisfies the most limiting of the prescribed operating criteria for a specified system configuration to ensure operation within acceptable reliability criteria. System operating limits are based upon certain operating criteria. These include, but are not limited to:

- Facility ratings (applicable pre- and post-contingency equipment or facility ratings)
- Transient stability ratings (applicable pre- and post-contingency stability limits)

- Voltage stability ratings (applicable pre- and post-contingency voltage stability)
- System voltage limits (applicable pre- and post-contingency voltage limits)

The SOLs are currently established based on offline planning and operation studies. Established SOLs account for reliability and transmission capacity of the system. Higher the limits are set, more transmission capacity is available. However, reliability is a primary concern and SOLs are capped by reliability constraints. System operating limits depend on operating conditions, and they are changing over the time. For that reason, system limits established in offline studies are conservative in order to account for study assumptions. In general, further we look into the future there is more uncertainty due to massive DERs. It includes load forecast, cost of the fuel, outages, weather conditions, outages, impact of neighbored conditions, and others.

Traditional Real-Time Contingency Analysis (RTCA) that is power flow with contingent element(s) out of service assures that these limits are not violated in real time. However, under certain conditions, RTCA cannot establish system operating limits in real time and cannot assure that previously established limits are safe for reliable operation. For example,

- RTCA assumes post-contingent system condition remains stable at a scheduled frequency of 50 or 60 Hz, which might not be a valid assumption.
- There are certain Remedial Action Schemes (RAS) models that are triggered based on voltage and frequency threshold violation over certain time period. For example, in the Western Interconnection, there is a RAS model that automatically inserts series capacitors in 500 kV lines based on a voltage dip threshold. This specific voltage threshold is set for voltage to drop below 525 kV for 0.85 s or more at a specific bus. Since RTCA has no way to account for the time delay, this RAS can only be partially modeled, e.g., "if voltage in steady state is below 525 kV, insert series capacitors."
- Under-frequency load shedding (UFLS) is an important mechanism for preventing frequency collapse. Since the system frequency is always assumed at 60 Hz flat in steady-state power flow, load shedding cannot be accurately modeled in RTCA. For a contingency that requires load shedding to prevent system collapse, RTCA would preserve the load and increase generation, thereby providing misleading results.
- Under certain conditions RTCA simply cannot solve, while those contingencies significantly change system conditions. The same phenomena can happen due to singularity or numerical instability. In such case it is difficult for operators to distinguish if an invalid solution is due to voltage collapse or numerical issues.

To overcome the limitations of RTCA on evaluating post-contingent SOL exceedances, it requires to implement real-time Transient Security Assessment Tool (TSAT) in the control rooms, because the TSAT can calculate and evaluate system operating limits by time domain dynamic simulation.

4.2   Basic Concept of Transient Stability

The concept of transient stability was well defined in many textbooks and IEEE literatures. A power system is transiently stable if it returns to equilibrium and reaches a new steady state following a system disturbance. Oftentimes, transient stability is referred to as angular stability or the ability of generation not to fall out of the step (lose synchronism). However, in practical implementation transient stability is a much broader concept as seen later in the chapter. The objective of this section is to provide the basic concept of transient stability. Consider a generator connected to the stiff system through a transformer and a line inductance jX as illustrated in Fig. 4.1.

Let the voltage phasor of the system be $\mathbf{E}=|\mathbf{E}|$ and generator external voltage $\mathbf{V}=|\mathbf{V}|e^{j\delta}$ where δ is a power angle between voltage phasors $\mathbf{E}$ and $\mathbf{V}$. In order to transfer real power P [MW] from generator to the system, generator voltage angle δ needs to lead voltage of the system $\mathbf{E}$.

In steady state:

$$P = \text{Re}\left\{\mathbf{VI}^*\right\}$$

$$\mathbf{V} = Ve^{j\delta}$$

$$\mathbf{E} = E$$

$$\mathbf{I} = \frac{\mathbf{V} - \mathbf{E}}{jX}$$

$$P = \text{Re}\left\{\mathbf{V}\left(\frac{\mathbf{V}-\mathbf{E}}{jX}\right)^*\right\} = \text{Re}\left\{Ve^{j\delta}\left(\frac{Ve^{j\delta}-E}{jX}\right)^*\right\} = \text{Re}\left\{Ve^{j\delta}\left(\frac{Ve^{j\delta}-E}{jX}\right)^*\right\} = \frac{VE}{X}\sin\delta$$

So the maximum power transfer in steady state occurs when angle delta is 90°. Any further increase in angle would result in the decrease of the real electrical power. In reality, this angle is much smaller than 90° due to transient stability concerns.

When a synchronous machine is connected to the grid, it must be synchronized with it. Excitation circuit current creates a rotor magnetic field, while stator currents

Fig. 4.1 Generator connected to the stiff system

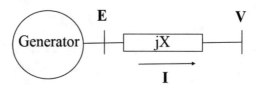

create a stator field. The synchronization process locks the rotor magnetic field with the stator magnetic field, and both rotate at a synchronous speed. If machine works as a generator, the internal voltage (open circuit voltage) leads the grid voltage. Rotor angle is the angle between the rotor axis and the stator magnetic field. This angle is zero during the synchronization process, but with increased loading of the generator, the angle increases together with the current through the generator stator. During steady-state operation, this angle is constant (Fig. 4.2).

The generator rotor is run by a turbine. During the process of energy conversion, mechanical power from turbine P_m is converted into the electrical power P_e through the intermediary of the electromagnetic field within the generator. During that process, electrical torque T_e is opposed to the mechanical torque T_m. In steady state, mechanical power supplied by the turbine is equal to electrical power consumed by the load plus the losses. If electrical load is suddenly changed due to a fault or some other reason than the mechanical power of turbine and electrical power does not match causing acceleration of the rotor and increasing rotor angle. When the fault cleared, the electrical power of the load is restored causing a deceleration of the rotor slowing down rotor angle increase. If the fault is not cleared fast enough and rotor angle has sufficiently increased, the generator rotor and the stator magnetic field can unlock causing generator to go out of the step or, what is called in operational practice, the loss of synchronism. In that case, the generator is said to become unstable, and it is removed from the system by an out-of-the-step protection relay. This simplified dynamic can be described by a swing equation and an equal area criterion.

Consider the turbine-generator example in Fig. 4.3, where J is the moment of inertia of rotor and turbine in kg²; theta is an angular position of rotor; w_s is the synchronous speed in rad/s; T_e and T_m are the electrical and mechanical torque, respectively; and H is normalized moment of inertia.

The equation of the motion can be written as:

$$J\frac{d^2\theta}{dt^2} = T_m - T_e = T_a$$

$$P_m = T_m\omega$$

$$P_e = T_e\omega$$

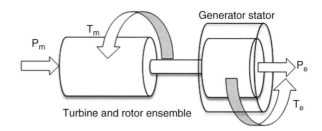

Fig. 4.2 Turbine-generator ensemble used for swing equation deduction

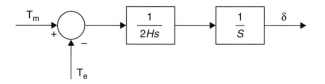

Fig. 4.3 Swing equation representation in S domain

$$\theta = \omega_s t + \delta$$

$$H = \frac{J\omega_s^2}{2S_{rated}}$$

$$J\frac{d^2\delta}{dt^2} = T_m - T_e = T_a$$

$$\frac{2H}{\omega_s^2}S_{rated}\frac{d^2\delta}{\partial t^2} = T_m - T_e = T_a$$

$$\frac{2H}{\omega_s^2}S_{rated}\frac{d^2\delta}{dt^2} = \frac{2H}{\omega_s}S_{rated}\frac{d}{dt}\left(\frac{\omega_r}{\omega_s}\right) = T_m - T_e = T_a$$

in p.u.

$$2H\frac{d^2\delta_{p.u.}}{dt^2} = T_{m\ p.u.} - T_{e\ p.u.} = T_{a\ p.u.}$$

or in space state form:

$$\frac{d\omega_{p.u.}}{dt} = \frac{1}{2H}\left(T_{m\ p.u.} - T_{e\ p.u.}\right), \omega_{p.u.} = \frac{d\delta_{p.u.}}{dt}$$

This leads to a representation in S domain:

Transient stability can be explained through the equal area criterion:

In the steady state, mechanical power P_m and electrical power P_e are equal (all losses are neglected). In that case accelerating power is equal to zero $d^2\delta/dt^2 = 0$, and the generator operates at the angle δ_0 on $P-\delta$ plain (see Fig. 4.4). In case of the fault, electrical power $P_e = 0$ and generator rotor will accelerate, since $2Hd^2\delta/dt^2 = P_m$, consequently increasing angle δ. A fault is typically cleared in a few cycles so mechanical power of the turbine will not change. When fault cleared after few cycles, angle already has increased from δ_0 to δ_1. In that moment, electrical power P_e will be larger than mechanical power since $\delta_1 > \delta_0$ and angle δ will continue to increase due to the rotor-turbine inertia. Since electrical power P_e is more than

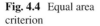

Fig. 4.4 Equal area criterion

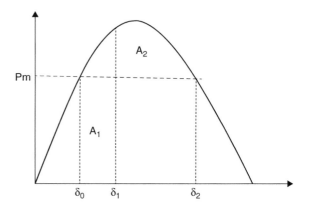

mechanical power P_m, it will deaccelerate rotor until the angle eventually reaches δ_2 on the same operating curve and the system will damp out oscillations of angle δ in case of the stable system or, in case of the unstable system angle, δ will keep increasing until the unit loses synchronism. Maximum angle δ_{MAX} is called critical angle. Areas A_1 and A_2 are called area of acceleration and area of deceleration. If $A_2 > A_1$, then the system is stable. Critical angle δ_{MAX} can be calculated from $A_2 = A_1$. It is worth noting that angle δ_1 depends on fault time clearing. Faster fault clearing time will result in smaller angle δ_1, smaller area A_1, and larger area A_2 making the generator more stable. Figure 4.4 depicts this relationship.

While equal area criterion is useful for understanding how generator can lose synchronism, in operation it is not a real concern if the single unit will go out of the step. Generators in the interconnected system have ride-through capability based on standards [2], and a generator should be able to survive a three-phase fault that's normally cleared on the generator interconnection with the grid without losing synchronism. Moreover, certain RAS can initiate intentionally large generation tripping that can result in more than 2000 MW of generation instantaneously disconnected from the system. For that reason, losing one or a few units from the system should not be a source of the significant operational concern. Such an issue should be addressed in the proper planning timeframe. In everyday operations the concern is with large-scale instabilities, uncontrolled islanding, and cascading that can lead to blackouts.

In the large interconnected system, such as the Western Interconnection, there are over 4000 generators. Each generating unit consists of multiple components such as a governor, a power system stabilizer (PSS), and an excitation circuit. The generators are of different types and different manufacturers. Each of these components needs to be modeled separately. That means described by set of differential equations and converted in S domain as it was previously done by the swing equation. It results in hundreds of different models including other component of power system such as shunts, loads, FACTS devices, and others. The dynamic models used in the Western Interconnection are described in the modeling tool user manuals [3, 4]. These models are standardized and implemented in various study packages such as GE PSLF, Powertech's TSAT, PowerWorld, PSSE/PTI, and others. Model parameters are based on individual unit being separately tested [5–7].

In real-time transient stability, as in most offline transient stability applications, the interest is not in fast transients but only on the electromechanical transients in range from 0.1 to 6 Hz. For that reason, a power network is not modeled using differential equations but rather using algebraic equations. This modeling is done as a set of differential-algebraic equations of the form:

$$\dot{x} = f(x, V)$$

$$I(x.V) = YV$$

where Y is the admittance matrix, V is the voltage vector, I is the current injection vector, x is the state vector, and f is a set of differential equation from the dynamic models. Figure 4.5a, b illustrates the abovementioned power system representation for dynamic simulations on small-scale IEEE 9 bus test system. Generators, some switching devices, and loads are represented using differential equations, while the transmission network is represented using algebraic equations. Each generator in the power flow has multiple components as shown in Fig. 4.5b.

Power flow is then solved and used to initialize dynamic simulation. Numerical integration is performed typically using an integration step of 0.25 or 0.5 cycle. Current vector **I** is updated using the voltages from dynamic models and the algebraic equations. More details are provided in the TSAT user's manual [8].

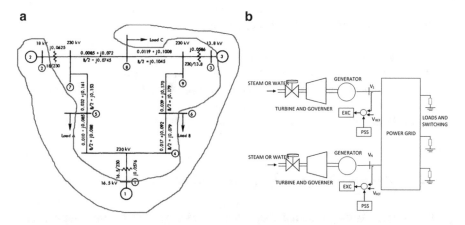

Fig. 4.5 Representation of the system for dynamic simulation. (**a**) Generators and some load and switching devices are represented using dynamic models, while transmission network is represented by Y matrix, and (**b**) each generator contains multiple blocks

4.3   Emerging Challenges and Industry Efforts

4.3.1   Emerging Operation Challenges

In the last decade, many renewable and inverter-based generation resources, such as solar PV and wind farm units, were continuously integrated into the WECC system. The renewable energy portion keeps growing rapidly. The 2018 summer study case shows approximately 11,000 MW of solar PV online, which represents 12% of the online generation. The penetration rate of overall renewable generations is expected to continuously increase for the next decade as the state of California is obligated to meet its 50% renewable resource mandate by 2030 or even sooner. High penetration of renewable energy resources is changing WI's operating patterns. The sustained declining of traditional generators with a large inertia mass requires RC to monitor system frequency response sufficiency in real time and assure for compliance of NERC's Interconnection Frequency Response Obligation (IFRO) persistently.

Since the 1996 blackouts, some original work was published in IEEE papers [4–7] for development of efficient real-time transient stability limit computation algorithms. There were a number of utilities to perform transient stability simulation study and operation risk assessment on planning basecases in a daily or routine basis. However, there is limited practice in major RCs under NERC to run real-time transient stability analysis in control rooms for operation decision. In WI, more recently, CAISO made the effort on implementing online TSAT tools on a regional system model [8] for computation of dynamic transfer limits. In 2014 Peak RC initiated the WSM-based online TSAT project for monitoring various stability issues across the entire Western Interconnection (WI):

- Loss of synchronism of cluster of generators and consequently their removal from the grid on a fault
- Unacceptable frequency response that can result in under-frequency load shedding or tripping of thermal and nuclear units
- Monitoring transient or dynamic RAS operation and evaluating RAS Gen drop impact to system frequency performance and cascading outage scenarios
- Real-time assessment of stability constrained SOLs and IROLs
- Uncontrollable oscillation in the system (damping criteria) under [N-1] or [N-k] contingencies
- Fast voltage collapses due to induction motor instability

4.3.2   Complexity of TSAT Application and Dynamic Models

Transient stability and dynamic phenomena spreads over interconnection and cannot be isolated. Frequency response of the interconnections depends on inertia of each generating unit that is online at that moment as well as on their governor response. At the same time, generation in different part of the interconnections interact against each other through various inter-area oscillation modes. For that reason, it is difficult to implement real-time transient stability if there is no available real-time steady-state model of the interconnections.

Transient stability, apart from the steady-state model, requires dynamic models of all components that are matched to the steady-state model of the observed power system. While a dynamic model might not be a challenge with a small test system with a few buses that are often used by the academic community to prove concepts, it is completely different with large-scale interconnection such as the Western Interconnection. In the Peak RC's West-wide System Model (WSM), there were nearly 4000 generating units modeled for State Estimator (SE) and Real-Time Contingency Analysis (RTCA) solutions. Typically, a dynamic model of each generator contains models of a generator with turbine, an excitation circuit, a PSS, and a governor.

It is important to note that:

- Different types of equipment will have different models based on unit fuel, manufacture, type of the component, and type of the control.
- All parameters and time constants need be populated in the model based on field tests.

The detailed model description that is used in commercial software is typically given in commercial software vendors user guides [3, 4]. The typical steady-state model of a generator contains only maximum/minimum generation outputs (MW and MVAR), a capability curve (when available), and a regulated bus so that the voltage can be adjusted to meet the voltage target. This model does not depend on type of generator, fuel, or manufacturer.

Figure 4.6 represents the Western Interconnection with existing WECC paths. A portion of WECC paths have been limited by different transient stability issues. Till date, all path limit ratings were calculated offline by individual path operators through the corresponding path operation procedures. The offline studied path limits tend to be conservative in general but could be underestimated under certain unplanned operation conditions, i.e., [N-k] contingencies. It's highly desirable to perform real-time assessment on the WECC paths subject to a stability issue.

The WECC system spreads over 39 balancing authorities (BAs) and 64 transmission operators. Each BA is responsible for balancing its own footprint. It is impossible for each BA to maintain real-time model of the interconnections due to lack of the modeling resources and data sharing issues. For that reason, only larger system operators may have enough capacity to maintain real-time state estimator model. Control room tools used for real-time decision-making by operators are subject to a variety of reliability standards. That can be very costly requiring dedicated IT infrastructure and 24/7 personnel support. It can be even more costly if tools lead operator to wrong decision or stop providing valid solution. Therefore, real-time applications have to be robust and, at the same time, provide good-quality trustworthy solution after each run so that operators can rely on. Robustness and accuracy need to compromise with each other for large system models, including steady state as well as dynamic, as those large models always contain some hidden errors.

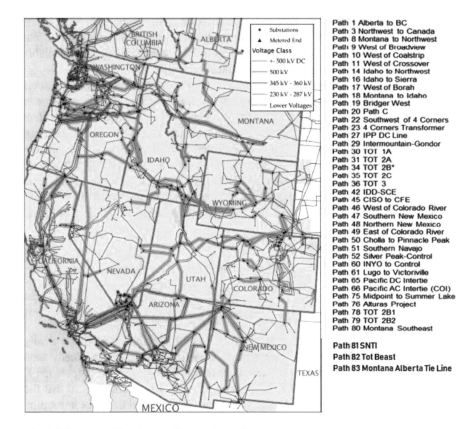

Path 1 Alberta to BC
Path 3 Northwest to Canada
Path 8 Montana to Northwest
Path 9 West of Broadview
Path 10 West of Coalstrip
Path 11 West of Crossover
Path 14 Idaho to Northwest
Path 16 Idaho to Sierra
Path 17 West of Borah
Path 18 Montana to Idaho
Path 19 Bridger West
Path 20 Path C
Path 22 Southwest of 4 Corners
Path 23 4 Corners Transformer
Path 27 IPP DC Line
Path 29 Intermountain-Gondor
Path 30 TOT 1A
Path 31 TOT 2A
Path 34 TOT 2B*
Path 35 TOT 2C
Path 36 TOT 3
Path 42 IDD-SCE
Path 45 CISO to CFE
Path 46 West of Colorado River
Path 47 Southern New Mexico
Path 48 Northern New Mexico
Path 49 East of Colorado River
Path 50 Cholla to Pinnacle Peak
Path 51 Southern Navajo
Path 52 Silver Peak-Control
Path 60 INYO to Control
Path 61 Lugo to Victorville
Path 65 Pacific DC Intertie
Path 66 Pacific AC Intertie (COI)
Path 75 Midpoint to Summer Lake
Path 76 Alturas Project
Path 78 TOT 2B1
Path 79 TOT 2B2
Path 80 Montana Southeast

Path 81 SNTI
Path 82 Tot Beast
Path 83 Montana Alberta Tie Line

Fig. 4.6 2019 WECC path map. (Source: WECC)

4.4   Peak RC Online TSTA Project Introduction

4.4.1   WSM-TSAT Overview

In the wake of the September 8, 2011, Pacific Southwest blackout, WECC Reliability Coordinator (the predecessor of Peak RC) started to develop real-time voltage stability and transient stability analysis tools on top of the WSM EMS applications including State Estimator (SE) and RTCA, in order for real-time assessment and monitoring of system operating limits (SOLs) and Interconnection Reliability Operating Limits (IROLs) [2, 3].

The WSM represented a large-scale and full western system operational model. Peak RC used to run the WSM-TSAT for basecase screening (cycled by 2–4 min) and power transfer analysis on the Western IROLs (cycled by 7–12 min) in parallel. The WSM-TSAT is characterized as follows:

- A bus-branch model basecase with 16,500 buses and 20,000 branches (dynamic). The basecase raw file not only follows PTI v30 format but also includes an EMS long name labeled for each network equipment.
- 3900 generating units with 96% online generation capacity mapped to WECC *.dyd file.
- 30 contingencies currently modeled active.
- Three transfer scenarios are currently modeled, i.e., SDGE Import IROL, SDGE/ CFE Import IROL, and Northwest Net Export/Oregon Net Export IROL.
- PDCI and Intermountain DC Ties modeled explicitly.
- 17 dynamic or transient RAS are modeled for Path 26, COI, PDCI, FACRI, MATL, BC Hydro RAS, and 11 other RAS selected for SDGE Import and SDGE/CFE Import IROL scenarios.
- About 6000 UFLS and UVLS models are built in to represent in lsdt1, lsdt2, lsdt3, and lsdt9.
- User-defined models (UDM) are used to model the RAS and consolidate unmapped small units between EMS model and WECC *.dyd file.

The underlying figure shows an overview of the WSM-TSAT system architecture for real-time implementation at Peak RC (Fig. 4.7).

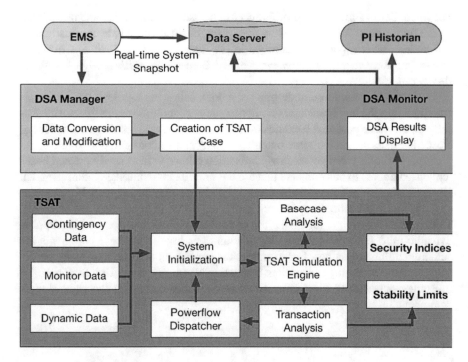

Fig. 4.7 WSM-TSAT system architecture and interface with EMS

The TSAT application is installed on four dedicated servers in two separate control rooms due to redundancy requirements. Only one site is active, and all others are in stand-by mode.

WSM-TSAT read in GE-EMS network model data in PTI v30 raw file format for every 5 min, containing last run SE basecase solution and EMS equipment IDs, unit D-curves, and Interface definition files exported from EMS. The EMS model files used for TSAT calculation are exported by Peak RC custom application called XTSAT. State Estimator (SE) is run on its own server's independent from transient stability servers. SE model representing snapshot of the Western Interconnection is exported every 5 min together with generator capability curves and interfaces definition. SE case is copied in power flow application and solved. Successfully solved power flow is exported in PSSE (PTI) format, and, during the process, bus numbers and equipment names are added into the case. During this process network model is converted from node-breaker to bus-branch model. The node-breaker model is typically used in SE and contains breakers so that topology can be changed by changing breaker states based on real-time status in the field that is mapped to the model and received by SE from SCADA system.

4.4.2   Node-Breaker Model vs. Bus-Branch Model

The figure below illustrates the differences in between node-breaker and bus-branch model for the same substation. More details on the differences between bus-branch and node-breaker model are described in [*R. Ramanathan, B. Tuck, S. Kincic, H, Zhang and D. Davies, "Equipment Naming Convention Methodology for Node/ Breaker Model"*].

During state estimation, node-breaker model, following topology processing, is collapsed into the bus-branch model. In that process, multiple nodes that are electrically on the same potential become collapsed into the single bus (e.g., both sides of closed breakers are on the same potential so they can be considered as single bus).

Export process from power flow application allows to map newly created buses (collapsed nodes) to bus numbers that are the same as in traditional planning model. However, due to modeling differences, not all buses may be mapped. For that reason, during export process equipment names are added in power flow as comments. These names are unique for each piece of equipment, and they are used in mapping dynamic models to power flow. Application does know that these comments are actually equipment names. For example, for generation unit the names are generated using the following convention: Substation Name_ kV level_Unit Id. Figure 4.8 illustrates power flow file format having equipment names on the right side as comments.

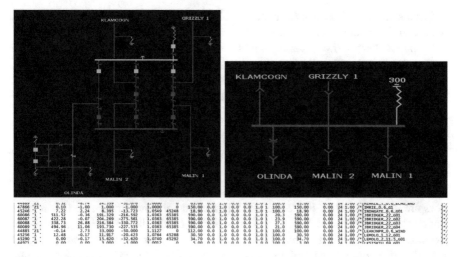

Fig. 4.8 Difference between node-breaker and bus-branch representation of the same substation

The cases are created every 5 min, represent system snapshots, and are saved on SE server automatically. The cases are then copied to TSA server where they are matched automatically with other files needed to perform accurate TSA. This file includes dynamic models of system components, RAS models, monitored variable, criteria file, and contingency and transfer definitions.

4.4.3   Power Flow Modification Files and Other Modeling Considerations

There are numerous other modifications that need to be performed on power flow "on the fly" in order to make simulation credible. This section illustrates some of these necessary modifications.

4.4.3.1   Distributed Slack Buses

Snapshot from State Estimator is bases for initialization of cases. Due to PTI export, State Estimator mismatches and swing bus chosen (first generator in PTI export file) sometime power flow was not able to solve or there was an excessive error in power flow solution. This issue was resolved by adding "Distributed Swing Buses" option. Peak RC added a new function in the power flow solver to use distributed swing buses when solving a basecase from the EMS. This helps the interface flows match up better with EMS, and it should also prevent much MW change at the swing bus. The disadvantage is that each unit's MW generation will change a little to account for losses in the system. This option causes all AGC generators in an AC island to help pick up the MW mismatch in the island. The MW mismatch is distributed according to the current MW output of each unit; i.e., units with higher MW output get a higher share of the percentage; in this way, the existing MW distribution would not be changed significantly. This process is performed on each island in the system. Slack is distributed proportionally to Pout. This option is used only for the first power flow solution as snapshot is passed from SE to TSAT. Each next power flow solution will use the default swing bus.

4.4.3.2   Base Load Flags

Some units have their governor blocked (nuclear, coal plants, etc.), and their governors are not responsive. Since we use PTI format for power flow and this format does not have base load flag, this flag is added on the fly each time a model is exported from the State Estimator. This flag is populated from WECC basecase, and this file needs occasional update. Consequence on not having correct units baseloaded will be wrong frequency response. This file simply lists units that are baseloaded.

4.4.3.3   Netting Small Generators Having Low MW Output

Due to lack of status signals and individual output MW measurements, for some generators it is not clear if they are online or offline. This is mostly the case with very small units or some renewable generation. These units can cause problem during dynamic simulation especially in case of wind farms modeled as asynchronous machine (types 1 and 2). These generators need to be automatically netted (dynamic model disabled). This is performed automatically using script on each real-time case.

4.4.3.4   HVDC Models

In the Western Interconnection, there are multiple HVDC links. The DC links are modeled in State Estimator. However, since the models are not standard power flow models, during export and conversion process from node-breaker to bus-branch model, in bus-branch model, they appear as sources and sinks (depending if poles operate as rectifier or inverter). In order to preserve dynamic HVDC links, one must match dynamic model to steady-state model. We need to reinsert HVDC steady-state model into the bus-branch model on the place of sources and sinks.

4.4.3.5   Static VAR Devices

In State Estimator model, many SVCs and STATCOMs are modeled as synchronous condensers. Some dynamic models representing them cannot be coupled with generator model in power flow, but instead they require SVC models in power flow.

For that reason, all the models have to be converted, on fly, into the SVC models. This is performed automatically using script.

4.4.3.6   Mapping Dynamic Models to Power Flow

There are nearly 4000 generators modeled in the WSM. Dynamic models are kept in WECC dynamic database. Generators are tested every 5 years based on WECC generation testing policy. Model parameters are based on test data. Dynamic models need to match one-to-one with steady-state model (power flow).

Figure 4.9 illustrates dynamic models for unit G_COULEE having id PG10 and connected at 13.8 kV. This specific unit also can work in pumping mode. Dynamic model of the specific unit consists of generator salient pole model gentpj, excitation model exst1, power system stabilizer model pssb, and governor model type hyg3. Generators in power flow that do not match one-to-one are netted (meaning that they will not have dynamic model and will be considered as MW injection). In our case some 5% to 9% of the units are netted.

Model mapping is maintained manually. Each time a new steady-state model (power flow) database is updated, one needs to check for changes on substation names containing generator, voltage level at the point of connection with generator, or generator Id. If changes are noted, they need to be transferred to file containing dynamic models. If new generators are added in steady-state model, they need also to be added in dynamic model. If new testing data are available for dynamic models, then dynamic models need to be updated. The approach is the same for other

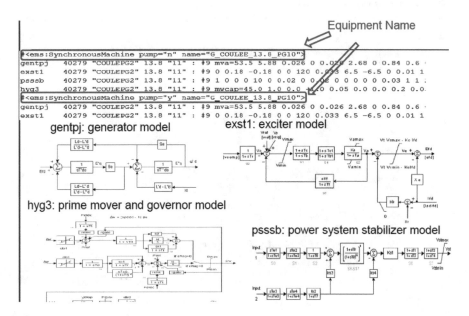

Fig. 4.9 Fig dynamic model for G_COLEE unit PG10

components having dynamic models such as loads, relays, and SVCs. In general, mapping is one-time effort and maintenance as incremental process that does not take too much time.

4.4.3.7 Contingency Definition

While in RTCA contingencies are defined by removing contingent element prior to power flow, calculation in dynamic simulation contingencies starts with contingent element in service. One needs to define the type of fault, fault position, fault imped- ance clearing time, integration method, and integration step. Typically, the integra- tion step used is 0.25 cycles, and the integration method is Runge-Kutta fourth order. Smaller integration time and larger order of integration yield more accurate results; however, it leads to slower execution time and more computation so com- promise in between two must be found. Below is an example of contingency defini- tion (Fig. 4.10).

If faulted line contains series capacitors, one needs to remember to bypass them during fault. In the field series, capacitors are protected against overvoltage and short-circuit currents by MOVs or some other gap device that is not necessarily modeled in dynamic model database. Our practice is to bypass series capacitors at the same time when fault is applied. If capacitor is not bypassed, simulation results will not be accurate. Effects of not bypassing series capacitors are decreased line impedance and increased short-circuit current leading to more serious fault in simu- lation than would be in reality. The effect of not bypassing series capacitors can be seen on bus voltages in Fig. 4.11.

```
Description MUC5L021 (BROADVIEW GARRISON 1#2 500 KV)
Simulation 10.000000 Seconds
Step Size 0.25 Cycles
Plot 3 Steps
Report 999 Steps
Integration RK4
At Time 30 Cycles
Modify Line ;'GARR_G1S1_1500_LN_A' ;'' ;    0 0.00011 0
Modify Line ;'GARR_G2S1_2500_LN_A' ;'' ;    0 0.00011 0
Three Phase Fault On Line ;'BRDV_GARR_1500_LN_A' ;'' ;    0
Three Phase Fault On Line ;'BRDV_GARR_2500_LN_A' ;'' ;    0
After 3 Cycles
Clear Three Phase Fault
/ opening of the line BROADVIEW GARRISON 1
Remove Line ;'BRDV_GARR_1500_LN_A' ;'' ;
Remove Line ;'BRDV_GARR_1500_ZBR_1Z' ;'' ;
Remove Line ;'BRDV_GARR_1500_ZBR_Z1' ;'' ;
Remove Line ;'GARR_G1S1_1500_LN_A' ;'' ;
Remove Line ;'BRDV_GARR_1500_LN_B' ;'' ;
Remove Line ;'BRDV_GARR_2500_LN_A' ;'' ;
Remove Line ;'BRDV_GARR_2500_ZBR_1Z' ;'' ;
Remove Line ;'BRDV_GARR_2500_ZBR_Z1' ;'' ;
Remove Line ;'GARR_G2S1_2500_LN_A' ;'' ;
Remove Line ;'BRDV_GARR_2500_LN_B' ;'' ;    |
Nomore
/
END
```

Fig. 4.10 Sample of a contingency definition in TSAT

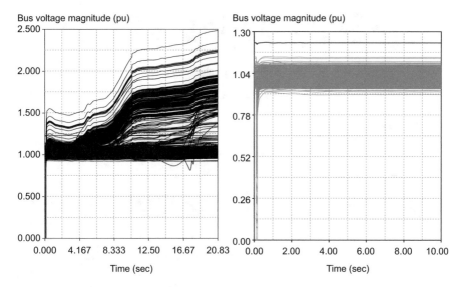

Fig. 4.11 Difference of TSAT simulation without and with bypassing in-series capacitor

4.4.3.8   RAS Modeling in TSAT

Remedial Action Scheme (RAS) senses abnormal or predetermined system conditions and takes corrective actions to maintain system reliability and/or security.
RAS supplement ordinary protection and control devices to prevent violations of
the NERC reliability standards and limit the impact of extreme events [3]. The most
common actions taken by RAS include changes in generation (MW and MVAR),
demand, or system configuration to maintain system stability, acceptable voltage, or
power flows [3]. As RAS are becoming more widely used in recent years [4], the
impact of these schemes and the potential interaction between RAS systems need to
be considered in real-time power system studies. Transient stability analysis focuses
on the power system behavior in the period of 10 s after the disturbance. Most RAS
schemes usually take action in this time frame and can affect the transient stability
limits in a given study area. Therefore, it is essential to understand and model the
effects of these automatic protection systems in real-time transient stability
assessments.

4.4.3.9   RAS Basic Structure

RAS basic structure, illustrated in Fig. 4.12, consists of four components: input
signals, RAS logic, actions, and in some cases arming tables. We used the following
procedure for implementing and testing WECC RAS schemes for online transient
stability analysis. First, based on the enabling and triggering conditions of a given
RAS, the logic diagram has been designed and built using user-defined models

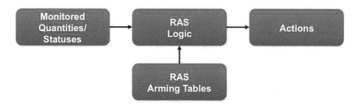

Fig. 4.12 RAS basic structure

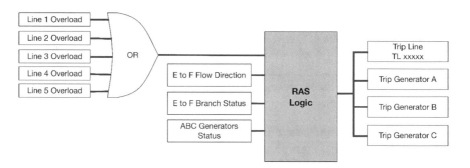

Fig. 4.13 Logic diagram of Path X RAS

(UDM). Once the UDM is created, its data file is included in the TSAT case (in the Dynamic Data section) to be used in computations. Then, a recent TSAT case has been selected from archive and tested to ensure its quality using a "no disturbance" test. In this analysis, a simulation is performed with no disturbance applied to the system. TSAT will list the generators with the largest speed and angle deviations from the simulation. Selected case is considered credible if all generator angle deviations after a 10s of flat run remain within $5°$. Finally, the functionality of the implemented RAS scheme is tested both offline and online for all enabling and triggering conditions using dynamic simulations and a pre-defined set of contingencies. During this study 11 RAS schemes in WECC were implemented and tested.

Below we will explain the implementation and testing procedure for one of the RAS schemes, named "Path X" RAS, in more details. The "Path X" RAS is a parameter-based RAS which is triggered based on overload on certain lines. Parameter-based RAS monitors variables for which a significant change confirms the occurrence of a critical event. Details on common RAS classifications can be found in [4]. Real names of the equipment and areas have been changed to maintain confidentiality. Figure 4.13 shows the logic diagram for this RAS mainly highlighting the required input signals and the actions. After designing the logic diagram, it is built using a graphical interface called UDM Editor. The TSAT UDM file is then added to the dynamic data of the case.

Table 4.1 RAS logic example

Enabling conditions	Triggering conditions			Actions
Always enabled	Loading on any of the 230 kV lines: Line 1, Line 2, Line 3, Line 4, Line 5 above 388 MVA	AND	Flow from bus E to F	In 2 s open TL xxxxx If the overload persists, the RAS will sequentially drop units at ABC generating station until the overload is alleviated.
			Flow from bus E to F	RAS will sequentially drop units at ABC generating station until the overload is alleviated.

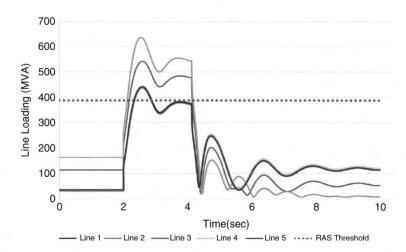

Fig. 4.14 Line loading and Path X RAS threshold

Table 4.1 shows the enabling and triggering conditions for Path X RAS and sum-marizes the actions. This RAS is always enabled and is designed to protect certain transmission lines from overload. It is triggered by detecting an overload above 388 MVA on any of the specified 230 kV lines. The RAS action depends on the direction of the power flow from bus E to F. It will trip a line and/or drop generating units until the overload is alleviated. For this RAS, all the actions should take place within 10 s which is during transient stability analysis time frame.

First, we simulated the no disturbance analysis to ensure the quality of the TSAT test cases. Initiation results proved proper quality of the testing cases as all genera-tor angle deviations after 10 s of flat run were within 5°. To test the functionality of the RAS, we used a double transmission line outage at 2 s. This contingency caused overload on all five 230 kV lines. In the selected case, the power flow was from E to F; therefore, a trip signal was sent in 2 s to open TLxxxxx line. Since the overload persisted after opening the line, RAS sequentially dropped units at ABC generating station. In this case all the three units were dropped. Figure 4.14 shows the line loading and the RAS threshold for this simulation. More simulations were performed to test all the triggering conditions for different E to F flow patterns.

In order to achieve practical results, Remedial Action Schemes (RAS) and system protection schemes (SPS) should be properly modeled in TSAT. In general, the RAS modeling in TSAT is more capable than RAS modeling such as in power flow-based tool such as STNET or PowerWorld in several aspects:

TSAT is able to model RAS associated with system transient behaviors. In the field, there are rate of change (ROC) relays and out-of-step relays. These relays cannot be accurately modeled in power flow-based software. The modeling of MATL SPS is a good example.

1. The triggering of the RAS/SPS based on post-contingency status does not depend on a feasible power flow solution. For example, a RAS is designed to trigger based on an overload of a line in the post-contingency status. If under a contingency there is no feasible power flow solution without this RAS in place, this RAS will not be triggered in RTCA/STNET. However, in TSAT, since it is a time domain simulation tool, even if this contingency is insecure without the RAS, the flow on the line after the contingency can still be calculated using numerical integration method. As long as the overflow threshold is met, the RAS will be triggered and will potentially mitigate this contingency.

The modeling of RAS in TSAT can be achieved using user-defined model (UDM). Also, in the WECC planning case, under-frequency/under-voltage load shedding functions are built in the dynamic file using standard PSLF dynamic modules. These two modeling practices will be introduced later in details.

In the TSAT case, there are ~6000 UFLS and UVLS models built in. These models are stored in dyn#.dat files in the TSAT case. UFLS/UVLS models are usually represented in lsdt1, lsdt2, lsdt3, and lsdt9. The use of such model is straightforward, and the references can be found in the PSLF user guide.

For more sophisticated RAS/SPS, user has to develop the schemes in UDM. Powertech has a graphical user interface called UDM Editor that can aid the modeling process. There were dozens of RAS being modeled in the WSM-TSAT at Peak RC.

4.4.3.10 Use of Real-Time RAS Arming Measurements

In the Western Interconnection, there is a lot of RAS intentionally tripping generation to provide temporally relief after contingency happens, until schedules are curtailed. Such RAS action may trip a few thousands of MW of generation. The amount of generation tripped is based on nomograms and various system studies. However, in real-time operation, different operators may choose more or less conservative approach for generation trip arming, based on system conditions and experience. Generation that is armed for tripping in control rooms across western control centers is sent to RC control room via ICCP link together with other real-time measurements and used in RTCA. The same approach is used with transient stability contingency calculation. The real-time arming data are sent from SE in text format together with real-time power flow snapshot and passed to TSAT RAS models every 5 min. It's essential to enhance accuracy of TSAT RAS models by use of real-time arming data from ICCP measurements. A new TSAT software enhancement implementation diagram and an example of RAS model by Peak RC are described as follows (Fig. 4.15).

4.4.3.11   Summary of Real-Time TSAT Implementation at Peak RC

Peak RC successfully implemented WSM-TSAT in control rooms in collaboration with vendor and entities. Development of WSM-TSAT involved tons of efforts on both software enhancements, system integration, model improvements, and validation against system events. After multi-year continuous efforts on use cases development and model validation, the tool has been cut off in control rooms to perform real-time monitoring on Colstrip ATR RAS and to back up Real-Time Voltage Stability Analysis (RT-VSA) tool for real-time assessment of IROLs. The online results of Colstrip ATR RAS monitoring are being shared with NWMT and other entities via Peak Web Portal www.peakrc.org to improve operation situational awareness across among stakeholders. Implementation experience and lessons learned from the WSM-TSAT project can be summarized as follows:

- WSM-TSAT is able to maximize model quality and to minimize uncertainties in order to increase transmission capacity utilization while maintaining reliability.
- Using real-time RAS Arming ICCP data for RAS modeling in TSAT is essential for accurate TSAT simulation results.
- Calculation of system inertia and frequency BIAS in real time provides an efficient way to monitor system frequency response capability.

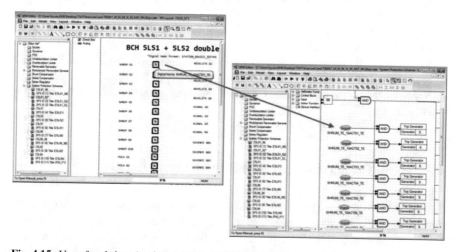

Fig. 4.15 Use of real-time Arming points in Path X RAS model

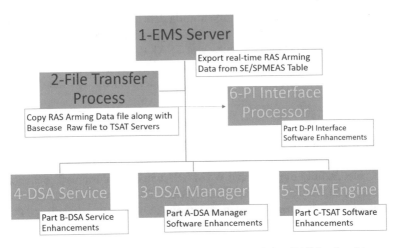

Fig. 4.16 TSAT software enhancement implementation by real-time RAS Arming data

- Setting of Unit Base Load Flags affects TSAT simulation performance of governor responses secondary control loop.
- Unit status availability and model accuracy matter.
- Lower-frequency dips in simulation might be caused by excess governor response from some units with incorrect $P_{\max}$ or base load flag settings.
- Modeling errors on wind farms and PV Solar plants are the ones usually causing TSAT solution issues.
- WSM-TSAT can back up RT-VSA for real-time assessment of the IROLs, particularly when the limiting factor is fast voltage collapse or angle instability.
- Parallel execution of basecase and transfer analysis phases in DSA Manager enables TSAT basecase screening to be completed as fast as RTCA and RT-VSA, i.e., cycling at 5-min interval. This feature allows TSAT to back up both RTCA and RT-VSA tools with no gap.
- Improve PV modeling capability in the software for more realistic simulation of PV Solar momentary cessation.
- Enhance ePMU simulation feature to allow large-scale adjustment of loads and generation MW for PMU training and modal analysis baselining.
- Improve power flow robustness against low bus voltages in basecase solution (Fig. 4.16).

4.5 Real-Time Transient Stability Study Criteria

Currently there are no commonly followed transient stability criteria that can be directly applied for real-time transient stability assessment. The purpose of this section is to establish practically applicable real-time transient stability guidelines that can be used for real-time transient stability analysis and determining transient stability-based transfer limits. Peak Reliability SOL Methodology (v8.0) requires the assessment of credible single and multiple contingencies. To ensure the reliability of the interconnection, these contingencies need to be assessed on a real-time basis to ensure that there are no transient instability risks.

4.5.1   Background for Transient Stability Criteria

Historically, WECC system performance criteria for frequency and voltages were defined in TPL-001-WECC-CRT-2.1 and utilized for planning studies and real-time assessment. These criteria are listed below in Table 4.2 and illustrated below in Fig. 4.17.

While these criteria could have been used as a starting point for discussion on real-time stability study criteria, they might not be suitable for real-time operation due to differences between system operations and planning. Moreover, these criteria have been replaced with new TPL-001-WECC-CRT-3 [12].

TPL-001-WECC-CRT-3 does not address frequency response of the system. Positive damping in stability analysis is demonstrated by showing that the amplitude of the power angle or the voltage magnitude oscillations after a minimum of 10 s is less than the initial post-contingency amplitude and transient voltage recovery criteria is less restrictive as illustrated in Fig. 4.17. The less restrictive voltage criteria are to accommodate for induction motor load and delayed voltage recovery due to induction motor stalling. Following fault clearing, the voltage shall recover

Table 4.2 Former WECC TPL-001-WECC-CRT-2.1 transient stability criteria for voltage and frequency

NERC and WECC Categories	Outage Frequency Associated with the Performance Category (outage/year)	Transient Voltage Dip Standard	Minimum Transient Frequency Standard	Post Transient Voltage Deviation Standard (See Note 3)
A	Not Applicable	Nothing in addition to NERC.		
B	≥ 0.33	Not to exceed 25% at load buses or 30% at non-load buses. Not to exceed 20% for more than 20 cycles at load buses.	Not below 59.6 Hz for 6 cycles or more at a load bus.	Not to exceed 5% at any bus.
C	0.033 – 0.33	Not to exceed 30% at any bus. Not to exceed 20% for more than 40 cycles at load buses.	Not below 59.0 Hz for 6 cycles or more at a load bus.	Not to exceed 10% at any bus.
D	< 0.033	Nothing in addition to NERC.		

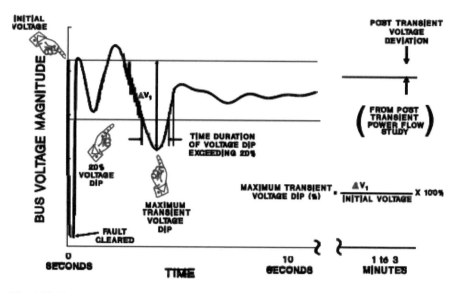

Fig. 4.17 Former voltage performance criterion WECC TPL-001-WECC-CRT-2.1

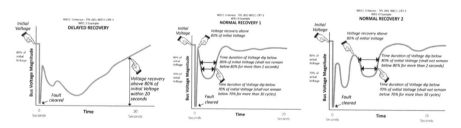

Fig. 4.18 Current TPL-001-WECC-CRT-3 voltage performance criterion WECC TPL-001-WECC-CRT-2.1

to 80% of the pre-contingency voltage within 20 s of the initiating event for all P1 through P7 events, for each applicable Bulk Electric System bus serving load as illustrated in Fig. 4.18.

However, these are planning criteria, and if they are violated, it does not mean that something ominous will happen as a consequence. In real-time operation, we need to define criteria and thresholds that, if violated, would require operator preventive action.

A step forward in definition of real-time transient stability criteria is the development of the new SOL methodology within WECC. The transient system performance table from the SOL methodology is given in Table 4.3.

Table 4.3 Transient system performance requirements

Transient system performance	Required for single P1 contingencies	Required for credible multiple contingencies
The system must demonstrate positive damping. The system is considered to demonstrate positive damping when there is a 53% reduction in the magnitude of the oscillation amplitude measured over any four cycles following the removal of fault or after the completion of any RAS action. A 53% reduction equates to 3.0% damping. The signals used generally include power angle, voltage, and/or frequency	Yes	Yes
The BES system must remain transiently stable and must not cascade or experience uncontrolled separation as described in the SOL methodology. System frequency in the interconnected system as a whole must not trigger UFLS. Any controlled islands formed must remain stable	Yes	Yes
Transient voltage or frequency dips and settling points shall not violate in magnitude and duration:	Yes	No
1. Generator ride-through capabilities as specified by PRC-024-2; no generating unit shall pull out of synchronism (or trip) in response to transient system performance; UFLS shall not be triggered		
2. Nuclear plant interface requirements		
3. Known BES equipment trip or failure levels, e.g., surge arrestors, transformer saturation levels, and generator over-excitation		

General notes:

1. UVLS or other automatic mitigation actions are permitted as specified within Peak Reliability's SOL/IROL methodology

2. A generator being disconnected from the System by Fault clearing action or by a Special Protection System is not considered pulling out of synchronism

3. If known BES equipment trip settings are exceeded, the appropriate actions must be modeled in the simulations

4. For generators that the GO or NPIR has identified as not being able to meet the PRC-024-2 requirements, either the unit must be tripped, or the Point of Interconnection (POI) frequency verified against the unit established trip values and the appropriate action taken

While the SOL methodology is defined for operational purposes, it is still intended for offline studies. The following statements apply:

- The SOL methodology is intended primarily for offline studies.
- Both system models (operation and planning) use same WECC dynamic data.
- SOL methodology requires addition of generic distance relay.
- No transient voltage criteria are in the proposed SOL methodology.
- The difference in dynamic load model is that offline simulation uses composite load model, while real-time model does not use composite load model yet.
- The SOL methodology only requires 10 s of simulation unless oscillations are seen.
- Real-time analysis must be robust.

The proposed real-time transient stability criteria are based on avoiding undesired behavior in the interconnection within the real-time operation timeframe so that such behavior can be prevented. Another consideration is that criteria need to be robust to avoid false alarming. These criteria then are in compliance with the objectives of the SOL methodology, PRC-024, and the WECC UFLS plan.

4.5.2 Monitoring for UFLS

UFLS settings are based on the WECC Off-Nominal Frequency Load Shedding and Restoration Plan [13]. This plan should be implemented for the real-time transient stability assessment. The first stage of UFLS is normally at 59.5 Hz in the Western Interconnection. The main reason to avoid a larger-frequency decay is preventing steam turbine tripping. Steam turbines are designed to rotate within a nominal speed range, or else generator components can be easily damaged, so under-frequency and over-frequency protection is used to keep the generator from operating outside the nominal speed range.

The load buses monitored for these criteria should also be the PMU telemetered buses. The rational for monitoring frequency on PMU telemetered buses is:

- This selection can allow the validation of the results since frequency events can be easily simulated and results then can be compared to the PMU measurements.
- The frequency performance should be similar for all buses on the same system.
- To avoid false alarming.

For those reasons the frequency threshold is set to 59.5 Hz for 0.1 s. However, some regions that can island from the interconnection have hydro only and have different frequency threshold since hydro turbines are not as sensitive to speed deviation.

4.5.3 Monitoring for System Frequency Response

As previously mentioned, there is no criterion in TPL-001 for frequency response. However, there is BAL-003 that requires sufficient frequency response from a BA [14]. However, BAL-003 has some deficiencies:

- It addresses generation loss only (does not address a WECC separation under surplus of renewable in California).
- Covers only the operating timeframe.
- Average over events, not required for every event (planners assume performance target each event).
- No requirement for category P2–P6 events.
- Does not address frequency dip (natural or inertia response).

Figure 4.19 illustrates typical frequency response in the Western Interconnection. A is the starting point; C is the lowest frequency point (nadir); and point B is the settling point following the primary frequency response. The capability to monitor the A–C or A–B frequency response measure in real time would provide situational awareness to operators on the deterioration of system frequency. A BA may or may not need to perform this analysis at the BA level. An analysis of the historical frequency response measured performance would reveal whether a real-time assessment is necessary.

4.5.4   Monitoring for Unit Frequency Ride-Through Capabilities

Unit ride-through capabilities are defined in PRC-024 [2], and the Peak RC SOL methodology for the Operations Horizon requires that these thresholds should not be violated as that would indicate a unit might pull out of synchronism. These ride-through capabilities can be modeled using a common generic relay model applicable for all units as a part of the dynamic data. In order for real-time transient stability assessment to be practical, a simulation time of 10 s is generally utilized. This simulation time may be increased if benchmarking indicates that unwanted unit trips may not be captured by running the simulation only for 10 s. These criteria are set as a warning only since an operator would not be able to act on it. However, it would help to point out a potential weakness in the generator buses during operation. The frequency ride-through criteria is shown in Fig. 4.20.

Fig. 4.19 Typical frequency response

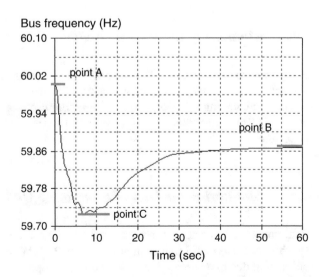

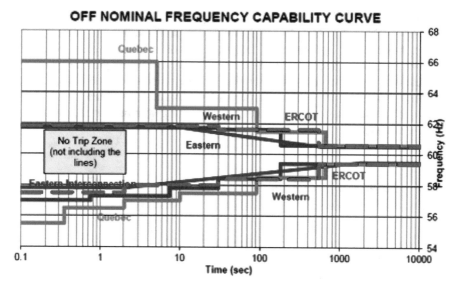

Fig. 4.20 Frequency ride-through criterion

4.5.5   Monitoring for Unit Voltage Ride-Through Capabilities

Unit ride-through capabilities are defined in PRC-024 [2] shown below, and the Peak RC SOL methodology for Operations Horizon requires that these thresholds should not be violated as that would indicate a unit pulling out of synchronism. These can be modeled using a common generic relay model applicable for all units as part of the dynamic data. The voltage ride-through criteria is shown in Fig. 4.21.

4.5.6   Monitoring for System Oscillation Damping

Per the revised Peak RC's SOL methodology for the Operations Horizon, evaluations need to be done on the power, angle, voltage, and/or frequency signals to ensure that damping is 3% or greater for any four cycles after all RAS actions have occurred. The same damping threshold is set for real-time transient stability application.

4.5.7   Monitoring for Loss of Generating Unit Synchronism

Real-time transient stability assessment should be set to monitor for units that lose synchronism. Non-BES units that are less than 25 MW in size and are observed to be tripping due to modeling issues should not be reported as tripping due to loss of

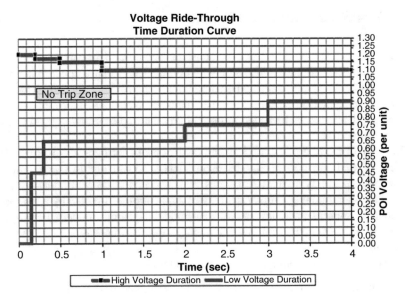

Fig. 4.21 Voltage ride-through criterion

synchronism. For determining transient stability-based path limits, the loss of one or more larger units may not cause reliability issues for the interconnection. In that scenario larger thresholds for the total amount of MWs tripped may need to be utilized for monitoring purposes. These thresholds are defined by geographical area based on dialog with area operator. The objective of these criteria would be to prevent cascading.

4.5.8 Monitoring for Additional Voltage Criteria on Custom Basis

Many TOPs have additional transmission voltage criteria specific to their footprint due to their local nature. These criteria are based on historical system events and "no return points" after system was not able to recover. An example of voltage criteria is provided below:

500-kV bus voltages along the California-Oregon Intertie:

– Shall not drop below 440 kV during a power swing
– Shall not stay below 460 kV for more than 30 cycles during a power swing

4.6   Model Validation and Tuning

All control room applications must be robust and accurate so that operators and supporting engineers can make a decision with confidence. A major concern in control room is, on the one hand, not to miss some important problem that can surface during real-time operation and, on the other hand, to avoid false alarming. For that reason, application must be tested thoroughly by the implementation team and end users continuously.

In order to move application into the control room for decision-making, engineering, IT, and operators need to have confidence. To gain and to keep confidence in accuracy of the dynamic assessments tool, it needs to be continuously benchmarked by operation engineers. Accuracy of application is affected by a number of components such as accuracy steady-state model, dynamic models, contingency definition, RAS models, and overall quality of real-time cases from the state estimator.

The objective of system model validation is twofold:

- First, make sure that there is no significant modeling error that affects results (for both steady-state and dynamic model).
- Second, identify specific study error margin for WECC major paths under different system stressing levels.

4.6.1   No Disturbance Test

The first test of quality of the cases is "no disturbance" 10-s test run. Each new case is submitted to this test run in real time, prior to contingencies run. Simulation results should be flat lines for all parameters since no system changes have been performed meaning no fault has been simulated and no single switching operation has been performed during that test. Normally the largest generation speed deviation and peak-to-peak angle should be small. The figure below illustrates the differences in between non-acceptable and acceptable case. In case that test is not passed, one needs to identify what causes oscillations in the system and to eliminate it. Usually it is related to dynamic model parameters. Generators having largest speed and angle deviations are listed and their dynamic models should be inspected.

Figure 4.22 shows the examples of two no disturbance test results: unacceptable and acceptable.

4.6.2   Event-Based System Model Validation

System model validation is performed by comparing results of simulation to PMU measurements for different system disturbances [20]. It includes frequency and non-frequency events. Since SE solution is saved each 5 min, the case just prior to

Rank	Bus No.	Bus Name	Gen. ID	Peak-Peak Angle Dev. (deg)
1	70334	PUB_DSLS	G1	102.128
2	70315	MANCHEF2	G2	13.578
3	70310	PAWNEE	C1	13.364
4	70589	RMEC2	G2	13.150
5	70588	RMEC1	G1	13.032
6	70501	QF_CPP3T	ST	12.781
7	70106	CHEROK4	C4	12.743
8	73449	FLATIRN2	1	12.708
9	70409	ST.VRAIN	G1	12.647
10	73418	RD_NIXON	1	12.626

Rank	Bus No.	Bus Name	Gen. ID	Peak-Peak Angle Dev. (deg)
1	46814	ROCKI U6	U6	1.452
2	46812	ROCKI U5	U5	1.311
3	35048	FRITOLAY	1	0.808
4	31782	CRBU 4-5	1	0.563
5	22487	MEF MR2	1	0.512
6	34610	HAAS	1	0.482
7	34610	HAAS	2	0.376
8	50438	KMO 13G2	8	0.308
9	46822	ROCKI U8	U8	0.306
10	50437	KMO 13G1	3	0.304

Fig. 4.22 Results of no disturbance test for non-acceptable and acceptable cases

disturbance is retrieved from the archive. This case is taken "as it is," and no adjustments are performed within power flow unless some other switching has happened in between the moment the case was archived and prior to system disturbance event. Figure 4.23 depicts steps in event-based system model validation. After system event happens, archived case just prior to event is taken. Precise event switching sequence is modeled as contingency and event is simulated. Simulation results are compared to PMU recording of the event [21–32].

Following example illustrates event-based system model validation:

A specific event happened on Thursday February 15, 2018, 20:53 PST. System frequency dropped to 59.877 Hz momentarily and became normal in 6 min. The event consists of large generation trip. Unit relayed while carrying 1334 MW. Remedial Action Scheme activated inserting series capacitors in remote 500 kV lines in the Pacific Northwest together with some shunt capacitors for voltage support. Event sequence has been obtained from system operator, and the event has been simulated using archived real-time case that has been saved a few minutes prior to the event itself. Figures below illustrate simulation results for frequency response, major bus voltages, and some line flows superposed to PMU measurements during the event. Figure 4.24 illustrates simulation results vs. available PMU measurements for major interface flows, bus voltages, and frequency.

4.6.3   Benchmarking Interface Flow

The Western Interconnection contains a number of interfaces also known as a "transmission paths." Paths are corridor that allows interchange in between balancing areas (BA). Single path typically consists of multiple lines. The

Fig. 4.23 System model
validation steps

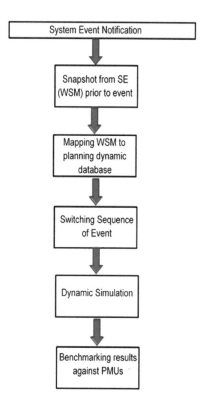

California-Oregon Interface (COI) is one of the major WECC transmission flow-gates (Fig. 4.25) consisting of three 500 kV lines allowing transfer of power from the Pacific Northwest to California or, occasionally, from California to the Pacific Northwest. System operating limit (SOL) imposed on each transmission path has to be respected during operation. SOLs are calculated in offline studies and can change with operating conditions. In general, they are set very conservatively to account for study error so there is a significant margin available. Performing frequent system model validation for given transmission path for different system operating condition will enable to set limits more accurately and to estimate this margin with more precision. Figure 4.26a–e illustrates results of benchmarking of the California-Oregon Interface (COI) flow for different operation patterns for multiple system events. One of the major reasons of the benchmarking is the assessment of error margin so that one knows what is the expected error from the simulation and studies.

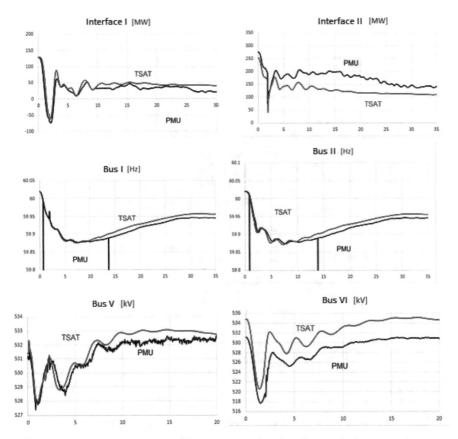

Fig. 4.24 Simulation results vs. available PMU measurements for major interface flows, bus voltages, and frequency

4.6.4   Validation of RAS Model

Remedial Action Scheme (RAS) modeling is crucial for accurate simulation. Example below illustrates how RAS models are validated. This specific RAS relay on measuring unit output in the plant and shaft speed and acceleration to choose, in real time, what, if any, unit will be tripped. For that reason, this specific RAS cannot be accurately modeled in power flow. Over the year there are multiple occasions when RAS model is triggered. These events are simulated as they happened in the field without tripping generators but waiting to see if correct tripping will be performed by modeled RAS action. Table 4.4 illustrates one such event and switching sequence. Table 4.5 shows simulation log demonstrating that correct units were tripped.

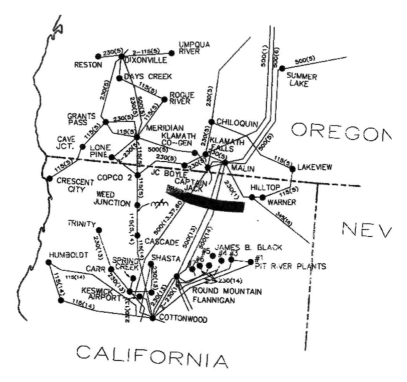

Fig. 4.25   California-Oregon interface

After the simulation is performed, multiple simulation results are compared to available PMUs to make sure that results are credible. Figures 4.27 and 4.28 show some of the results.

4.6.5   Contingency vs. Event

As shown in the previous sections, transient stability application validation and testing is performed on multiple levels using archived cases from real time as a starting point. Continuous testing is done by benchmarking results of simulation to PMU measurements for simulated system events mimicking the event. The other way of looking into it is that, if the field event has been modeled as contingency to be run in real time, to compare predicted results to PMU traces acquired during the event, after the event happened. While it is not possible to predict the event completely (type of fault, position of fault, fault impedance), simulation results should be close enough (Fig. 4.29).

On October 16, 2017, British Columbia and Alberta separated from the Western Interconnection due to contingency and formed separate islands. Similar event has

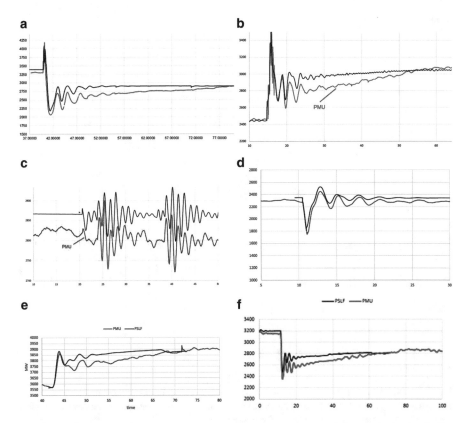

Fig. 4.26 (**a**) Interface flow for simulated event (measurement vs. real-time node-breaker WSM model for 2826 MW generation drop). (**b**) Interface flow for simulated event (measurement vs. real-time node-breaker WSM model for RAS event resulting in 1708 MW generation drop). (**c**) Interface flow (measurement vs. real-time node-breaker WSM model for double reclosing and tripping of the Garrison-Taft line). (**d**) Interface flow (measurement vs. real-time node-breaker WSM model for brake insertion into the system). (**e**) Interface flow (measurement vs. real-time node-breaker WSM model for loss of large generation unit). (**f**) Interface flow (measurement vs. real-time node-breaker WSM model for loss of two large generation units)

Table 4.4 Event logs

Time	Event
01:00:58.941	A-Phase fault on Colstrip - Broadview 500 kV Line A; Z=0+ j0.05981 pu
01:00:58.983	Fault cleared by opening Colstrip - Broadview 500 kV Line A
01:01:06.724	A-Phase fault on Colstrip - Broadview Line B; Z=0+ j0.05981 pu
01:01:06.766	Fault cleared by opening Colstrip - Broadview 500 kV Line B
01:01:06.908	Colstrip Unit 4 Tripped
01:01:07.000	Colstrip Unit 1 Tripped
01:01:07.180	Colstrip Unit 2 Tripped
01:01:31.000	NAVAJO Unit 3 Trip

Table 4.5 Simulation log

Switching Event Report - Basecase 1 2017_01_20_01_01_50 - Part 1 : COLSTRIP SINGLE PHASE FAULT

No.	Time (Seconds)	Description	Additional Information
1	1.000	LINE TO GROUND FAULT	BUS : 97134 COLSTRP 500. ZERO SEQ. : R = 0.00000 X = 0.05981 P.U. NEG. SEQ. : R = 0.00000 X = 0.00000 P.U.
2	1.042	LINE TO GROUND FAULT CLEARED	BUS : 97134 COLSTRP 500.
3	1.042	LINE REMOVED	FROM BUS : 97118 BROADVU 500. TO BUS : 97134 COLSTRP 500. CKT : 1 EQ. NAME: BRDV_COLS_1500_LN_B
4	1.070	LINE TO GROUND FAULT	BUS : 97127 COLSTRP 500. ZERO SEQ. : R = 0.00000 X = 0.05981 P.U. NEG. SEQ. : R = 0.00000 X = 0.00000 P.U.
5	1.112	LINE TO GROUND FAULT CLEARED	BUS : 97127 COLSTRP 500.
6	1.112	LINE REMOVED	FROM BUS : 97119 BROADVU 500. TO BUS : 97127 COLSTRP 500. CKT : 1 EQ. NAME: BRDV_COLS_2500_LN_B
7	1.184	GENERATOR DISCONNECTED BY SPS	GENERATOR : 623504 COLSTP 4 26.0 ID : 1 EQ. NAME: COLSTRIP_26_G04
8	1.212	GENERATOR DISCONNECTED BY SPS	GENERATOR : 623501 COLSTP 1 22.0 ID : 1 EQ. NAME: COLSTRIP_22_G01
9	1.212	GENERATOR DISCONNECTED BY SPS	GENERATOR : 623502 COLSTP 2 22.0 ID : 1 EQ. NAME: COLSTRIP_22_G02
10	1.220	IMPEDANCE ADDED	BUS : 623107 COLSTRP 230. AMOUNT : R = 0.83333 X = 0.00000 P.U.
11	25.000	GENERATOR DISCONNECTED	GENERATOR : 15983 NAVAJO 3 26.0 ID : 1 EQ. NAME: NAVAJGEN_26_U3_NET

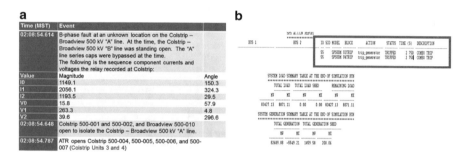

Fig. 4.27 (**a**) Event sequence as it happened. (**b**) Simulation log. Event is simulated as it happened without tripping units. Units are tripped by RAS model. Simulation log shows that RAS model has tripped the same units as it happened in the field

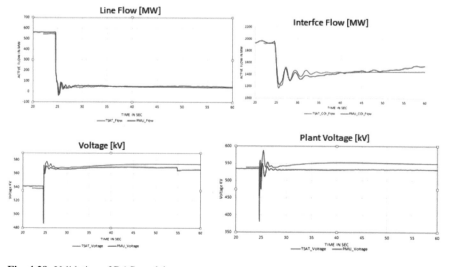

Fig. 4.28 Validation of RAS model

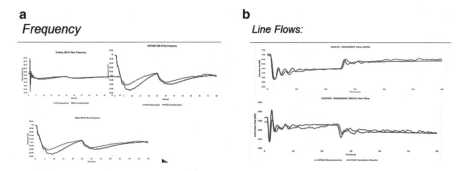

Fig. 4.29 (**a**) Frequency response of the system. (**b**) Line flows

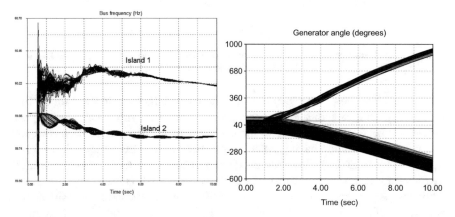

Fig. 4.30 (**a**) Frequency of interconnection before and after separation. (**b**) Rotor angles of generators

been modeled as a contingency in online stability application. Real-time transient stability results before the event show that contingency is safe and that there is no instability of any kind.

Following event that occurred in October 2017 has been modeled as N-2 contingency. In reality, one of the 500 kV lines linking US part to Canada part of the interconnection was out of service for maintenance. Second 500 kV line trips triggering multiple RAS actions leading to separation of interconnection in two islands. RAS actions were modeled as in field, as UDM, and they are triggered based on conditions. Generation arming was from real time, based on field setting. Simulation results show the frequency of the two islands in Fig. 4.30a and the generator angles in Fig. 4.30b. It can be clearly seen from the two figures the separation of interconnection into two islands and that frequency settles down above 60 Hz for one island and below 60 Hz for the second island, depending on over- and under-generation. Figure 4.30b indicates that generator angles stay together within the two separate

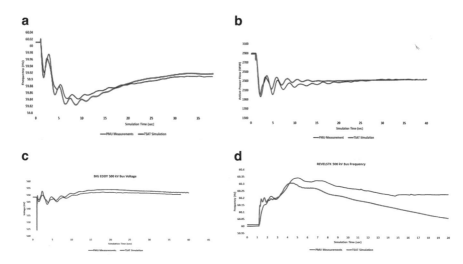

Fig. 4.31 (**a**) Frequency response PMU vs. simulation. (**b**) The California-Oregon Interface flow. (**c**) Voltage on one 500 kV bus. (**d**) Frequency response for BC Hydro footprint

groups. Flat line shows the generation that was tripped. Increase in angle for the first group of generators indicates that this group is rotating with synchronous speed having frequency above 60 Hz, so the angle seems to increase, while the second group indicates rotation having synchronous speed having frequency below 60 Hz. Voltages show no voltage instabilities. Results clearly show that there are no stability issues. Further, simulation results are compared to PMUs to see how close simulation results correspond to the event itself. Difference between event and simulation is timeline of RAS triggered as well as type of the fault and position of the fault. Figure 4.30 shows just a few comparison results of simulation vs. PMU measurements.

Results indicate high level of correspondence in between predicted results and what actually happened in the field. There is small frequency mismatch in Fig. 4.31d. This mismatch was due to modeling issue of missing voltage limiter on synchronous condenser, as it can be seen in Fig. 4.32a. Increase in voltage leads to increase of the load consequently decreasing frequency. Once the limiter was added, the frequency response matched.

4.7   TSAT Use Cases Development at Peak RC [51]

4.7.1   Acceleration Trend Relay Monitoring in Real Time

Colstrip is one large 2380 MW coal-fired power plant consisting of four generating units and operated by Northwestern Energy from Montana. Plant is integrated into the Pacific Northwest system through long 500 kV double transmission line.

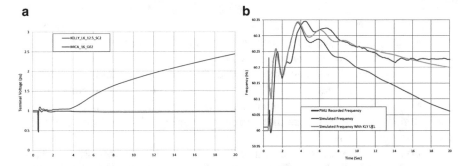

Fig. 4.32 (**a**) Terminal voltage on problematic synchronous condenser (blue trace). (**b**) Frequency response after OEL was added

Acceleration trend relay (ATC) is a special RAS action that is designed to protect the grid in Montana and Wyoming from larger-scale instability during fault and line outage on 500 kV system. The ATR will trip Colstrip generators to return system to stable condition while minimizing amount of generation tripped. The device itself is a computer-based relay which monitors speed (tooth wheel counting), acceleration, and angle of four Colstrip units. This information is used to detect unstable event in progress, on flay. The ATR use 11 different algorithms to detect instability. After instability detected trip, logic chooses which units, if any, need to be tripped. More details on ATR is given in [https://www.wecc.org/Reliability/Acceleration_Trend_Relay_Model_Specification_DRAFT.PDF C.A. Stigers; C.S. Woods; J.R. Smith; R.D. Setterstrom "The acceleration trend relay for generator stabilization at Colstrip" IEEE Trans on Power Delivery, Vol12 Issue 3 June 1997.

While ATR functions as designed, it represents a major challenge for Northwestern operators and Reliability Coordinators:

- They do not know if and how many generation units will be tripped until the event happens.
- ATR is designed to prevent instability but not overload of lines.
- Since the amount of generation tripped is based on speed and acceleration measurement, ATR cannot be modeled in RTCA that is a power flow with contingent element(s) out of service.

As system operators and Reliability Coordinators are required to run RTCA for N-1 and credible N-2 and to operate the transmission facilities within system limits. As ATR cannot be modeled in RTCA accurately and there is no advanced knowledge on what units would be tripped for a given contingency that is expected to trigger ATR, there is a set of the contingencies that indicates overload or contingencies simply cannot be solved. Each such contingency sends an alarm to the operator. Consequently, the operator must investigate if alarm is false or there is a real violation. In this regard, Reliability Coordinators communicate the violation to Northwestern Energy operators. The operator communicates back that ATR would resolve the violation, but it is an assumption. As ATR objective is to prevent instability and minimize generation tripping, it does not necessarily miti-gate thermal limit overload of underlying lower voltage grid. TSAT reports which units are tripped at Colstrip for each contingency. If ATR action does not prevent overload, then generation output must be reduced so that the system operates without violation for N-1 and credible N-2. Reducing unit output has economic impact as well as unit unnecessary unit tripping. Therefore, it is crucial to model accurately ATR in real-time transient stability assessment tool and to send real-time arming into the RTCA so that correct generation units are tripped in RTCA for specific contingency.

Overview of the process is shown in Fig. 4.33. State Estimator provides base-cases for both RTCA and TSAT every 5 min. TSAT has all information to evaluate what contingencies would activate ATR and how many units, if any, would be tripped. Results are sent, in real time, to historian and displays are shared with Northwestern in real time. At the same time, generation arming for each contin-gency are converted into ICCP points and sent back to RTCA so that RTCA has an accurate real-time information on how many units each contingency will be tripped. Figure 4.34 gives an example to show how an TSAT calculated RAS gen trip arming point is sent to EMS/SCADA and used in RTCA for RAS modeling.

In case violation for N-1 and credible N-2 contingencies is noticed, Northwestern operator needs to runback generation output to acceptable level to prevent potential violations. A PI display shown in Fig. 4.24 was built for RC to obtain ATR RAS operation situational awareness and shared with impacted entities.

For each of the pre-selected N-1 or N-2 contingencies, ATR RAS Arming point for each Colstrip unit trip is updated by WSM-TSAT every 5 min. The RAS Arming data is transferred to SCADA via ICCP and next passed onto RTCA via SE base-case. Since individual ATR RAS Arming points in SCADA are used for modeling those relevant RAS in RTCA, it's accurate and deterministic for RTCA to drop appropriate Colstrip units or now under real-time assessment. Below is such an example.

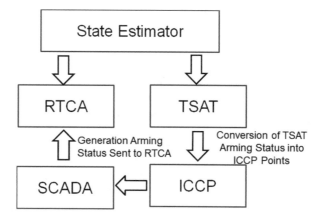

Fig. 4.33 RTCA-TSAT integration design

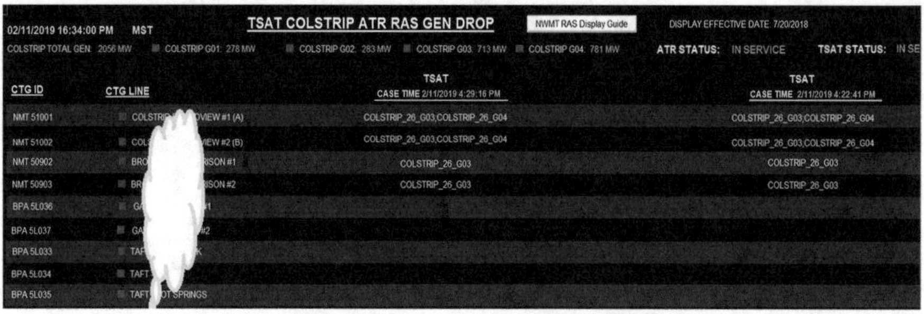

Fig. 4.34   ATR RAS operation situational awareness PI display

Such combination of real-time transient stability and real-time steady-state tool provides system operators with valid information to verify if Colstrip ATR actions resolve thermal violations leading to operational certainty and allows to maximize usage of the resources. The following example demonstrates application of ATR RAS monitoring in TSAT.

In December 2018 both 500 kV Broadview-Garrison lines were tripped. This N-2 contingency is disabled in RTCA leading to a large number of violations without knowing how many units would be tripped by ATR. Assumption based on offline studies was that ATR will mitigate overload. After the event happened, ATR tripped unit 3 only which resulted in more than 30 major branch violations which consequently resulted in failed cascade test and potential IROL violation. Mitigation included curtailment of Glacier Wind 1 and 2, Rimrock East and West, and derate of remaining Colstrip units. When the same N-2 was run in TSAT using pre-disturbance case, only unit three tripped, same as in the field. The TSAT simulation results are benchmarked against the PMU recording, as shown in Fig. 4.35, including frequencies, voltages, and COI flow. The closeness between the two confirms the quality of the TSAT results. It's still unknown what really happened (type of the fault, position of the fault, etc.). In the TSAT contingency definition, it's assumed three phase faults on two lines at the end of each line.

In the existing practice for WECC path rating definition and RAS Arming determination, nomograms are created ahead of time to predict a safe region whereby operating inside the region would be expected to result in acceptable pre- and post-contingency system performance. They is a mechanism to describe interaction between elements or paths with the objective of ensuring that the system is operated in a safe and reliable state, while the use of related elements or paths is simultaneously maximized. Nomograms may be used to provide system operators with helpful guidance as part of an operating plan; however, they are not considered to be SOLs unless the nomogram represents a region of stability (i.e., the nomogram defines a stability limit). If both real-time VSA and TSAT tools are implemented on a full WI/EI/TI operation model like WSM, empowered by super computers, it will be possible and feasible to calculate various stability limits for given system paths, internal paths, and IROL Cutplanes and evaluate RAS Gen drop arming levels in near real-time horizon. Peak RC has proved this concept by real-time assessment of

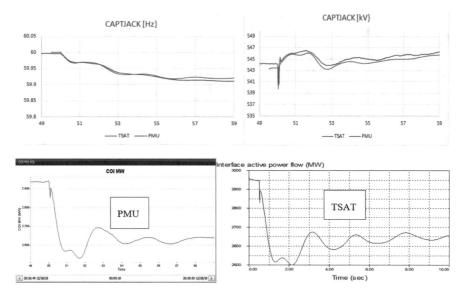

Fig. 4.35 TSAT model validation on ATR RAS trip event in December 2018

the IROLs using the RT-VSA tool and evaluation of ATR RAS unit drop using the online TSAT. The NAERM project can expand such success story to a larger model scale and broader scope of use cases.

4.7.2   Monitoring Island Frequency Response and Wide-Area System Instability

The following example in Fig. 4.36 illustrates control room implementation of the transient stability assessment tool in the prevention of frequency collapse. Usually, during spring conditions, there is abundance of the water in the Pacific Northwest due to snow melt and rain. Coupled with large wind resources and lack of water storage capacity, it drives electricity price down. During that time power flows from the USA to Canada through WECC Path 3. Path 3 is tied in between BPA and BC Hydro (BC) consisting of two 500 kV and two 230 kV lines. Two 500 kV lines are adjacent, and their loss is considered as credible N-2 contingency. During flow from North to South (the USA to Canada), loss of these two lines would trigger Northern Intertie Separation RAS islanding BC Hydro system.

In case of heavy import and light loading condition during the night hours, BC Hydro would rely on under-frequency load shedding and inertia in its own island to prevent frequency collapse. Under such conditions, frequency can go as low as 58 Hz before it starts recovering. Frequency decline has to be rapidly arrested by a combination of load shedding, inertia, and governor response. Since BC Hydro

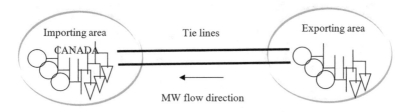

Fig. 4.36 Conceptual illustration on US-Canada Ties

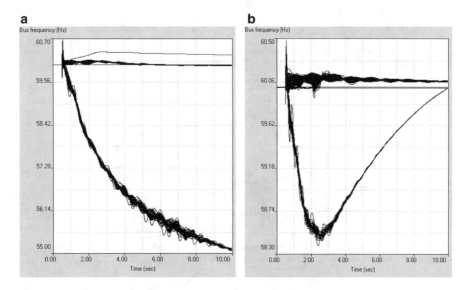

Fig. 4.37 BC island frequency collapse vs. BC island frequency recovery

generation fleet consist of hydro plants, they are not susceptible to under-frequency tripping as steam turbines are. Typically, hydro turbine would sustain frequency of 57 Hz for 30 s or more.

Figure 4.37 illustrates two BC islanded cases without and with sufficient combination of load shedding, inertia, and droop response. In this case frequency threshold for transient stability tool is set at 58 Hz at major buses in BC Hydro region vs. 59.5 Hz for the rest of the interconnections. If the threshold is violated during contingency analysis, the alarm will be sent to the operator. In order to prevent frequency collapse, the operator needs to add additional generation to spin to increase system inertia and improve frequency response.

Similar situation can occur on a larger scale. In case of the activation of NE/SE Western Interconnection separation scheme (FC Separation Scheme impacted area is plotted in Fig. 4.38). This RAS is armed if flow on COI is North to South and larger than 600 MW. In case that COI becomes separated, power is forced onto eastern lower voltage lines leading to overload, low voltage, and instability. NE/SE

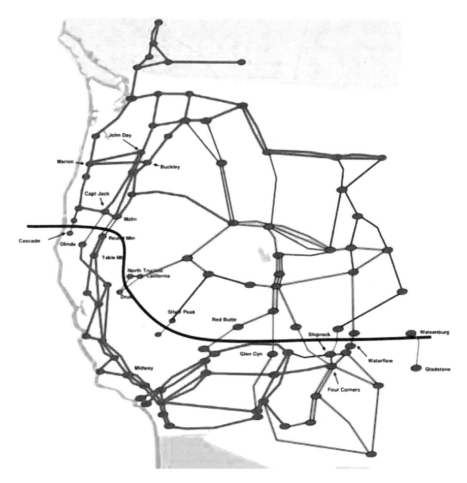

Fig. 4.38 WECC-1 RAS Separation Scenario

protection schema would trip multiple lines initializing controlled separation of NE portion of the system from the SE portion of the system. As most of the system inertia is situated within the Pacific Northwest and due to large penetration of renewable generation in SE part of the system, frequency response and even island frequency stability might be a problem with the next decade or two. It's recommended to perform preliminary case studies using the WSM or similar WI-wide operation model for at least three scenarios: (1) current high DER penetration cases, (2) 50% renewable generation in California, and (3) 100% renewable generation in California.

During BC Hydro heavy export from Canada to the USA, for credible N-2 ID-CT contingency, BC Hydro relays on generation tripping to prevent over-frequency and instability. Real-time transient stability indicates if quantity of generation is armed for tripping in case separation of the USA and Canada is sufficient to prevent system instability in British Columbia, Canada.

During August 16th 20xx export from British Columbia to USE was steadily increasing from 0 MW to over 2100 MW over a few hours' interval. With increase in export, the amount of generation that needs to be tripped increases as well, and operator in the control room would manually arm generation to be tripped in case of N-2 contingency that would result in separation. Real-time arming points are sent via ICCP link from source BA to Peak RC control room and automatically sent to real-time transient stability so that the right amount of generation is tripped during real-time transient stability run for a given contingency.

If sufficient amount of generation is not armed, real-time transient stability indicates instability, and alarm is sent to the operator. Following the alarm, the operator increases the amount of generation that needs to be tripped, and the system becomes stable against the given contingency. Both unstable and stable case results are shown in Fig. 4.39 for comparison side by side.

4.7.3   Monitoring Pacific DC Intertie (PDCI)

Pacific DC Intertie or Path 65 is a HVDC line spanning from the Pacific Northwest (Celilo Converter Station) to Los Angeles area (Sylmar Converter Station) which is a 846-mile-long DC line shown in Fig. 4.40. The project was completed in 1970 and upgraded multiple times. Originally mercury-arc valves were used, and, later, it was replaced by thyristor-based converters. Capacity is 3200 MW North to South and 2200 MW South to North. The original purpose of PDCI was to bring cheaper electricity from Columbia River to L.A. area. However, now with a large number of renewables in California, it becomes more frequent that power flows opposite direction. DC line operates at 500 kV. More detail on PDCI can be found in https://placesjournal.org/article/the-infrastructural-city/.

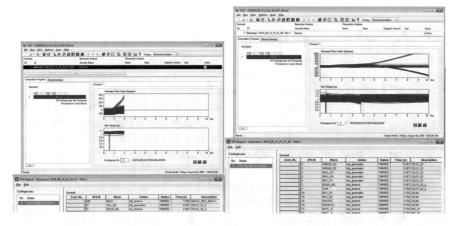

Fig. 4.39 Insecure vs. secure cases due to insufficient vs. sufficient gen drops

Fig. 4.40 PDCI vs. COI
Transmission Corridors

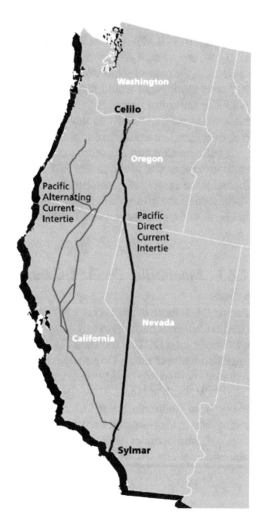

Loss of PDCI is a credible contingency and, depending on PDCI loading and neighbor paths and line loading and line in service, can result in overloading of major 500 kV lines, system instability, and voltage collapse. Based on the parameters (PDCI loading and neighbor paths and line loading and line in service), real-time operators in the Bonneville Power Administration (BPA) control center will arm up to 2800 MW of generation to instantaneously trip, if PDCI is lost, to prevent instability, uncontrolled separation, and cascading. The amount of generation armed is based on real-time conditions, offline computed nomograms, and individual operator. Operators are conservative so they tend to trip more generation to be on the safe side. If PDCI is tripped, RAS will shed armed generation, and AGC will be suspended across the Pacific Northwest until schedules are curtailed with the objective to prevent loading pickup of parallel 500 kV lines in the neighbored of the PDCI. Such large disturbance can affect the interconnection as well as frequency response. Real-time dynamic simulation helps to optimize the amount of generation tripped and, at the same time, help prevent instability and cascading.

In April 2017, PDCI flow was around 3000 MW. Simulation indicated that the amount of generation tripped by RAS action was 1023 MW. Generation arming in real-time dynamic simulation is automatic, based on real-time generation armed in BPA control center and sent via ICCP link. The same amount of generation was in RTCA and the PDCI outage contingency was unable to solve.

Dynamic simulation indicated wide-area instability, cascading, and voltage collapse. However, in reality amount of generation armed was 2900 MW and dynamic simulation indicate stable case. FACRI was triggered as well in this case. The problem was found – some RAS ICCP points were missing, and it was immediately corrected. It demonstrates how real-time transient stability can indicate possible instability and help operator to prevent it in real time. On the other hand, further TSAT simulation found if 2000 MW of generation tripped instead of 2900 MW the system remains stable. This indicates that real-time dynamic assessment can help optimize or minimize amount of generation tripped while maintaining integrity and reliability of the system.

Online or real-time TSAT is a viable tool to identify abnormally low-frequency excursions caused by massive generation drop and high renewable generation penetration (e.g., low real-time system inertia). As loss of PDCI occasionally happens, it allows validation of the WSM and TSAT model. For the example displayed in Fig. 4.41, PDCI was tripped by the RAS due to N-1-1 outages twice in 2014. It dropped over 2200 and 2500 MW generation, respectively. The first trip event caused the system frequency to drop down to 59.75 Hz, yet it recovered rapidly, because the total wind generation in BPA was close to 0 at the event time. Otherwise, if the event occurred at several hours later, when BPA wind generation went up to 3000 MW, the system frequency could drop lower significantly and recover much slowly. Such risk of frequency dip or even instability will be emerging more frequently and severely when California system embraces 100% renewable generation penetration in the upcoming two decades. To prepare for emerging challenges, we need to develop the larger operational model for real-time transient stability analysis to predict and prevent the uncontrolled frequency drop issues.

4.7.4 IFRM and FRM Real-Time Calculation

As NERC defined, one calculates *FRM* by:

$$FRM_{\text{Interconnection}} = \left(P_{\text{GenLoss}} - P_{\text{LoadLoss}} \right) / 10 \left(f_{\text{B}} - f_{\text{A}} \right),$$

where P_{GenLoss} is an interconnection generation loss and P_{LoadLoss} is an interconnection load loss.

$$BA\,FRM = \left(P_{\text{Interchange A}} - P_{\text{Interchange B}} + P_{\text{outBA}} \right) / \left(f_{\text{A}} - f_{\text{B}} \right)$$

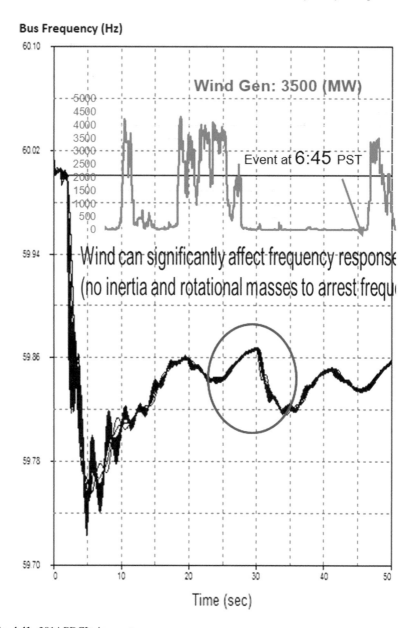

Fig. 4.41 2014 PDCI trip event

For Point B, TSAT will take the last frequency point if the simulation is less than 20 s. From 20 s to 52 s, it calculates an average. Based on a recent 2PV case, it seems 30 s are necessary, at least. However, I really think this is something the user should customize in the case setup, rather than TSAT forcing the case to run 60 s.

For the plot in Fig. 4.42, TSAT would have reported the following values for Point B frequency.

1. 10 s contingency: 59.735 Hz
2. 20 s contingency: 59.8128 Hz
3. 30 s contingency: 59.8323 Hz
4. 60 s contingency: 59.8538 Hz

WSM-TSAT software has been enhanced to calculate system inertia and IFRO Measure A-C on WECC system or individual BA level. Figure 4.43 shows PI trend on frequency BIAS of WECC, BPA, and SCE systems between February 20, 2018, and September 5, 2018. It's interesting to see the patterns of WECC system and two BAs change noticeably on a seasonable change: spring to summer.

Figure 4.44 gives a 24 h WECC system FRM curves calculated by online TSAT. As one can see, the system frequency responsive sensitivity (i.e., equivalent to frequency BIAS) changes within a range of 750–1150 MW/0.1 Hz. For operation planning study, WECC currently recommends for a value of 840 MW/0.1 Hz. A fixed or constant frequency BIAS value may not provide an accurate signal for RC/ Power Pool and/or Market Operator to control system frequency by securing adequate yet economy spinning reserves. Real-time calculation of system inertia and frequency response measures will be essential for monitoring impact of high penetration of renewable generation in the future.

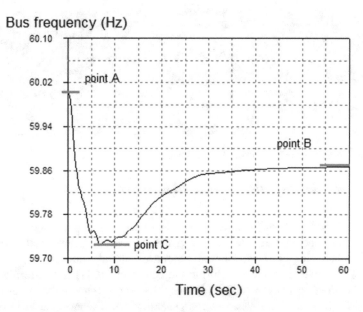

Fig. 4.42 Frequency responsive measure points. (*For further details, refer to NERC BAL-003-1 Frequency Response and Frequency Bias Setting Reliability Standard," Atlanta, GA, 2015)

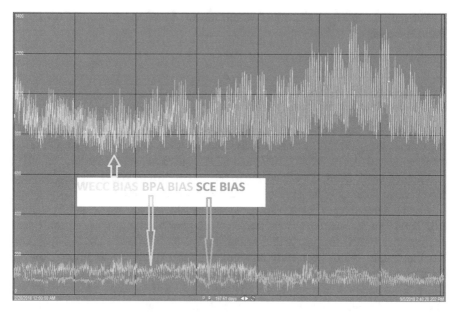

Fig. 4.43 Seasonal variation of frequency responsive sensitivity

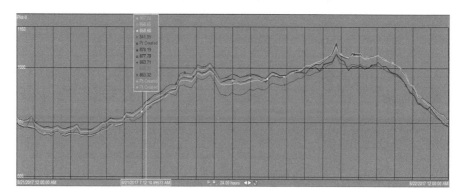

Fig. 4.44 24-h System FRM Curve. (Source: Peak Reliability)

4.7.5 *Monitoring Inverter-Based Resources*

California's 50 % renewable policy by 2030 and decrease in the cost of solar panels contribute to increase in the rate of penetration of inverter-based generators. Dynamic behavior of inverter-based generators already impacts BES, and this impact is expected to increase. During the Blue Cut fire [1] in California, a fault resulted in the frequency in WECC to reach 59.87 Hz and caught the power industry by surprise because there was no large generator that disconnected from the grid. Investigations revealed that several PV generators in the area temporarily stopped injecting power into the network due to a short voltage dip on the transmission system caused by the fault. This temporary phenomenon observed across a large number of generators has been termed as "momentary cessation." This was not an isolated incident. There have been multiple frequency events attributed to momentary cessation. This behavior of inverter-based resources has the potential of cascading loss of generation leading to blackout, and they need to be accounted in all system studies.

Current solar plant models in official WECC database and used in all system studies (planning and operation) use default parameters so as such do not illustrate "momentary cessation" behavior. For that reason, such events were not anticipated in planning timeframe during generation integration studies or during operation reliability studies that are routinely performed to account for reliability of the operation. This behavior of inverter-based resources has the potential of cascading loss of generation leading to blackout, and they need to be accounted in all system studies. The main contribution of this section is tuning current model parameters, based on event recordings so that actual model behavior corresponds to reality. Further, such tuned models are implemented in real-time transient stability application for the Western Interconnection that Peak Reliability runs in real time as Reliability Coordinator for the majority Western Interconnection.

Momentary cessation is triggered when the terminal voltage is beyond a pre-specified threshold. The generator stops injecting power, and after a time delay, generator output is ramped up at a constant rate. This voltage threshold, time delay, and ramping rate vary across manufacturers of power converters. Therefore, the user must be able to define these quantities in the model. The PV generator model in power system simulation software must have the following properties:

- The model should be able to inject zero current for the specified time period when the voltage at the terminal goes beyond the user-defined threshold.
- The model should be able to ramp up to the pre-fault levels of power injection at the ramp rate that can be specified.
- The models should have independent real and reactive current control.
- The model should be able to restrict the reactive power injection during the fault condition.

Over the years, several generic models have been developed for converter-based generators that are used for transient stability, voltage stability, and frequency stability studies of the BES [4]. NERC recommends that the momentary cessation be modeled using the REGC_A and REEC_A model. The voltage threshold and time delay are modeled using REEC_A model, and the ramp rate is modeled using REGC_A model.

A frequency drop event that was attributed to the momentary cessation was observed on April 20, 2018. There was a line to line fault on a major 500 kV line in California. The fault was cleared in less than four cycles (Fig. 4.45). In spite of the short fault clearing time, many of the PV plants entered into momentary cessation mode resulting in significant frequency drop in WECC system. The frequency dropped to 59.87 Hz (Fig. 4.46). This frequency drop indicated a generation loss of

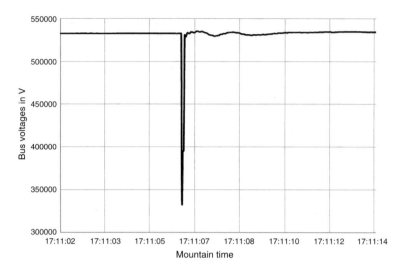

Fig. 4.45 Bus voltage drop at the PV momentary cessation

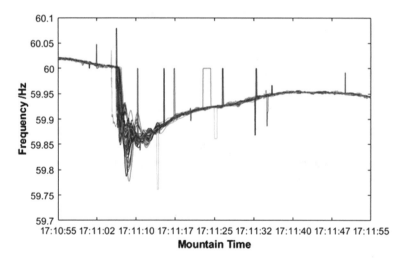

Fig. 4.46 Frequency excursion at the PV momentary cessation

about 1200 MW. Data from SCADA, which is collected every 10 s, indicated that a total of 600 MW of PV generation dropped after the event. However, in order to capture momentary cessation accurately, a more precise data set was needed. The DFR recording at several PV plants is used. This data is collected 30 times a second. It is not available at all PV plants points of connection. Therefore, it is not possible to replicate the frequency response of the system. It is also almost impossible to accurately represent plants behind meters. Therefore, PMU data available at many various locations on major transmission buses is used to further improve other PV plant models so that the measured and simulated values align as much as possible. This approach allows us to estimate the actual power drop due to PV plants entering into the momentary cessation mode including the ones in the distribution system.

The simulation is done using the TSAT software and the WSM model. The base case is a bus-branch model with 16,000 buses and 20,000 branches (dynamic) and 3800 generating units. The PV generators are modeled using the REEC_A and REGC_A models. Archived case used is the case recorded just prior to the event. The fault was applied using the details given in the NERC event report.

Initially, the default parameters for modeling momentary cessation that are used are as follows: voltage threshold of 0.9 p.u., time delay of 0 s, and a ramp rate of 10. There is very little impact of momentary cessation on the frequency response. Another simulation with time delay of 10 s was done, and the response shows frequency drops to 59.6 Hz. This large difference in the impact of momentary cessation between the two simulations clearly illustrates the need for accurate modeling of parameters for simulation studies. The TSAT simulation results are compared in Fig. 4.47a, b.

In order to accurately model momentary cessation, each PV plant is individually tuned to replicate the event recording from data fault recorders (DFRs). Two such generators are discussed in detail. DFR measurement and the final tuned model simulation of the PV generators are shown in Figs. 4.48 and 4.49.

Based on these DFR measurements, there was a reduction of almost 1500 MW momentarily, and generators producing 500 MW of power remained disconnected. As not all measurements were available at all PV stations and several PV generators

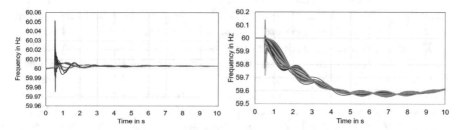

Fig. 4.47 (a) Frequency response using default parameters during the April 20, 2018, event. (b) Frequency response using a time delay of 10 s during the April 20, 2018, event

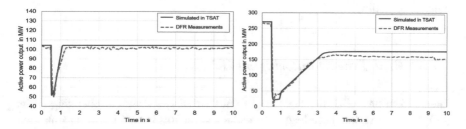

Fig. 4.48 Comparison of PV real power responses between DFR and TSAT simulation

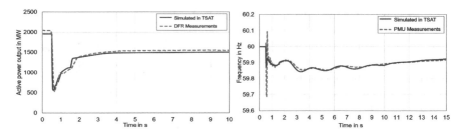

Fig. 4.49 Comparison of PV real power responses between DFR and TSAT simulation

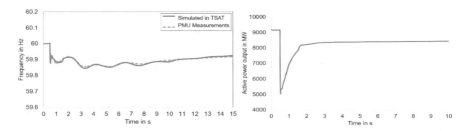

Fig. 4.50 Comparison of PV total gen drop between DFR and TSAT simulation

connected to the distribution system might have entered momentary cessation and there is no method of measuring them, PMU measurements available at certain buses the event is used to tune the model further. The final tuned model response at a few buses with the PMU measurements at those buses is shown in Fig. 4.50.

The total generation loss for this case is shown in Fig. 4.50 too. A total of 900 MW of generator remained disconnected even after 10 s. However, there was a reduction of almost 4000 MW momentarily. This number might not be all from the PV generators in transmission line; there might have been several generators in the distribution system as well. The verified PV modeling parameters were applied to WSM-TSAT in April 2019. Since then, Peak RC has been collecting real-time TSAT solution results for further validation study. With those settings implemented in real-time WSM-TSAT, it's anticipated that validating and tuning a future PV momentary cessation event, if occurring in WECC footprint, will be completed in a much shorter time.

4.7.6   Real-Time Monitoring on IROLs

Interconnection Reliability Operating Limits (IROLs) are system operating limits (SOLs) that if violated can lead to cascading, wide-area instability, or uncontrolled system separation that, consequently, can lead to blackouts. Presently a few voltage stability IROLs are identified within Interconnection: San Diego Import and Pacific

Northwest or Oregon Export. They are primarily monitored in real time using volt-age stability application since voltage stability is of main concern. However, tran-sient stability during transfer increase would first solve power flow prior to dynamic simulation, and, if the solution cannot be met for given level of the transfer, it will indicate voltage collapse.

For example, the Oregon Net Export (also called Northwest Net Export) IROL Cutplane consists three WECC paths including COI (California and Oregon Intertie) that dominates the IROL. This IROL is normally voltage stability limited, but under certain 500 kV line outages, the limiting factor will be angle instability. From Fig. 4.51, WSM-TSAT still solves the VSA limit for the IROL when RT-VSA diverges in basecase power flow or solves lower IROL limit. Hence, WSM-TSAT can back up the RT-VSA tool in case of power flow diverges or RAS not fired in RT-VSA.

Moreover, transient stability can identify fast voltage collapse due to induction motor modeling, while traditional voltage stability cannot do it. In the Western Interconnection, motor load can reach significant percent of the load. Figure 4.52 illustrates behavior of motor load on voltage recovery following fault on transmis-sion line.

Due to fault all induction motors seeing voltage dip will slow down, and, after fault cleared, they accelerate all at the same time. During acceleration they draw larger current and, consequently, induce additional voltage drop. That behavior can lead to fast voltage collapse during hot summer days when a large number of resi-dential air-conditioning devices are active. The figures below illustrate voltage on distribution bus following fault on transmission bus for different percentages of motor load, motor currents, and voltage on transmission bus following three phase faults cleared within 0.1 s. In case of longer fault clearing time due to stuck breaker, this impact is more perilous (Fig. 4.53).

The other reason for using real-time transient stability to monitor voltage stabil-ity IROLs is simply to have backup application. The definition of transfer and con-tingencies is the same. The deference is in solving mechanism. Results of both applications are closely compared against each other and also compared against CAISO voltage stability application. The figure below illustrates three calculated limits (Fig. 4.54).

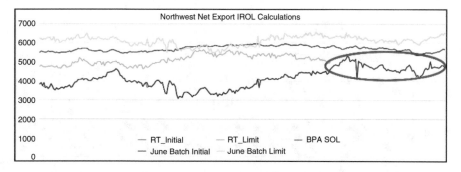

Fig. 4.51 Comparison of Northwest Net Export IROL values solved by different tools

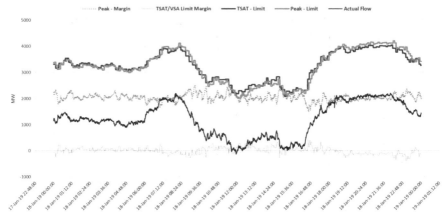

Fig. 4.52 Comparison of Oregon Net Export IROL values solved by different tools. (Source: Peak Reliabillity)

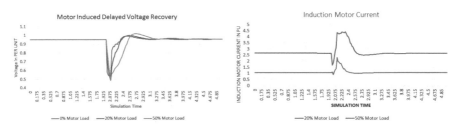

Fig. 4.53 TSAT simulation results with 20% motor loads applied

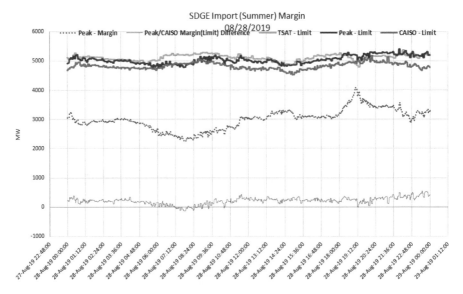

Fig. 4.54 Comparison of SDG&E Import (summer) margins solved by different tools

4.7.7   Wide-Area Angle Separation Monitoring

Angle monitoring in interconnected power system can indicate a level of the stress of the system and improve wide-area situational awareness [42–48]. In the Peak control room, we have implemented "virtual bus angle" concept in real-time transient stability tool [48]. First, there is brief explanation of the concept of "virtual bus angle." This implementation is for baselining purposes and no alarms are generated and no operator action is required.

For a single transmission line, the power transferred depends on the voltage phasors of the sending and receiving ends. Under the DC power flow assumption, the real power flow is linear to the angle difference between the sending end voltage and receiving end voltage as expressed:

$$P \sim \frac{1}{X}\left(\theta_S - \theta_R\right)$$

From the DC power flow approximation, real power transferred is closely correlated with the angle difference. In real world, a transfer path or a transfer boundary usually consists of more than one transmission lines and has multiple sending and receiving ends. The correlation between the transferred power and the bus voltage angles is not straightforward. Figure 4.55 shows a transfer path that consists of L transmission lines, LN_1 to LN_L; M sending end buses, S_1 to S_M; and N receiving end buses, R_1 to R_N.

Under the DC power flow assumption, the real power transferred over this path can be expressed as the summation of real power over the L lines:

$$P_{path} = \sum_{i=1}^{L} P_i \approx \sum_{i=1}^{L} \frac{1}{X_i}\left(\theta_{Smi} - \theta_{Rni}\right) = \sum_{i=1}^{L} \frac{1}{X_i}\theta_{Smi} - \sum_{i=1}^{L} \frac{1}{X_i}\theta_{Rni}$$

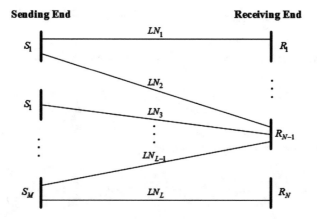

Fig. 4.55 Illustration of a transfer path

where X_i is the reactance of LN_i and S_{mi} and R_{ni} are the sending end and receiving end of LN_i, respectively.

Define equivalent reactance X_{eq} as:

$$\frac{1}{X_{eq}} = \sum_{i=1}^{l} \frac{1}{X_i}$$

Then, the path flow can be represented by:

$$P_{path} \approx \frac{1}{X_{eq}} \sum_{i=1}^{L} \frac{X_{eq}}{X_i} \theta_{Smi} - \frac{1}{X_{eq}} \sum_{i=1}^{L} \frac{X_{eq}}{X_i} \theta_{Rni} = \frac{1}{X_{eq}} (\theta_S - \theta_R)$$

where

$$\theta_S = \sum_{i=1}^{L} \frac{X_{eq}}{X_i} \theta_{Smi} \text{ and } \theta_R = \sum_{i=1}^{L} \frac{X_{eq}}{X_i} \theta_{Rni}$$

Define θ_S as the *virtual sending bus angle* and θ_R the *virtual receiving bus angle*. Equation (4) serves as the theoretical evidence of the high correlation between the path flow and the angle difference of the virtual sending angle and receiving angle. Moreover, by introducing the virtual angle concept, the flow on a complex path is now analogous to a simple expression in (1).

Grouping the angles by sending or receiving end buses, virtual bus angle θ_S and θ_R can be expressed by:

$$\theta_S = \sum_{m=1}^{M} \left(\sum_{j \in A_m} \left(\frac{X_{eq}}{X_j} \right) \theta_{Sm} \right), A_m = \{j : L_j \text{ connects to } S_m\}$$

$$\theta_R = \sum_{n=1}^{N} \left(\sum_{j \in B_n} \left(\frac{X_{eq}}{X_j} \right) \theta_{Rn} \right), B_n = \{j : L_j \text{ connects to } R_n\}$$

Define the *weighting factors* for sending or receiving buses as:

$$w_{Sm} = X_{eq} \cdot \sum_{j \in A_m} \left(\frac{1}{X_j} \right), \text{and } w_{Rn} = X_{eq} \cdot \sum_{j \in B_n} \left(\frac{1}{X_j} \right)$$

The path flow now can be expressed as:

$$P_{path} \approx \frac{1}{X_{eq}} (\theta_S - \theta_R) = \frac{1}{X_{eq}} \left(\sum_{m=1}^{M} (w_{Sm} \cdot \theta_{sm}) - \sum_{n=1}^{N} (w_{Rn} \cdot \theta_{Rn}) \right)$$

Through equation, the correlation between the real power flow over the path and the individual sending and receiving end bus angles can be expressed by the weighting factors w_{Sm} and w_{Rn} derived in equation. A larger weighting factor means a larger correlation between the path flow and the bus angle.

In daily operation, outages of transmission lines and series capacitors or resistors actions can largely affect the weighting factors. This is especially true in the Western Interconnection where a good amount of series capacitors is installed to increase the transfer capacity. A physical angle pair that have high correlation to path flow may become less correlated when operating condition shifts. One advantage of the proposed VBPA is that it can dynamically adjust the weighting factors and maintain a high level of correlation.

While implementation of "virtual bus angle" uses PMUs, one needs first to fine limiting angles and perform baselining. Virtual bus angle difference of the current operation condition alone is not sufficient in indicating system stress. The limit of the angle difference is needed to inform the operators about operating margin of the system. With this motivation, the limit calculation is implemented in Peak RC's online TSAT engine.

The virtual bus angle definition is built in the online engine. After each transfer analysis run, the virtual bus angle of the critical operating point is calculated and archived in PI database. Figure 4.56 is a screenshot of the transfer analysis display of the online TSAT tool. The speedometer chart for "COI_VBP" is the display for the virtual angle. The virtual angle difference in base case is 4.5°. The angle difference limit is 8.0° which is also the boundary between the green and red zones.

The virtual bus angle limit is archived in PI database and can be easily trended as shown in Fig. 4.57.

The performance of the virtual bus angle during two recent system events in the Western Interconnection is analyzed in this section. Initial findings on the proposed approach are documented.

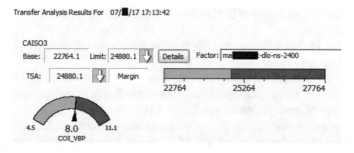

Fig. 4.56 Display of virtual angle concept in online TSAT

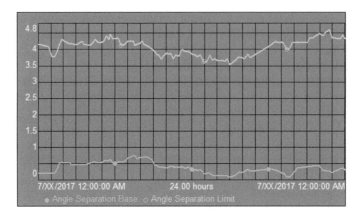

Fig. 4.57 Twenty-four hour trending of the virtual angle difference and limit

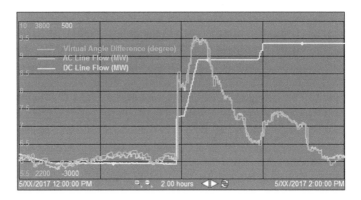

Fig. 4.58 DC line tripping event

4.7.7.1 Single Pole Loss on the DC Line

During this event, one pole of the DC line is lost at 12:57 pm, and the DC line flow suddenly dropped from 2673 MW to 1605 MW as shown in Fig. 4.58 (yellow curve). From 1:00 to 1:07 pm, the flow on the remaining DC pole further reduced to 359 MW.

Flow on the parallel AC path (blue curve) suddenly increases from 2433 MW to 3093 MW following the single pole loss. Correspondingly, the virtual bus angle difference of the AC path jumps from 6.23° to 8.16°.

In this event, because of the loss of transmission capacity on the DC line, north to south flow flooded onto the parallel AC path and causes increased system stress. The proposed virtual bus angle difference clearly indicated the increased system stress.

4.7.7.2   Forced Outage on Transfer Boundary

In this event, forced outage happened on one of the 500 kV lines on the transfer boundary at 6:48 am. There is a spike of the path flow at the same time of the outage as shown in Fig. 4.59. However, the path flow remains at almost the same level following the spike. This is because the flow on the outage 500 kV line shifted onto the remaining lines on the boundary. Loss of the 500 kV line will stress the system significantly although the overall path flow did not increase.

On the other hand, following the outage, virtual angle difference increases from 0.93° to 2.91° and stays around the same level after the event. During this event, the virtual angle difference clearly indicates the increased system stress following the forced outage.

RT-VSA (Real-Time Voltage Stability Analysis) transfer limit is also shown in Fig. 4.48 to justify the increased system stress. The VSA run following the event gives a much lower limit at 3304 MW compared to 4732 MW limit of the previous run. However, the indication from VSA limit is delayed more than 10 min due to the interval between limit update.

During this event, it is demonstrated that the virtual angle difference provides early indication of system stress, while path flow following the outage did not provide similar indication.

4.7.8   Real-Time Assessment of COI Limits

Main objectives of power system operation are reliability and economy of operation. System operating limits (SOLs) account for both. Higher limits mean more available transmission capacity, but it might affect inadvertently reliability of the

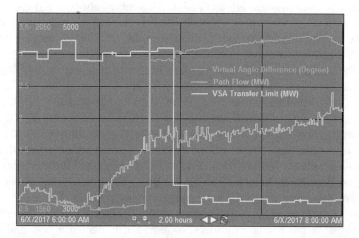

Fig. 4.59 Forced outage on transfer boundary

system. It is very important to be able to accurately calculate system operating limits. Normally SOLs for transmission corridors are assessed in the planning timeframe and then re-adjusted on a seasonal basis, in advance of planned outages, and they are re-evaluated day ahead and hour ahead based on the actual system conditions. However, it is very difficult and time-consuming to adjust system model to desired operating conditions. It might take multiple days to adjust the model. Moreover, due to a large number of Transmission System Operators within the interconnection, one operator may not know operation conditions of other system operators, and transient stability disturbances can affect large portions of the interconnection. For that reason, most utilities in the Western Interconnection use multidimensional nomograms trying to account for various system conditions.

During real-time operation, we do not want to violate pre-established SOLs for N-1 and credible N-2. Current practice of seasonal calculation of the limits leaves a lot of margin and unused system capacity on the one side and, on the other side, does not account for all possible operating conditions.

Multiple transmission corridors in the Western Interconnection are impacted by transient stability constraints. While basecase analysis provides a means to assess if the system is stable under current operating conditions for modeled contingencies, it does not allow to calculate margin to instability (system limits) and available transmission capacity. We want to recalculate these limits using real-time cases. This approach allows dynamic calculation of SOLs and consequently helps unlock available transmission capacity while maintaining reliability of the grid.

System operating limits (SOLs) are estimated in offline studies. It is known that SOLs depend on operating conditions and offline SOLs are very conservative locking available transmission capacity. While thermal limits can be dynamically estimated based on environmental temperature, current flow through the line and wind speed stability limits requires simulation.

In order to calculate system limits in real time, one needs to stress interface by increasing flow over it in steps. This is performed by defining source within sending area and sinks within receiving area. After each increment N-1 and credible N-2 contingencies are applied in time domain. Limit is reached when some of the stability or thermal criteria is violated. Used criteria are same as in basecase analysis. Table 4.6 illustrates how system limit for specific interface changes based on real-time operating conditions. Specific interface limit was locked at 4800 MW for over the decade. Based on real-time simulation, this limit can be significantly increased unlocking additional transmission capacity. Results below show an example of heavily stressed real-time cases from August 31, 2017, to September 1, 2017, afternoon showing how limits are changing with operating conditions and what is the limiting reason. In the example below, two most severe contingencies are applied for both cases. One is loss of PDCI (Pacific DC Intertie), and the second is loss of two largest Western Interconnection units in the south of the system. Both contingency stresses significantly COI increasing flow on it. As PDCI is in parallel with COI, all power from PDCI is transferred on COI activating large generation trip (RAS action) in the Pacific Northwest to temporarily decrease flow and prevent

Table 4.6 August 31 COI limit estimates

Contingency	Actual flow COI MW	Date	New limit COI MW	Limiting reason	RAS generation trip MW
HVDC	4260		6346	Voltage collapse	2774
Large generators N-2	4260		5478	Voltage drop	0
HVDC	5105		6377	Voltage collapse	2836
Large generators N-2	5105		5955	Voltage drop	0

overload and additional line tripping. The amount of generation tripped is limited due to frequency response. From examples it can be clearly seen that the limit at that time could have been set at 5400 MW or at 5900 MW for the second case. Contingency yielding lower limit is considered limiting contingency. Voltage criteria applied on 500 kV buses are that voltage shall not drop below 440 kV during power swing after the fault cleared and that voltage shall not stay below 460 kV for more than 30 cycles. This criterion is experience based. Damping threshold was set at 3%. Figures show post-contingent voltages and COI flows. Please note that limiting flows refer to pre-contingent flows.

Figure 4.60 shows interface flows and major 500 kV voltages for limiting system conditions for August 31 cases.

4.8 Lessoned Learned

4.8.1 Enable Parallel Execution of Basecase and Transfer

Normally State Estimator runs every minute and provides cases for the all downstream applications. RTCA runs every 5 min as well as real-time voltage stability. Case for real-time transient stability is provided also every 5 min. As time domain simulation is more computational, it is expensive to complete both basecases, and transfer takes more time, depending on the number of the contingencies modeled for N-1 and N-2 as well as on the number of the transfer scenarios modeled. Typically transfer scenarios take much more time to execute, and, originally, results are updated only after both basecase and all transfers are computed.

The TSAT software changes need be performed so that basecase and transfer scenarios are executed and updated in parallel, independent of each other. The changes were accomplished by reserving certain number of the servers only for basecase analysis (N-1 and N-2).

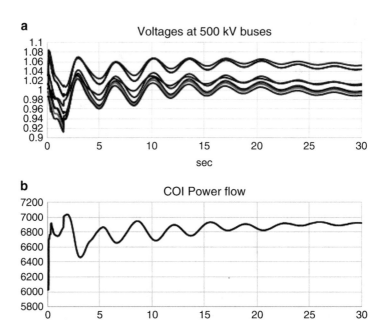

Fig. 4.60 (**a**) August 31th major 500 kV voltages for HVDC trip. Simulation results for limiting flow. (**b**) August 31th interface flow for limiting conditions and HVDC trip. (**c**) August 31 major 500 kV voltages for two large units' trip and limiting flow. (**d**) August 31 interface flow for two large units' trip and limiting flow

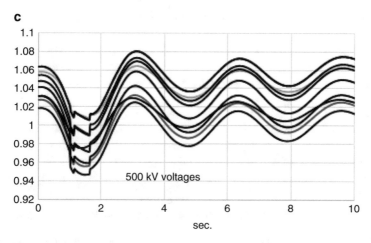

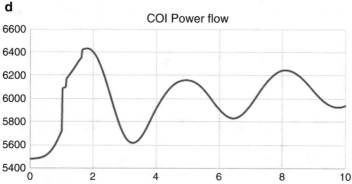

Fig. 4.60 (continued)

4.8.2   Impact of AGC and Generation Ramping-Up on Results

Traditionally, dynamic simulation duration is 10–20 s. Typically, AGC action is not simulated. During frequency events (loss of generation) for runs longer than 20 s, AGC will act to try to restore ACE. Simulation results might not match with PMU measurements, and it is important to recognize this mismatch. The following example illustrates the abovementioned remark: Simulation is performed for loss of BC Hydro 525 MW unit loss. Figure 4.61 illustrates currents through one of the WECC major 500 kV lines. Within first 20 s (initial and primary response), current waveform from simulation and PMU measurements match remarkably. After 20 s current through the line builds up due to generation secondary response (AGC action). In simulation current becomes constant since there is no AGC action modeled.

Figure 4.61 shows voltage profile on buses at the end of these lines. First 20 s voltage profile of simulation and measurements match remarkably. After 20 s

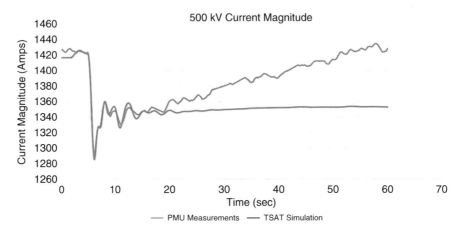

Fig. 4.61 Impact of AGC action is reflected in increasing current over the line (blue trace). AGC action is not modeled, and in simulation, after primary response current becomes constant (orange trace)

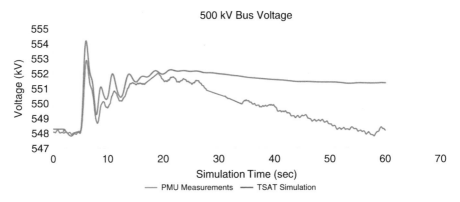

Fig. 4.62 Increasing current over the line sag voltage (blue trace). Simulation does not model AGC action (no current increase over the line) so voltage drop remains constant (orange trace)

voltage profile in simulation becomes constant, but in reality, voltage depresses due to increase in current through the line (AGC action) (Fig. 4.62).

Figure 4.63 illustrates another point: simulation is always started from steady state, and some movements in the system cannot be captured. The plots show flow over the California-Oregon Interface. Simulation starts with flat line. However, it can be seen from PMU trace that prior to the event, flow over interface was ramping down probably due to unit schedule change and, after disturbance, this trend continues (after 45 s of simulation up to 48 s of simulation). This mismatch cannot be considered as a modeling error.

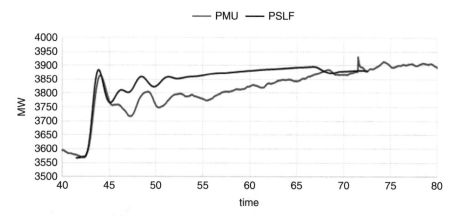

Fig. 4.63 Illustration on how flow decreases due to units ramping down (red trace from PMU measurement) vs. simulation results where simulation starts from flat line (black)

4.8.3   Impact of Archived Case Timestamp on System Model Validation

NERC standards require planning coordinators to perform system model validation, for their footprint, by simulating a specific event and comparing results to high-resolution measurements such as PMUs or DFRs (MOD-33). In order to perform comparison, offline model needs to be adjusted on pre-disturbance condition before performing simulation of the event. Offline model and real-time model are different. It is very time-consuming to adjust offline model that needs to be validated. Overall interconnection model needs to be adjusted to pre-disturbance condition in order to match frequency response and oscillatory behavior of the system. Single planning coordinator has available real-time data only from its own footprint and needs to request measurements that would be used to adjust power flow case for the specific time from all other operators. In large interconnection such as the Western Interconnection, it might take a lot of time to obtain the required measurements. In order to avoid this delay in the Western Interconnection, we use real-time dynamic simulation snapshot from the archive closest to the event since that case is already adjusted on pre-event operating conditions and this case is used to transfer initial conditions to offline model. As DSA application run every 5 min, there might be a time gap in between event and archived case that might be up to 5 min. In 5 min, a lot of things can change as illustrated in the following example:

On September 7 13;10 MW of generation were tripped in Arizona. Consequently, COI flow increased from 3800 MW to 4200 MW in 10 s and to 4300 MW in 50 s after disturbance, based on PMU data. Archived case closest just before the disturbance was taken and simulation is performed. Figure 4.64 illustrates simulation

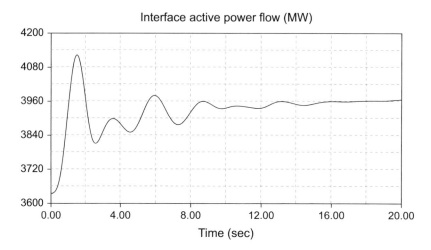

Fig. 4.64 Interface flow-simulation results

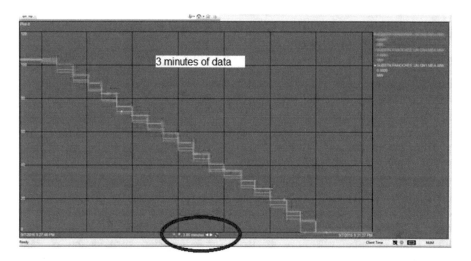

Fig. 4.65 Three units ramped down after snapshot from SE has been taken but before the event itself

results. It can be seen that in simulation, interface flow increased from 3636 MW to 3966 MW. PMU traces show flow increased from 3634 MW to 4200 MW.

The reason for the discrepancy was the fact that the archived case was saved 4 min prior to the disturbance and, in the meantime, three units in California ramped down as shown by historian (Fig. 4.65). Consequently, units in the Pacific Northwest picked up additional load increasing flow on interface. After these units' output adjusted to zero, simulation result becomes more accurate as it can be seen from graphs in Fig. 4.66.

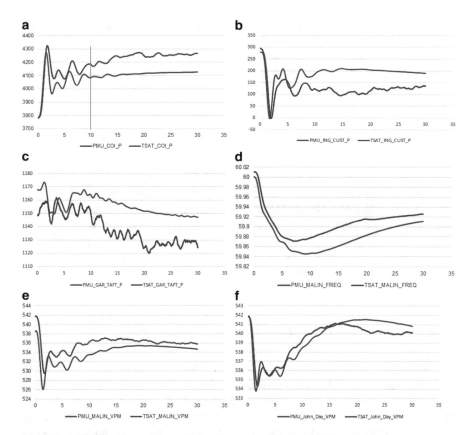

Fig. 4.66 MW flows, frequency, and bus voltages simulation vs. PMU measurements

- Comparison between TSAT simulation and PMU trace shows:

  - ~100 MW gap at 10 s event resulted that can be due to:

    Modeling issues
    Difference between SE basecase and PMU recording

  - Additional COI flow change is mainly driven by AGC of multi-operating BAs, such as BPA, CAISO, etc.

  Following steps are undertaken in order to improve modeling:

- Verify the load model accuracy for frequency response under frequency dip events against PMU (need PMU at load centers, e.g., PRC-002/2).
- Check TSAT simulated primary response of conventional steam plants against PMU and PI data.
- Verify if each steam turbine of a combined cycle facility on low-frequency events is blocked.
- Verify units not having dynamic models.

- Verify if there is any TSAT tool issue resulting in basecase solved differently than SE input case.
- Investigate impact of SE pseudo injections.
- Difference in between SE solution, SCADA, and PMU (e.g., some bus kV typically estimated 3–5 kV lower than PMU).

  - 1% bus kV off could reduce COI flow by 30–40 MW.

- Continue to improve unit mapping between WSM and *.dyd files (hundreds of units not mapped yet).

In order to produce better cases, closer to the event, we use Dispatcher Training Simulator (DTS) replay mode. Energy Management System records all SCADA values in Historical Data Record (HDR) files in real time. DTS "Replay" function can replay these HDR files in training environment. It enables creating State Estimator case from replay so snapshot from SE up to a few seconds prior to event can be created and used as basis for dynamic simulation. That way we can capture system state just prior to event.

4.8.4 Impact of SE Pseudo Injections (Mismatches)

It is interesting to investigate impact of pseudo injection. In order to solve, State Estimator injects pseudo loads (pseudo injection) all over the system. Without pseudo loads SE would not be able to provide valid solution for large system. Pseudo injections (loads) typically point at places where measurements do not match SE solution. Pseudo loads may be positive or negative MW and MVAR. They are consequence of modeling errors, topology error, measurement error, and wrong weight assignment to measurements, wrong position of the specific measurements, or something else. There is a threshold in MW and MVAR for pseudo loads that, if violated, would lead to "invalid solution." In our case these thresholds are set at 50 MW and 50 MVAR. Most of the time, pseudo loads are way below that threshold; however, they add up and create artificial sources and sinks. Example below illustrates the point:

Figure 4.67a represents the sum of pseudo loads for one specific snapshot for different area within the Western Interconnection. In this case the largest individual mismatch in PG&E was 16.45 MW which is way below the threshold of 50 MW. Table shows that total sum of pseudo loads in PG&E area which adds up to 480 MW. As it can be expected, there is additional flow from area to area as a consequence of pseudo loads. Figure 4.67b illustrates additional flows, described as follows:

- Ellipse bubbles are control areas.
- Line between the areas represents the SUM of tie-line flows between the areas.
- The flow arrows below represent the *additional* line flows that appear as a result of these "pseudo source/sinks."

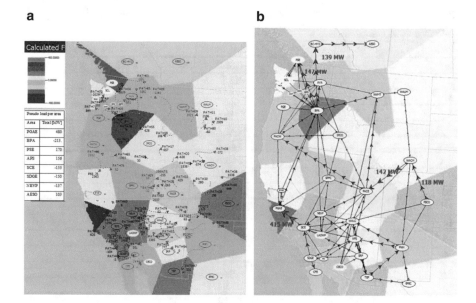

Fig. 4.67 (**a**) Sum of pseudo loads. (**b**) Additional flow in between area

- Pink numbers label the incremental flows higher than 100 MW.
- As expected, flows tend to leave red/orange regions and go toward the blue regions.

However, due to those pseudo loads, measurement of flows through the interfaces (original) and state estimator solution (solution) matches very well as it can be seen on Table 4.7.

This pseudo loads are not scaled during transfer studies, and they do not have dynamic counterparts. During dynamic simulation they are treated as default load model (constant current for MW and constant impedance MVAR). They can be converted to be constant impedance for active power as well depending on user preference.

4.8.5 Modeling Voltage Droop

Another issue is when two or more units share the same bus and they are on AVR. It can lead to oscillations in between units.

Units sharing the same bus would need either:

- "xcomp" negative as part of generator model record
- "ccomp" model (John Day, The Dalles, Chief Joseph 17–27)

Table 4.7 Interface flows for specific snapshot SE solution (solution) vs. measurements (original)

Interface flows comparisons			
Path	Solution	Original	Difference
1	−159.4	−159.2	−0.2
2	−75.1	−75.1	0
3	−119.7	−120.1	0.4
4	5609.1	5608.7	0.4
5	3353.4	3351.7	1.7
6	1224.8	1225.2	−0.4
8	1340.7	1340.6	0
9	2021.4	2020.7	0.7
10	2008.5	2008.6	−0.1
11	2107.9	2106.2	1.6
14	76.9	76.1	0.8
15	868.3	861.3	7
16	30.2	30.1	0.1
17	657.9	657.2	0.8
18	206.5	205.9	0.6
19	1537.3	1537.8	−0.5
20	−636.9	−637.7	0.8
22	1446.4	1447.4	−1
23	64.7	65.4	−0.7
24	14.3	14.3	0
25	−2.2	−2.1	0
26	1298.7	1295.9	2.8
27	2036.9	2036.9	0
28	−264.8	−264.7	−0.1
29	9.9	9.9	0
30	279.9	279.9	0
31	−11.2	−10.6	−0.6
32	−32.2	−32	−0.1
33	428.4	428.5	−0.1
35	−155.9	−156.4	0.5
36	1037.9	1038.3	−0.4
37	−573.2	−573.5	0.4

CCOMP is implemented as analog circuit in old voltage regulators or as software with cross-unit telecom for digital AVRs. A cross-current compensation model (CCOMP) is added to enable stable reactive power sharing among paralleled units.

Figures 4.68 and 4.69 illustrate improvements with implementation of CCOMP model.

In WECC dynamic database file, many CCOMP models are missing, the reason being when units are adjusted on the same output, they do not have this problem. In real-time simulation, it cannot be done since unit output is adjusted by real-time measurements.

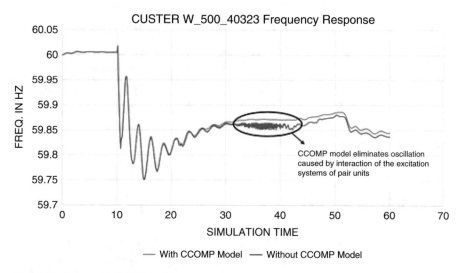

Fig. 4.68 Impact of CCOMP model on frequency response. It can be clearly seen that implementation of CCOMP eliminates oscillatory response

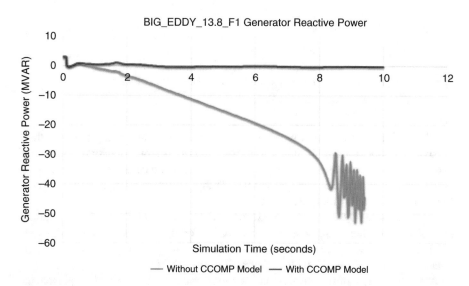

Fig. 4.69 Impact of CCOMP on reactive output of generator. Without CCOMP unit becomes unstable

CCOMP model is added as one model for pair of units sharing the same bus. The figure below illustrates how to add CCOMP model for two BIG_EDDY units G01 and G02 on same bus (Fig. 4.70).

```
#<ems:SynchronousMachine name="BIG_EDDY_13.8_G01">
gentpj    44041   "TDA 0102"  13.80  "01" : #9  mva=94.42   4.096  0.024   0.0   0.036  :
exstl     44041   "TDA 0102"  13.80  "01" : #9  0.0    99  -99   1.0   2.5   250   0.0   4.1
hyg3      44041   "TDA 0102"  13.80  "01" : #9   1   0   1   0.05   0   0.05   0.1   0.1   0.1
pss2a     44041   "TDA 0102"  13.80  "01" : #9   1   0   3   0   10   10   10   0.0   0.0   10
#<ems:SynchronousMachine name="BIG_EDDY_13.8_G02">
gentpj    44041   "TDA 0102"  13.80  "02" : #9  mva=94.42   4.096  0.024   0.0   0.036  :
exstl     44041   "TDA 0102"  13.80  "02" : #9  0.0    99  -99   1.0   2.5   250   0.0   4.1
hyg3      44041   "TDA 0102"  13.80  "02" : #9   1   0   1   0.05   0   0.05   0.1   0.1   0.1
pss2a     44041   "TDA 0102"  13.80  "02" : #9   1   0   3   0   10   10   10   0.0   0.0   10
#<ems:SynchronousMachine name="BIG_EDDY_13.8_G01">
ccomp     44041   "TDA 0102"  13.80  "01" : #9   0.0  -0.120000  0.0  0.0  0.0  1.000000
#
```

Fig. 4.70 Implementation of CCOMP model for two units

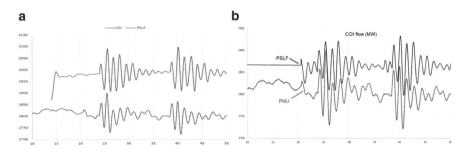

Fig. 4.71 MW flow over the interface

4.8.6   Steady-State and Dynamic Models Do Not Have the Same Rating

Another common problem is when the steady-state generator rating is larger than the dynamic model rating. If unit output from SE case is larger than dynamic model allows generator controller will decrease unit output in the beginning of the simulation. Consequently, it can increase output of other units and change flows on lines. Below is an example that illustrates the point.

By simulating system event and comparing results to PMU measurements (Fig. 4.71), we see additional pickup on COI for dynamic simulation although power flow is close to measurement data at the beginning of the simulation (orange trace). Blue trace shows PMU measurement of the MW flow.

After the review of the steady-state and dynamic model, the problem is isolated and corrected. There were four units modeled in this specific solar farm in the planning case with the following MVA bases:

- 233 MVA/48 MVA/36 MVA/245 MVA

In the SE snapshot, these four units were dispatched to 156 MV, 122 MW, 126 MW, and 154 MW as illustrated in Figure 4.72.

- 114.5/159.1/152.1/126.5 MW

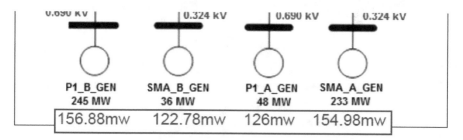

Fig. 4.72 PV plants' steady-state rating in WECC basecase (black) vs. unit output from real-time case (red)

The correct output is from SE case since it is based on measurements. The problem was that real-time steady-state and planning steady-state models were not matching each other. As dynamic model rating was based on WECC planning steady-state case, dynamic model ramped down unit outputs. Consequently, units on the other part of the system compensated with their governor response which increased the flow through the interface. After contacting owner and model update simulation were much closer to measurements.

4.8.7 Frequency Response Benchmarking and Tuning

Frequency response depends on each unit in the system and load model. In the online TSTA, we are interested in the initial response (due to system inertia) and primary or governor response. Secondary response (AGC) is not modeled. Following factors affect frequency response:

- Setting of Unit Base Load Flags.
- Secondary control loop.
- Lower-frequency dips in simulation might be caused by excess governor response from some units with incorrect P_{max}.
- Unit status availability and accuracy (online vs. offline).
- Dead band on governors.
- Hydrological conditions.
- Generator mode of the operation (motoring or generating).

Normally, frequency response of the system is validated and analyzed after system events comparing simulated frequency response to PMU measurements. Examples in Fig. 4.73 illustrate errors in frequency response. Figure 4.74 shows events where discrepancy in between simulated and measured frequency response was large. Results were analyzed with the objective to further increase accuracy. In most cases the problem was due to withdrawal of units due to secondary control loop or erroneous base load flag setting. Analysis was performed using snapshots

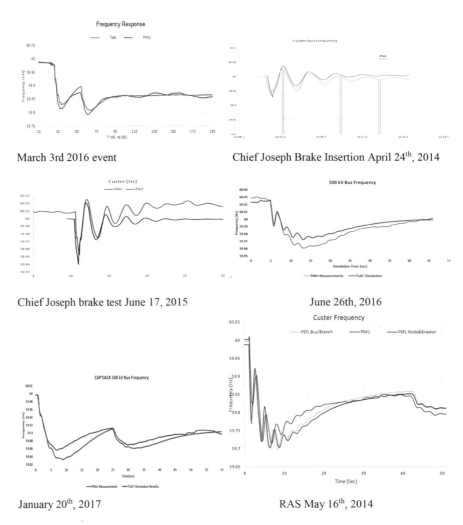

Fig. 4.73 Frequency response for various system events (simulation vs. PMU measurements)

from SE after the event and comparing generator output before and after the event. It was noted that some generator output after the event was lower than before that was consequence of secondary control loop.

4.8.8   Units in Pumping Mode

Some generators can work as pumps (in pumping mode) becoming synchronous motors drawing power from the grid. During that mode of operation, they start online at zero MW. Gate is typically 15% open. Then gates are closed completely;

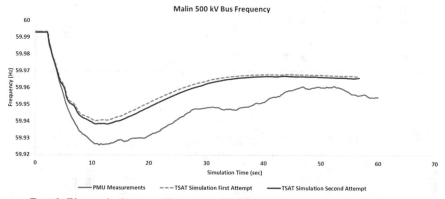

Turned off large units that were at low output (<10 MW)
Real-time case data had units at low power output that likely should have been off

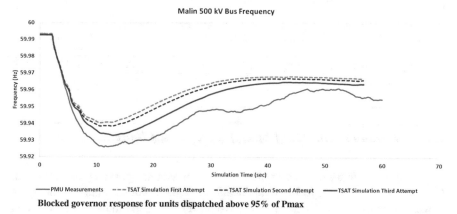

Blocked governor response for units dispatched above 95% of Pmax

Figure 4.74 Frequency response for various system events (simulation vs. PMU measurements)

the generator starts motoring with the turbine submerged in water consuming MW to cover rotational losses. Unit inertia does not change. However, there are some modeling changes that need to be performed:

– Generator parameters stay the same.
– Governor model is removed.
– PSS should be disabled.
– AVR gains should be adjusted to PSS OFF settings were available.

Real-time software, based on MW flows (out of the unit or in the unit), must detect automatically the mode of the operation and, in case of motoring, disable PSS and governor. The figure below illustrates how it is resolved in a dynamic model file.

In a unit at 37309 in pumping mode but in dynamic model, this option is not given. You need to add pumping option.

Before:

```
# Retest data for SMUD McClellan GTF Facility tested by Edison ESI 3/9/13, submitted by Dave Blevin
#<ems:SynchronousMachine name="MCLELN_13.8_GN1">
enrou    37309  "MCCLELLN"   13.80 "1 " : #9 mva=82.440   8.680  0.050  0.40  0.050  6.440  0.0  2.090
xac8b    37309  "MCCLELLN"   13.80 "1 " : #9  0.0  200.0  25.0  50.0  0.080  0.10  0.0  35.0  -35.0  .
ggov1    37309  "MCCLELLN"   13.80 "1 " : #9  mwcap=75.950  0.060  1.0  1.0  0.050  -0.050  10.0  2.0
ss2a     37309  "MCCLELLN"   13.80 "1 " : #9  1  0  3  0  15.0  15.0  15.0  0.0  0.0  15.0  1.1640  1  0
# #tlin1 records added to trip HELMS units on underfrequency.  Updated with new breaker times submit
# Rick Padilla (PG&E) confirmed that the Helms pumps are set to trip at 59.5 Hz, changed from 59.65 t
#tlin1    30820 ! !  "1 "   34600 "HELMS 1 "  18.00  "1 " 1 : #9  0 0 2 34600 59.50 0.1 0 10000  34600
#tlin1    30820 ! !  "1 "   34602 "HELMS 2 "  18.00  "1 " 1 : #9  0 0 2 34602 59.50 0.1 0.10000  34602
```

After:

```
exst1    37314  "ROBBS PK"   13.80 "1 " : #9 0.01  0.1  -0.1 1.4     16.0     260 0.0   3.39    -
eeegs    37314  "ROBBS PK"   13.80  1   : #9 0.2 0.2 0.01 -0.005 1.0  0 0.055 0.52  15  2.
# Retest data for SMUD McClellan GTF Facility tested by Edison ESI 3/9/13, submitted by  a
#<ems:SynchronousMachine pump = "n" name="MCLELN_13.8_GN1">
enrou    37309  "MCCLELLN"   13.80 "1 " : #9 mva=82.440   8.680  0.050  0.40  0.050  6.440   0
xac8b    37309  "MCCLELLN"   13.80 "1 " : #9  0.0  200.0  25.0  50.0  0.080  0.10  0.0  35.0  0
ggov1    37309  "MCCLELLN"   13.80 "1 " : #9  mwcap=75.950  0.060  1.0  1.0  0.050  -0.050
ss2a     37309  "MCCLELLN"   13.80 "1 " : #9  1  0  3  0  15.0  15.0  15.0  0.0  0.0  15.0

#<ems:SynchronousMachine pump = "y" name="MCLELN_13.8_GN1">
enrou    37309  "MCCLELLN"   13.80 "1 " : #9 mva=82.440   8.680  0.050  0.40  0.050  6.440   0
xac8b    37309  "MCCLELLN"   13.80 "1 " : #9  0.0  200.0  25.0  50.0  0.080  0.10  0.0  35.  0
# #tlin1 records added to trip HELMS units on underfrequency.  Updated with new breaker ti
# Rick Padilla (PG&E) confirmed that the Helms pumps are set to trip at 59.5 Hz, changed f
```

Every unit that can have negative MW output in steady-state model should have motor option enabled.

4.8.9   Impact of the Load Model on System Limits

This section illustrates the impact of load model on system operating limits (SOLs). Load model used for the real-time application of DSA is constant current for active power and constant impedance for reactive power with 20% induction motor load penetration. Induction motor dynamic model is mapped to the load buses. Around 5000 motor loads are mapped. In order to show the impact of different percentages of induction motor load penetration on SOLs, we use offline studies and cases having 0%, 20%, and 50% of motor load. Figure 4.65 shows such an example analyzed in an offline study for SDGE scenario. From Fig. 4.75 it can be seen that for that specific snapshot from real-time SOLs significantly decreases for increased induction motor penetration: for 0%, 20%, and 50%, motor load penetration SOL is set at 4209 MW, 4028 MW, and 3692 MW, respectively, and voltage collapse is always case for reaching limiting conditions. Composite load model is not used for real-time application.

4.8.10   Simulation of PMU Signals

Over the last decade, a large number of PMUs have been installed within the Western Interconnection [49]. It can be very useful to be able to simulate PMU signals. That way it is possible, based on simulation, to choose best positioning of PMUs within

Transfer Limit Summary

Show Delta Values

0% Motor Load

Cont. No.	Independent Variable (MW)	Dependent Variable (MW)	Interface1 [SDGE] (MW)	Interface2 [ECO_MNG]_1500_L] (MW)	Interface3 [SAML_SO] NG_1230_L] (MW)	Interface4 [SAML_SO] NG_2230_L] (MW)	Interface5 [SAML_SO] NG_3230_L] (MW)	Interface6 [SONG_TA] LE_1230_L] (MW)	Interface7 [SONG_TA] LE_2230_L] (MW)	Interface8 [OMES_TJ] UA_1230_L] (MW)	Interface9 [OCOT_SU] NC_1500_L] (MW)	Transient Stability Index (PU)	Volt. Drop Duration Index (Sec)	Volt. Rise Duration Index (Sec)	Freq. Drop Duration Index (Sec)	Freq. Rise Duration Index (Sec)	Status	Reason for Limitation	Contingency Description
1	9591.19	4830.2	4209.95	1676.24	135.51	190.38	140.00	216.52	216.52	421.61	1212.26	23.950	0.650	0.000	0.000	0.000	Secure	Voltage Drop Duration	eco-miguel-sio
2	9591.19	4830.2	4209.95	1676.24	135.51	190.38	140.00	216.52	216.52	421.61	1212.26	79.260	0.000	0.000	0.000	0.000	Secure		SCE 500KV WROLOM 3 PHASE BUS FAULT
3	9591.19	4830.2	4209.95	1676.24	135.51	190.38	140.00	216.52	216.52	421.01	1212.26	15.680	0.000	0.000	0.000	0.000	Secure		w-ngla

Transfer Limit Summary

Show Delta Values

20% Motor Load

Cont. No.	Independent Variable (MW)	Dependent Variable (MW)	Interface1 [SDGE] (MW)	Interface2 [ECO_MNG]_1500_L] (MW)	Interface3 [SAML_SO] NG_1230_L] (MW)	Interface4 [SAML_SO] NG_2230_L] (MW)	Interface5 [SAML_SO] NG_3230_L] (MW)	Interface6 [SONG_TA] LE_1230_L] (MW)	Interface7 [SONG_TA] LE_2230_L] (MW)	Interface8 [OMES_TJ] UA_1230_L] (MW)	Interface9 [OCOT_SU] NC_1500_L] (MW)	Transient Stability Index (PU)	Volt. Drop Duration Index (Sec)	Volt. Rise Duration Index (Sec)	Freq. Drop Duration Index (Sec)	Freq. Rise Duration Index (Sec)	Status	Reason for Limitation	Contingency Description
1	9714.19	4853.2	4228.38	1643.65	109.44	153.62	113.07	200.87	203.87	412.47	1180.39	26.160	0.650	0.000	0.000	0.000	Secure	Voltage Drop Duration	eco-miguel-sio
2	9714.19	4853.2	4228.38	1643.65	109.44	153.62	113.07	200.87	203.87	412.47	1180.39	98.370	0.000	0.000	0.000	0.000	Secure		SCE 500KV WROLOM 3 PHASE BUS FAULT
3	9714.19	4853.2	4228.38	1643.65	109.44	153.62	113.07	200.87	203.87	412.47	1180.39	9.910	0.000	0.000	0.000	0.000	Secure		w-ngla

Transfer Limit Summary

Show Delta Values

50% Motor Load

Cont. No.	Independent Variable (MW)	Dependent Variable (MW)	Interface1 [SDGE] (MW)	Interface2 [ECO_MNG]_1500_L] (MW)	Interface3 [SAML_SO] NG_1230_L] (MW)	Interface4 [SAML_SO] NG_2230_L] (MW)	Interface5 [SAML_SO] NG_3230_L] (MW)	Interface6 [SONG_TA] LE_1230_L] (MW)	Interface7 [SONG_TA] LE_2230_L] (MW)	Interface8 [OMES_TJ] UA_1230_L] (MW)	Interface9 [OCOT_SU] NC_1500_L] (MW)	Transient Stability Index (PU)	Volt. Drop Duration Index (Sec)	Volt. Rise Duration Index (Sec)	Freq. Drop Duration Index (Sec)	Freq. Rise Duration Index (Sec)	Status	Reason for Limitation	Contingency Description
1	9303.19	4322.2	3692.28	1500.27	61.11	85.47	63.13	180.22	180.22	395.94	1143.72	27.040	0.650	0.000	0.000	0.000	Secure	Voltage Drop Duration	eco-miguel-sio
2	9303.19	4322.2	3692.28	1500.27	61.11	85.47	63.13	180.22	180.22	395.94	1143.72	42.800	0.025	0.000	0.000	0.000	Secure		SCE 500KV WROLOM 3 PHASE BUS FAULT
3	9303.19	4322.2	3692.28	1500.27	61.11	85.47	63.13	180.22	180.22	395.94	1143.72	90.900	0.000	0.000	0.000	0.000	Secure		w-ngla

Fig. 4.75 Impact of induction motor load on SOLs for 0%, 20%, and 50% of induction motor penetrations

the system in regard to different applications. ePMU is a module added to the Powertech TSAT software and can be considered as a connection between the TSAT simulations and the synchrophasor applications such as FFDD-Online and MAS tools. A detailed description on ePMU settings can be found in the ePMU user manual [50]. The aim of this section is to provide our further experiences with ePMU module.

The ePMU module uses the ordinary TSAT simulation results and downsamples the results from the simulated rate (usually 0.25 or 0.5 cycle) to 30 samples per second (2 cycles). It can also be used to provide the data as a .CSV file (for offline usage) or stream the signals to the PDC (for online usage) based on the IEEE C37.118 format.

Usage Restrictions: Since the ePMU is supposed to stream the simulated signals in a real-time manner to PDC, at each time, only one ePMU simulation can be done. Because multiple ePMU simulations could interfere as they all would, stream the data by the same communication port. This is in contrast with TSAT ordinary simulation where as many as needed simulations can be done in parallel by opening multiple TSAT programs. You can perform one ePMU simulation along with multiple TSAT simulations. The ePMU simulation is designed to be performed as fast as real time. However, since the WECC model is very huge, its speed is about 6 times slower than real time. In other words, it takes ePMU about 6 s to perform 1 s simulation for the WSM. For small systems, the ePMU module performs the simulation in real time, although it can do much faster than real time.

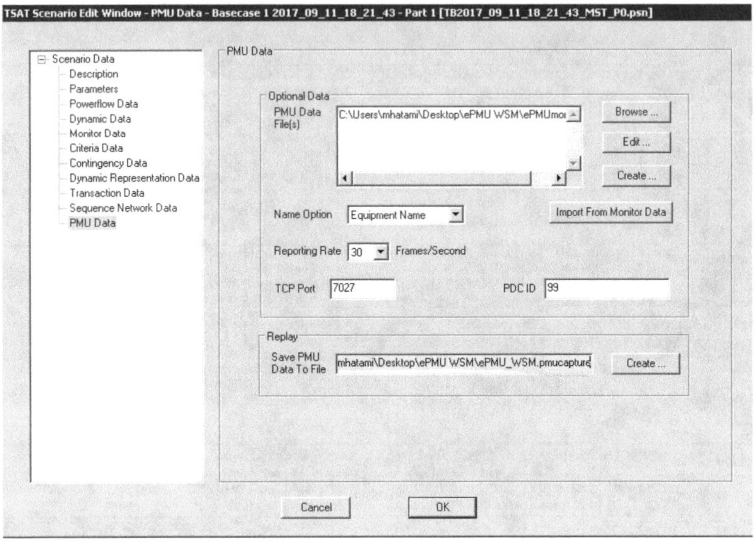

Fig. 4.76 Saving simulated PMU data

There are two modes of operation:

We can perform the ePMU simulation in two "offline" and "online" modes.

Offline: As shown in Fig. 4.76, we can specify a ".pmucapture" file in the "Replay" section to save the ePMU data. This file can be later played back by PMU Connection Tester to save the data in .CSV format (will be explained in detail later). If we do so, the data cannot be streamed to PMU Connection Tester or Open PDC in real time; rather, they will be streamed after the ePMU simulation is done. Saving the ePMU simulation as the pmucapture file makes it possible to have access to the simulation results later when the simulation is done.

Online: In this mode of simulation, we are interested in streaming the ePMU simulations to the PMU Connection Tester or Open PDC in an online manner, where the simulation results can be monitored in PMU Connection Tester or Open PDC while doing the simulations. To perform the online mode, we should leave the "Replay" section blank (i.e., no file should be saved). The disadvantage of this mode is that after the simulation is done, we will not have access to the simulated ePMU data.

4.8.11   Future Use Case: Application Composite Load Model in Real-Time TSAT

The goal of this use case is to implement composite load model in real-time tools using real-time cases and to perform continuous validation of composite load model used in offline and online studies. Since composition of the load is continuously changing with time, the implementation of composite load model in real-time dynamic simulation requires change of composition of the load on fly. This work will leverage and integrate work performed over the years by BPA, WECC/MVWG, and LMTF in the development of composite load model and load composition and work performed by Peak Reliability in the implementation of real-time dynamic simulation using cases from EMS SE. Since real-time dynamic simulation is performed every 15 min, a large number of the cases having different load compositions are created and archived allowing easy validation of the load model against system measurements for system disturbances for different times and for different operating conditions. Upon completion of the project, major software will be able to generate composite load model, with load composition generated on fly, together with real-time snapshots for real-time and offline dynamic studies. Also, at the same time, this project will provide major platform for validation of composite load model.

- The developed methodology and software tool will provide several major benefits for the industry.
- Platform for validation of composite load model for different times and different operating conditions.
- Capability of the major software to change load composition and generate load dynamic profile based on given time the dynamic study is performed for.
- Increase confidence in composite load model.
- Evaluate the impact of composite load model on system behavior through a large number of real-time simulations and benchmarking against system disturbances.
- More efficient (accurate) use of full topology EMS cases for dynamic simulation.
- Help with compliance with MOD-33.
- Increase the accuracy of real-time transient stability security assessment currently performed by RC office since composite load model will replace ZIP load model that is currently used.
- Better use of real-time cases in offline tools.
- Cross-validate results of real-time modal analysis tool against real-time simulations.

4.9 Summary

Peak RC has successfully implemented WSM-TSAT in RC control room in collaboration with vendor and entities. Development of WSM-TSAT involved tons of efforts on both software enhancements, system integration, model improvements, and validation against system events. Till date WSM-TSAT remains the only online TSAT tool running with a full western system operational model and a dozen of accurate RAS model representations.

After multi-year continuous efforts on use case development and model validation, the tool is operated in near real time for monitoring on Colstrip ATR RAS. The online TSAT results of Colstrip ATR RAS calculations are shared with NWMT and other entities via Peak Web Portal www.peakrc.org to improve operation situational awareness across among stakeholders.

Practical experience and lessons learned from the WSM-TSAT implementation project are summarized below:

- WSM-TSAT allows us to maximize model quality and to minimize uncertainties in order to increase transmission capacity utilization while maintaining reliability.
- Using real-time RAS Arming ICCP data for RAS modeling in TSAT is essential for accurate TSAT simulation results.
- Calculation of system inertia and frequency BIAS in real time provides an efficient way to monitor system frequency response capability.
- Setting of Unit Base Load Flags affects TSAT simulation performance of governor responses' secondary control loop.
- Unit status availability and model accuracy matter.
- Lower-frequency dips in simulation might be caused by excess governor response from some units with incorrect P_{max} or base load flag settings.
- Modeling errors on wind farms and PV Solar plants are the ones usually causing TSAT solution issues.
- WSM-TSAT can back up RT-VSA for real-time assessment of the IROLs, particularly when the limiting factor is fast voltage collapse or angle instability.
- WSM-TSAT can visualize power transfer stress in voltage phase angles and/or angle pair difference for user-selected buses or virtual bus angle pairs. WSM-TSAT solved angle limits have the potential for wide-area angle separation monitoring in the future [14].
- Enable parallel execution of basecase and transfer analysis phases in DSA Manager. The feature is desired because the results of basecase analysis will be exported to PI once basecase analysis is finished.

Future works for the TSAT enhancements include:

- Improve PV modeling capability to do more realistic simulation for PV Solar momentary cessation events.
- Enhance ePMU simulation feature to allow large-scale adjustment of loads and generation MW for PMU training and modal analysis baselining.
- Improve power flow robustness against low bus voltages in basecase solution.
- Implement accurate SVC models in replacement of generator equivalent SVCs.

References

1. Kundur, P. *Power system stability and control.*
2. NERC. *Standard PRC-024-2 generator frequency and voltage protective relay settings* (Online). https://www.nerc.com/pa/Stand/Reliability%20Standards/PRC-024-2.pdf
3. "PSLF User's Manual". General Electric International Inc.
4. "DSA Tools User-Defined Model Manual". Powertech Labs Inc.

5. NERC. *Standard MOD-027-1 – Verification of models and data for turbine/Governor and load control or active power/Frequency control functions* (Online). Available: http://www.nerc.com/pa/Stand/Project%20200709%20%20Generator%20Verification%20%20PRC0241/MOD-027-1_redline_to_initial_ballot_2012Feb23_rev.pdf

6. NERC. *Standard MOD-026-1 – Verification of models and data for generator excitation control system or plant Volt/Var control functions* (Online). Available: http://www.nerc.com/pa/Stand/Project%20200709%20%20Generator%20Verification%20%20PRC0241/MOD-026-1_clean_2012Dec11.pdf

7. WECC Generating Unit Model Validation Policy (Online). https://www.wecc.org/Reliability/WECC%20Generating%20Unit%20Model%20Validation%20Policy.pdf

8. TSAT User Manual. Powertech, November, 2018.

9. 1,200 MW Fault Induced Solar Photovoltaic Resource Interruption Disturbance Report. NERC, Atlanta, Georgia, USA. June 2017. Available (Online): https://www.nerc.com/pa/rrm/ea/1200_MW_Fault_Induced_Solar_Photovoltaic_Resource_/1200_MW_Fault_Induced_Solar_Photovoltaic_Resource_Interruption_Final.pdf

10. 900 MW Fault Induced Solar Photovoltaic Resource Interruption Disturbance Report. NERC, Atlanta, Georgia, USA. June 2017. Available (Online): https://www.nerc.com/pa/rrm/ea/October%209%202017%20Canyon%202%20Fire%20Disturbance%20Report/900%20MW%20Solar%20Photovoltaic%20Resource%20Interruption%20Disturbance%20Report.pdf

11. Arizona-Southern California Outages on September 8, 2011: Causes and Recommendations. http://www.ferc.gov/legal/staff-reports/04-27-2012-ferc-nerc-report.pdf

12. (Online). Available: https://www.wecc.org/Reliability/TPL-001-WECC-CRT-3.1.pdf

13. (Online). Available: https://www.wecc.org/Reliability/Off-Nominal%20Frequency%20Load%20Shedding%20Plan.pdf

14. NERC BAL-003-1Frequency Response and Frequency Bias Setting Reliability Standard, Atlanta, GA, 2015 (Online). Available: https://www.nerc.com/_layouts/15/PrintStandard.aspx?standardnumber=BAL-003-1.1&title=Frequency%20Response%20and%20Frequency%20Bias%20Setting&jurisdiction=United%20States

15. DSA Manager User Manual. Powertech

16. Ramanathan, R., Tuck, B., Kincic, S., Zhang, H., & Davies, D. (2017). *Equipment naming convention methodology for node/breaker model.* UPEC.

17. Special Protection Systems (SPS) and Remedial Action Schemes (RAS): Assessment of Definition, Regional Practices and Applications of Related Standards. North American Reliability Corporation (NERC) Technical report, April 2013.

18. Remedial Action Scheme Design Guide. Western Electricity Coordination Council (WECC), Remedial Action Scheme Subcommittee, Technical report, 2006.

19. Mahmoudi, M, Kincic, S., Zhang, H., & Tomsovic, K. (2017). *Implementation and testing of remedial action schemes for real-time transient stability studies.* PES General meeting.

20. NERC. *Standard MOD-033-1, steady state and dynamic system model validation* (Online). Available: http://www.nerc.com/_layouts/PrintStandard.aspx?standardnumber=MOD-033-1&title=Steady%E2%80%90State%20and%20Dynamic%20System%20Model%20Validation&jurisdiction=null

21. Lu, Y., Kincic, S., Zhang, H., & Tomsovic, K. (2017). *Validation of real-time system model in Western interconnection.* PES GM 2017 Chicago, IL.

22. Mahmoudi, M., Kincic, S., Zhang, H., & Tomsovic, K. (2018). *Model enhancements for real-time transient stability assessment in Western interconnection.* T&D conference 2018 Denver, CO.

23. Overholt, P., Kosterev, D., Eto, J., Yang, S., & Lesieutre, B. (2014). Improving reliability through better models: Using synchrophasor data to validate power plant models. *IEEE Power & Energy, 12*(3).

24. Huang, R., Diao, R., Li, Y., Sanchez-Gasca, J., Huang, Z., Thomas, B., Etingov, P., Kincic, S., Wang, S., Fan, R., Matthews, G., Kosterev, D., & Yang, S. (2017). *Calibrating parameters of*

power system stability models using advanced ensemble Kalman Filter. Transaction on Power Systems.

25. Huang, Z., Du, P., Kosterev, D., & Yang, S. (2013). Generator dynamic model validation and parameter calibration using phasor measurements at the point of connection. *IEEE Transactions on Power Systems, 28*(2), 1939–1949.

26. Pacific Northwest National Laboratory. *Power Plant Model Validation Tool (PPMV)* (Online). Available: https://svn.pnl.gov/PPMV

27. Li, Y., Diao, R., Huang, R., Sanchez-Gasca, J., et al. (2017). *An innovative software tool suite for power plant model validation and parameter calibration using PMU measurements*. IEEE PES General meeting 2017, Chicago, IL.

28. Quint, R. D. (2016). *Power plant model verification guideline* (Online). Available: https://www. wecc.biz/Administrative/Power%20Plant%20Model%20Verification%20-%20Ryan%20 D.%20Quint.pdf

29. Kosterev, D. N., Taylor, C. W., & Mittelstadt, W. A. (1999). Model validation for the August 10, 1996 WSCC system outage. *Power Systems, IEEE Transactions, 14*, 967–979.

30. Kosterev, D., & Davies, D. (2010). *System model validation studies in WECC*. PES GM June 2010, Minnesota.

31. Thomas, B., Kincic, S., Davies, D., Zhang, H., & Sanchez-Gasca, J. (2016). *A new framework to facilitate the use of node-breaker operations model for system model validation in WECC*. PES-GM 2016, Boston, MA.

32. Kincic, S., Davies, D., Kosterev, D., Zhang, H., Thomas, B., Vaiman, M., Weber, J., & Ramanathan, R. (2016). *Bridging the gap in between operation and planning in WECC*. PES-GM 2016, Boston, MA.

33. (Online). Available: https://www.wecc.org/Reliability/Acceleration_Trend_Relay_Model_ Specification_DRAFT.PDF

34. Stigers, C. A., Woods, C. S., Smith, J. R., & Setterstrom, R. D. (1997). The acceleration trend relay for generator stabilization at Colstrip. *IEEE Trans on Power Delivery, 12*(3).

35. (Online). https://placesjournal.org/article/the-infrastructural-city/

36. Quint, R. D., Etingov, P. V., Zhou, D., & Kosterev, D. N. (2016). *Frequency response analysis using automated tools and synchronized measurements*. PES GM 2016 Boston, MA

37. Federal Energy Regulatory Commission. (2014). *Frequency response and frequency bias setting reliability standard*. Docket NoRM13-11-000, Order No. 794, January 2014.

38. WECC Renewable Energy Modeling Task Force. (2014). *WECC PV power plant dynamic modeling guide*. WECC, Salt Lake City, UT, USA. April 2014. Available: https://www.wecc. biz/Reliability/WECC%20Solar%20Plant%20Dynamic%20Modeling%20Guidelines.pdf

39. NERC. (2018). *Modeling notification recommended practices for modeling momentary cessation initial distribution*. NERC, Atlanta, GA, USA. February 2018. Available: https://www. nerc.com/comm/PC/NERCModelingNotifications/Modeling_Notification_-_Modeling_ Momentary_Cessation_-_2018-02-27.pdf

40. NERC Reliability Guideline. (2018). *Power plant model verification for inverter-based resources*. NERC, Atlanta, GA, USA. June 2018. Available: https://www.nerc. com/comm/PC/Documents/4.b_Reliability_Guideline_-_PPMV_for_Inverter-Based_ Resources_-_2018-05-17.pdf

41. NERC Inverter-Based Resource Performance Task Force (IRPTF). (2018). *Resource loss protection criteria assessment*. NERC, Atlanta, GA, USA. February 2018. Available: https://www. nerc.com/comm/PC/InverterBased%20Resource%20Performance%20Task%20Force%20 IRPT/IRPTF_RLPC_Assessment.pdf

42. NERC. (2016). *Phase angle monitoring: Industry experience following the 2011 Pacific Southwest outage recommendation 27*. June 2016 (Online). Available: https://www.naspi.org/ node/517

43. NASPI Control Room Solutions Task Team. (2016). *Using synchrophasor data for phase angle monitoring*. May 2016 (Online). Available: https://www.naspi.org/sites/default/files/reference_documents/0.pdf

44. Midcontinent ISO. (2014). *Whitepaper MISO phase angle differences monitoring and control*. 2014 (Online). Available: https://www.misoenergy.org/_layouts/MISO/ECM/Redirect. aspx?ID=182446

45. Bhargava, B. (2013). *Eastern interconnection phase angle base lining study*. June 2013 (Online). Available: https://energy.gov/sites/prod/files/2013/07/f2/2013TRR-1Bhargava.pdf

46. Martíne, E., et al. (2006). *Using synchronized phasor angle difference for wide-area protection and control*. In Proceedings of the 33rd Annual Western Protective Relay Conference, Spokane, WA.

47. Wu, Y., Musavi, M., & Lerley, P. (2016). Synchrophasor-based monitoring of critical generator buses for transient stability. *IEEE Transactions on Power Systems, 31*(1), 287–295.

48. Yuan, H., Zhang, H., & Lu, Y. (2018). *Virtual bus angle for phase angle monitoring and its implementation in the Western Interconnection*. T&D Conference 2018 Denver, CO.

49. Kincic, S., et al. (2012). *Impact of massive deployment of PMUs on reliability coordination and reporting*. PES GM San Diego, July 2012.

50. ePMU_User_Manual. Powertech

51. Kincic, S., Zhang, H., & Howell, F. (2019). Monitoring RAS operation by online transient stability analysis tool: Implementation and lessons learned. *IEEE Power & Energy Society General Meeting (PES-GM)*, August 5–8, Atlanta.

Chapter 5
Implementing a Advanced DTS Tool for Large-Scale Operation Training

Sherrill Edwards and Hongming Zhang

Organization Acronyms

AESO	Alberta Electric System Operator
BPA	Bonneville Power Administration
CAISO	California Independent System Operator
CENACE	Centro Nacional de Control de Energia
CFE	Comisión Federal de Electricidad
FERC	Federal Energy Regulatory Commission
GE	General Electric
IPCO	Idaho Power Company
NERC	North American Electric Reliability Corporation
PG&E	Pacific Gas and Electric
PowerTech	PowerTech Research Labs Inc
SCE	Southern California Edison
SDG&E	San Diego Gas and Electric
V&R	V&R Energy System Research Laboratory
WECC	Western Electricity Coordinating Council

NERC/WECC Acronyms

ACE	Area Control Error
AGC	Automatic Generation Control
AVR	Automated voltage regulator
BA	Balancing authority
BES	Bulk Electric System
COI	California-Oregon Intertie
DCS	Disturbance Control Standard
DTS	Dispatcher Training System
EMS	Energy Management System
FACRI	Fast AC Reactive Insertion RAS
FTL	Frequency Threshold Limit

© Springer Nature Switzerland AG 2021
H. Zhang et al., *Advanced Power Applications for System Reliability Monitoring*, Power Systems, https://doi.org/10.1007/978-3-030-44544-7_5

GOP   Generation Operator
ICCP   Inter-Control Center Communications Protocol
IROL   Interconnection Reliability Operating Limit
LSE   Load Serving Entity
LTC   Load Tap-Changing (Transformer)
PDCI   Pacific DC Intertie (PDCI)
RAS   Remedial Action Scheme
RC   Reliability Coordinator
RCSO   Reliability Coordinator System Operator
RTCA   Real-Time Contingency Analysis
SCADA   Supervisory Control and Data Acquisition
SOL   System Operating Limit
SE   State Estimator
SPS   Special Protection Scheme
SVC   Static VAR Compensator
TOP   Transmission Operator
UFLS   Under-Frequency Load Shed
ULTC   Under Loading Tap Changer
UVLS   Under-Voltage Load Shed
WI   Western Interconnection
WSM   West-wide System Model

Other Acronyms

CB   Circuit Breaker
CEH   (NERC's) Continued Education Hours
DoD   Dynamic of Disturbance
DPF   Decoupled Power Flow
DSA   Dynamic Stability Assessment
DTSPSM   (GE's) DTS Power System Model database
GridEx   (NERC's) Grid Security Exercise
HDR   (GE's EMS) Historical Data Recording application
MUC   Multiple Outages Contingency
NETMOM   (GE's EMS) Network Model database
NPF   Newton Power Flow
NSD   Non-simulation Data
NSSCDB   (GE's DTS) Non-simulated database
OAG   (GE's EMS) Open Access Gateway application
PI   (OSIsoft's) Plant Information platform
POM   (V&R's) Physical and Operational Margin application
PCS   Peak RC Cloud Services
QKNET   (GE's EMS) Quick Network Topology Estimator
ROE   Real-Time Operation Engineer
ROSE   (V&R's) Region of Stability Existence application
RTDYN   (GE's EMS) Real-Time Dynamic Ratings application

RTGEN   (GE's EMS) Real-Time Generation control (i.e. AGC) application
RTNET   (GE's EMS) Real-Time Network Analysis application (i.e. SE)
RT-VSA   (Peak RC's) Real-Time Voltage Stability Analysis tool
SBO   Select Before Operate
STNET   (GE's EMS) Study Network Analysis application
TSAT   (PowerTech's) Transient Stability Analysis Tool
VSA   Voltage Stability Analysis

5.1 Introduction

5.1.1 Objectives

This chapter goes through description, design, and use of a power system simulator which utilizes the energy management software as a basis to provide a learning environment for training operations personnel.

The main objective of this chapter is to provide guidance and knowledge for simulation engineer and operations trainers in configuring and operating the power system simulation application based on the authors' experience with the Peak Reliability (Peak RC) implementation of the Dispatcher Training System (DTS) for the entire Western Interconnection.

The DTS is an excellent tool when training operators for blackstart situations, partial restorations, situation reenactments, or basic training.

- DTS offers a complexity that can closely resemble the dynamics of the electrical grid in a real-life situation. This complexity can be layered to offer increasing difficulty to the trainee.
- These layers can be used to tailor the simulation to the level of the trainees and their experience.
- This presentation examines the aspects of simulation development and reviews the GE/Peak RC tools/techniques developed over the last 5 years for the simulator operation during blackstart/restoration training.

Other uses for the DTS include:

- Testing new software
- Testing new databases
- Testing system patches
- Testing model philosophy
- Reviewing new/revised procedures
- Event analysis
- Proof of concept

5.1.2   Motivation

The Western Interconnection has experienced a few major blackout events over the last 32 years. Twice in the summer of 1996, a voltage collapse broke the Interconnection into multiple islands [1, 2] with some areas experiencing a blackout. These blackouts ranged from 2 million to 7.5 million customers being out of service for a period of time that ranged from several minutes to 6 h. An additional voltage collapse blackout event occurred in the early fall 2011 in the southwest region of WECC which impacted 2.7 million customers across Arizona, California, and Baja California for a period of time that ranged from 2 to 12 h [1–4].

The event findings report for these events provided several recommendations which have led to the establishment of large-scale restoration trainings in the Western Interconnection. Among the key elements to be included in the restoration training are the following: utilizing the Reliability Coordinator System Operator (RSCO) as the key role in orchestrating the restoration on any multi-entity event; practicing communication between different entities on the restoration plans that will be implemented; having periodic checkpoints where progress, setbacks, and next significant actions are discussed; and the coordination of entity to entity island syncing by the RSCO [2].

A study method can allow a snapshot of a current system to be used with possible changes, and a single step solution can determine the new state of the system based upon the changes made, but the study method can't provide a dynamic response that a simulation can.

This led to the establishment of large-scale restoration trainings in the Western Interconnection. The Key elements to be included in the restoration training are: utilizing the RSCO as the key role in orchestrating the restoration on any multi-entity event; practicing communication between different entities on the restoration plans that will be implemented; having periodic checkpoints where progress, setbacks, and next significant actions are discussed; and the coordination of Entity to Entity island syncing by the RSCO. [5]

NERC requires use of realistic simulators to:

- Verify restoration procedure (EOP-005, EOP-006)

  – Demonstrate blackstart units and cranking paths can perform

- Systematically train on Reliability-Related Tasks (PER-005)

  – Many times simulation hours are expected
  – Provide emergency operations training
  – Verify each of the participating system operator's capabilities to perform each assigned task

5.1.3   Peak RC DTS History

The Peak RC DTS system went through an evolution lasting nearly a decade. The group responsible for the DTS remained the same over that time though the players in the group change numerous times. The DTS system started with WECC but then transitioned to Peak RC when WECC was bifurcated. The DTS system software also went through a transition of ownership several times. The original system was developed by ESCA which then transitioned to Cegelec. The system then transitioned to Alstom before going to Areva and then back to Alstom again before becoming the current developer General Electric (GE). As with Peak RC, the group responsible remained the same over this time with numerous personnel changes over the years.

Peak RC's DTS support team grew from a single individual to a maximum of four before Peak RC shutdown.

The DTS system started with the 2.5 **e-terra***platform* software with certain customs for the operation of a reliability control area. As there was no generation model in this EMS system, the DTS generation model was created through a series of programs using the data in the alarm model, network model, SCADA model, and several flat files. The simulation model was then created from the alarm, generation, network, and SCADA models using another series of programs. The **e-terra***source* model management software is only used to maintain the network and alarm models. The need to handle the numerous changes coming in from over 44 entities within the Western Interconnection in a timely manner (every 6 weeks on average) lead to the WECC/Peak RC model management scheme to come up with regular updates to the WSM placed on the online EMS.

The DTS system started with two systems, one for development of the system and simulations and the second for training the operations personnel. The system started with a VPN connectivity but transitioned to a Cloud connectivity capability. The DTS system was then upgraded to the 2.6 **e-terra***platform* software. OSIsoft PI data history systems were then added to the DTS system environment. A third DTS system was added and tasked for development, and the original development system was re-tasked to training. Peak RC worked with Alstom/GE to add enhancements to the DTS specific software to better handle island frequency control and better tracking of trainee operations during the simulation. See Fig. 5.1 for the timeline of the Peak RC DTS evolution.

5.2   Understanding DTS Simulation

DTS simulation utilizes the model that has been created for a power grid to calculate how the electrical power flows through this model as power is created in a Source and flows to a Sink or load. This model includes the topology of the power grid with the electrical characteristics of the interconnections, the switching devices which can be used to direct how the power can be forced to flow, the protection elements

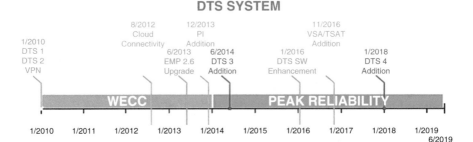

Fig. 5.1 Peak RC DTS system evolution

to make decisions on if and how switching devices can direct flow, the generating sources providing both the real and imaginary power to the grid, and the loads of the grid which change depending upon the time and conditions of the grid.

DTS simulation also provides a means for the training of operators to real-time operational problems through the manipulation of the model to simulate possible or previous events on the power grid. These manipulations are controlled by the instructor to challenge the operator into an experience that can be remembered for actual operations on how and how not to handle a problem situation.

5.2.1 Key Needs to Simulation

DTS simulation has several keys needs. First and foremost is a good model with the system components of the electrical grid that the operator will be managing. Next is the personnel to utilize the simulator for training the operators. The simulation environment then needs to match the operator environment.

5.2.2 Personnel Roles to Support a DTS Training Session

The personnel needed to support a simulation involves a varied skill set. This skill set will usually be spread between a team of individuals. Peak RC organized the basic simulation support team with a trainer and a simulation engineer to provide a simulation for a trainee. For special needs with the simulation, a model/display specialist may be involved.

Some simulations may have all the roles in a single person, while others may require multiple members of the same role to accomplish the simulation (Fig. 5.2).

The DTS engineer is responsible for maintaining the DTS system and that the associated tools functional properly, developing various DTS training simulation

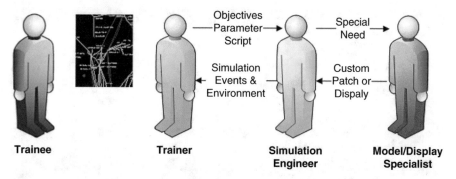

Fig. 5.2 Simulation team relationship

scenario cases upon trainer's requirements, and providing engineering support to trainers and DTS users internally and externally during a training session.

The modeling engineer provides the means to meet special needs for a simulation. While most special configurations of the DTS can be accomplished by the simulation engineer, some may require the talents that the specialist can bring to address the special needs. An example can be a new display that currently doesn't exist in the DTS display files. Another example could be a new substation that doesn't exist in the current model but has been built for a future update to the model and can be incorporated into the current DTS model for use in training on the substation before it goes online.

5.2.3 System Components

The simulator incorporates models of the power system components, including generating units, loads, lines, transformers, capacitors/reactors, DC lines, and protective relays. These models provide a valid representation of the effects of small and large deviations in voltage and frequency which can occur during emergency operations.

A DTS environment contains the components to functionally replicate the trainee operational environment. While the entire operational environment may not be duplicated, the key aspects of normal operations need to be present for the trainee to get best training experience from operational challenges to be simulated.

The Peak RC DTS environment includes:

- A simulator system (DTS)
- A PI data historian system
- Remote training capability
- RC applicable applications
- Adequate storage

Fig. 5.3 Peak RC DTS environment

The Peak RC DTS environment also includes the same type of phone system and a subset of the trainee's workstation to allow for appropriate working environment (Fig. 5.3).

5.2.4   Simulator System

The simulator portion of the DTS environment incorporates models of the power system components, including generating units, loads, lines, transformers, capacitors/reactors, DC lines, and protective relays. These models provide a valid representation of the effects of large deviations in voltage and frequency, which can occur during emergency operation.

Application software is made up of three major components:

1. Power system simulation: provides the power system dynamic simulation functions
2. Energy management System: a replica of an EMS system
3. Instructor control: used for setting up and controlling the simulation scenarios, reviewing the dispatcher's performance, and teaching the dispatcher

Figure 5.4 shows the components of the simulator and how they interact with each other. At the core is the EMS subsystem with SCADA bringing in the grid data and providing data to the State Estimator (RTNET) and prompting the Alarm Processor (ALARM) with violations and status changes to be provided to the trainee. The State Estimator provides the estimated snapshot of the Grid to Contingency Analysis (RTCA) and to the Study Application (STNET).

The normal grid data source for the EMS subsystem would be through a grid communications network of Remote Terminal Units (RTUs) or grid data

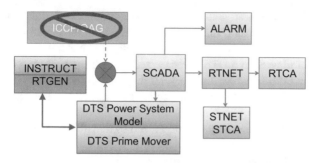

Fig. 5.4 Peak RC simulator system

concentrators bringing the data in through the Inter-Control Center Communications Protocol (ICCP). Peak RC doesn't use RTUs but instead gets all the grid data through ICCP.

The simulator system is not using the actual real-time grid data but instead uses simulated data which becomes the source for SCADA.

This simulated data is created using a prime mover software to solve the problem of what a generator can do given the controls it has received and the capability that the generator is configured for. The generator output then flows into the simulation, and the resulting calculations on how that flow affects the components are seen.

The simulator allows the instructor to automatically create an EVENTS scenario based on the changes recorded within the available Historical Data Recording (HDR) files on the DTS Replay directory.

5.2.5 Storage

The DTS system utilized two hard drives:

- A 279 GB Windows drive which kept the Windows operating system and application on it (WINDOWS disk)
- A 1.09 TB hard drive for the simulation software and the rest of the storage needs for the complete system (DTS data disk)

The drives were backed up regularly off-system to ensure continuity from a hardware failure.

The storage for the DTS data disk was set up in key directories for management of the system. Some of the structure was required by the DTS software, but additional structure was set up to help with the management of the system as the system evolved.

A key additional was the USERS directory which provided the work storage areas for the DTS engineers and trainers on each DTS system. Within the USERS directory was a sub-directory for saving the older DTS savecases and appropriately named DTS savecases. The main DTS savecase directory for running savecases can cause difficulties when it gets over a certain number of files within it. The DTSCASES task can crash if the number of files within the directory (including all the ones in sub-directories) exceeds the parameter that has been configured for the max number of files. So if a DTS savecase set hasn't been used for a while, move it out of the main DTS savecase directory. (If the Main DTS savecase directory is not known, check the Windows environment variables for the value of HABITAT_CASES.)

The DTS savecases directory was organized so that all files were within a folder. The folders were organized so that each database release had a folder for all saved cases (e.g., DB99). However, if a significant event was simulated, a separate folder would include those files (e.g., blackstart). Any simulation development-related file would also be saved under that folder. Several of these folders would also include the HDR files from when the simulation was run (e.g., GridEx). This structure proved useful in coming back to a simulation years after it had been run and have the data to easily recreate the simulation with the newest system database.

The OSIsoft PI historical data management program is connected to the DTS using the same interface used for the EMS. The **e-terra***habitat* SAMPLER application sends the historical data from the DTS to the DTS connected PI System.

SAMPLER is pointed at the SCADAMOM database of the SCADA application in the DTS family. The default sample rate is 5 s.

The PI data is stored in the chronological order of when it is shipped out from the DTS and not by the simulation time running in the DTS. This is an area that could use further investigation as to sending a time setup command with the current simulation time and have the PI system record the data using that time frame.

5.2.6 Simulations Developed and Simulated DTS Events

Table 5.1 lists a set of primary DTS simulation cases Peak RC developed for operations training programs and other training classes since 2012.

Table 5.1 List of major DTS training simulation cases

2017 Eclispse Replay with "What-ifs"	Natural Gas Shortage Restrictions
BA Islanding	North-South WECC Seperation
Frequency Events Recreations	Nuclear Plant Generation Loss
Frequency Oscillation	Operator Readiness Evals for R/T & Study
General Basecase	Recreation of 1996 WECC Breakup
Generation Restriction	Replay of 2011 Blackout
GridEx	R/T Supercase with 5 Event Sets
IROL Examples	Study Supercase with 4 Event Sets
Major Line Loss Impacting Path Ratings	Undervoltage Area
Major Substation Loss due to fire	Wild Fire Impacts
Restoration Training (Complete Blackout & Partial System Restoration)	

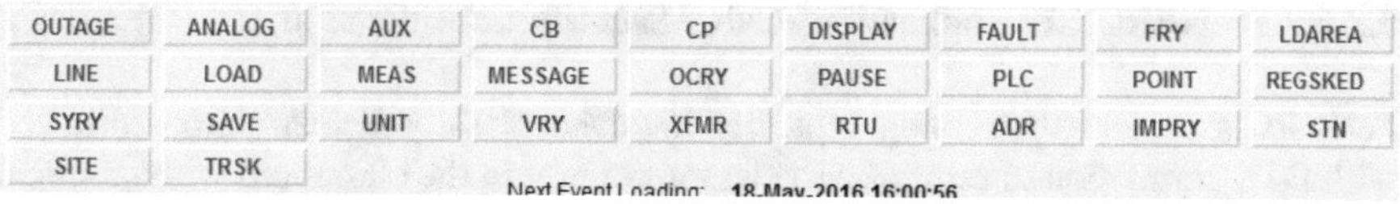

OUTAGE	ANALOG	AUX	CB	CP	DISPLAY	FAULT	FRY	LDAREA
LINE	LOAD	MEAS	MESSAGE	OCRY	PAUSE	PLC	POINT	REGSKED
SYRY	SAVE	UNIT	VRY	XFMR	RTU	ADR	IMPRY	STN
SITE	TRSK							

Next Event Loading: 18-May-2016 16:00:56

Fig. 5.5 EVENT action selection choices

5.2.6.1 DTS Event Types and Event Actions

A simulated event is a collection of actions to achieve a change in the simulation that produces the desired system conditions to occur. Particularly,

- **Directed action** – known as Deterministic, this event type results in an action determined by scenario that happen at a set time or manually trigger the action.
- **Conditional** – this event type is determined by specific set conditions that trigger the action.
- **"Random"** – known as Probabilistic, this event type is determined by a selectable probability that triggers the action when that probability occurs.
- **Event actions**
    - **Measurement based**
    - **Relay based**
    - **Specific action**
    - **Internal DTS –**
    - **Communications**
    - **Schedule changes**

Figure 5.5 lists all events the DTS tool can simulate in the Peak RC training environment.

5.2.6.2 Other Things in Consideration

5.2.6.2.1 Communications Failures

Simulations can have communications failures that occur through the use of the RTU and SITE event action commands. Since Peak RC received the system information completely through ICCP, the RTU action was not used in simulations, but the SITE action provided a means to simulate a ICCP site failure.

5.2.6.2.2 Transaction Schedules

Peak RC worked with the vendor to develop a means of changing the transaction schedule through an event action. Peak RC simulations can have a transaction schedule change made through the TRSK event action. This allowed a schedule to be cut or increased by a defined amount.

5.2.6.2.3   NERC CIP Implications/Cyber Security Consideration

Peak RC's system of training typically involved classroom instruction followed with DTS simulation to reinforce the lessons taught in the classroom with practical experience. The Peak RC trainings could be for just an internal audience or could be expanded to WECC-wide training with the entities.

5.2.6.2.4   Job Function

Peak RC utilizes the DTS to provide operational experience training for the Peak RC and other entities personnel such as system operators, system engineers, study engineers, and planning engineers.

5.3   About Peak RC DTS

5.3.1   The DTS Architecture with Cloud Service

Peak RC restoration training participation has been over a thousand people annually, and each training session might include 60–200 participants. The total travel cost would have been significant if all trainees had to attend the training in person. Further, the Peak RC training rooms at both offices could only accommodate a total of 60 people.

The current WSM-DTS was designed to allow for remote login and access to participants outside Peak RC through a Cloud connectivity by means of integrating Microsoft Remote Desktop Services with the DTS systems [6]. Figure 5.6 is the DTS-Cloud architecture diagram utilizing the Remote Desktop Services.

The Cloud service allowed easy access to multiple DTS servers and required only Internet Explorer 11 be available versus installing a separate program. Windows 7 and 10 were supported. The hardware and bandwidth have been configured to be able to handle a maximum of 300 users concurrently in a training session. Each user could have a dual-headed workstation with each monitor being 1900 × 1200 pixels.

A Cloud client guide was provided to each entity for use by their IT organization to meet the requirements for the Cloud connectivity. Several weeks before the start of the actual drills, each entity is given opportunities to test the connectivity to the Cloud and to review that year's restoration model for final changes. This ensures that each entity IT organization has the time to make any changes necessary in their security. It also ensures that the entity trainers have a model with the challenges they expect for their operators to experience during the drill.

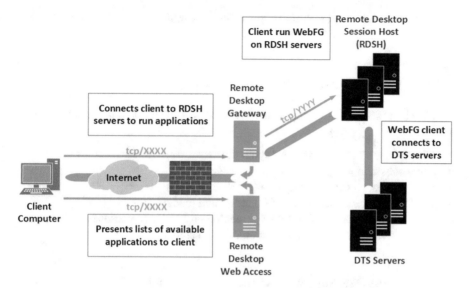

Peak Reliability DTS Cloud Connectivity

XXXX and YYYY are Assigned Port Numbers

Fig. 5.6 Peak RC DTS architecture with cloud service

5.3.2 Overview of the WSM-DTS System

Peak RC used GE Grid Solutions e-terra EMS/DTS (EMP2.6 version) product with many custom features. Apart from common DTS features presented in earlier IEEE papers [7–11], the **e-terra***simulator* software has several unique implementations published in two IEEE transaction papers: (1) an AGC Implementation for System Islanding and Restoration Conditions [6] and (2) an Online Dispatcher Training Simulator Function for Real-time Analysis and Training [12].

5.3.2.1 DTS Operating Modes

The **e-terra***simulator* supports two operating modes: Simulation and Replay. Switching between modes is allowed:

- Switching from simulation mode to replay mode can be performed when **e-terra***simulator* is paused or down
- Switching from replay mode to simulation mode can be performed only when **e-terra***simulator* is paused.

Peak RC used both DTS modes for blackout simulation and system restoration training.

5.3.2.2 WSM-DTS Model Scales

The WSM-DTS is built on top of the WSM EMS models. The latest WSM statistics were:

- 16,000 buses/20,000 branches (dynamic)
- 8,900 substations, 3,900 units
- 13,900 lines, 5,600 transformers (45 phase shifters)
- 100,800 breakers/switches, 11,000 individual loads
- 8,000 contingencies enabled RAS screening on 515 RAS records modeled active
- 177,000 SCADA points including analog and status

5.3.2.3 WSM-DTS-Supported Applications

The WSM-DTS was intended for replicating a full EMS tool suite, other vendor's real-time applications, and custom RC monitoring tools being deployed in Peak RC's control rooms [13, 14]:

- SCADA/alarms/permission management
- OAG (Open Access Gateway)/ICCP application
- Fast Network Analysis (QKNET) enhanced with reactive reserve monitoring and custom Forced Outage Detection (FOD) functions
- Dynamic Ratings (custom version)
- Advanced Network Applications, i.e., SE, RTCA, etc.
- RC Workbook visualization platform (in house tool)
- PI ProcessBook (custom configured)
- V&R Real-Time Voltage Stability Analysis (Peak RC ROSE RT-VSA) tool
- PowerTech DSA Manager/Transient Stability Analysis Tool (custom online TSAT)

The architecture of integrating RT-VSA, TSAT, and PI external applications into DTS is given in Fig. 5.7.

The WSM-DTS also utilizes the e-terra base product AGC features (that's not implemented in Peak RC Production EMS) to mimic balancing authority (BA) function for each operating entity and three regional reserve sharing groups on WECC footprint:

- AGC function capable for normal and islanding conditions
- Reserves Monitoring and Reserves Sharing Group
- Interchange Scheduling (automated by in-house developed interchange transaction processor)

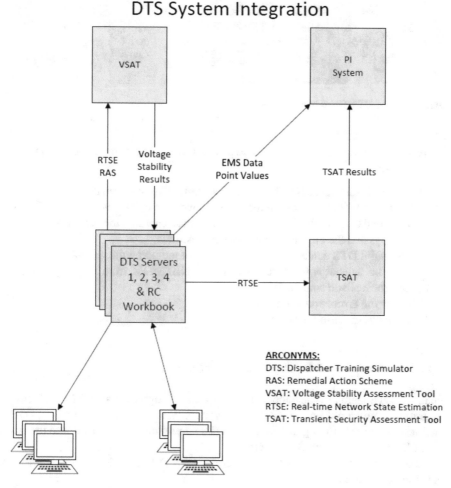

Fig. 5.7 Architecture of WSM-DTS System Integration

5.3.2.4 Peak RC Custom Implementations with the DTS

5.3.2.4.1 Automated DTS Model Build

To make the DTS model to be built more efficiently and consistently, Peak RC DTS team developed a custom C++ program to build a full DTS database automatically from current or historical EMS savecases and the relevant unit fuel type data file. The auto DTS model build process enables creating a new DTS basecase and having it run in DTS simulation mode within a couple of hours. A flow chart of DTS auto build process is illustrated as follows (Fig. 5.8).

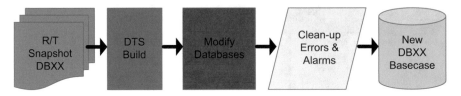

Fig. 5.8 WSM-DTS model build workflow

To automate this process, the first step was to break down the process. The process included:

- Grab a real-time savecase of the EMS system. This was usually done a week after the new build had gone online. This allowed for any patches that were found to be needed to be applied. The real-time savecase captured the current values for the analogs and the current manual switch positions.
- Create initial DTS databases from the snapshot including the generation model.
- Modify the databases for parameters not in NETMOM from the snapshot but needed for realistic simulation.
- Clean-up the errors and alarms.
- Create new basecase for DBxx.

5.3.2.4.2   Automated DTS Case Migration Process

Peak RC DTS team also developed a C++ software program to automate DTS case migration from an old EMS model to a new one. With limited manual effort, DTS engineers were able to migrate dozens of DTS simulation library cases into the latest EMS models for every 3–6 months.

Take current online EMS DBxx, and create DTS DBxx version as a basecase to build specific simulations.

- Simulation transition from DByz to DBxz.

Take functioning simulation on DBxy, and move to new DBxz environment to keep it up to date.

- Simulation creation

Create simulation based upon specific circumstances to provide a training event for exercising a RC to gain experience in handling a new problem or an old problem with a twist.

By leveraging an automation process, DTS engineers currently create a blackstart case much quickly and more consistently than before. The following steps are performed for creating a blackstart simulation case as shown below (Fig. 5.9):

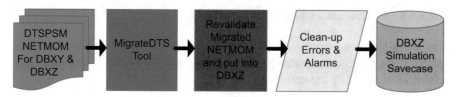

Fig. 5.9 Migrate simulation to new database DTS basecase

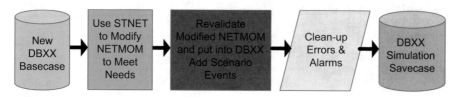

Fig. 5.10 Create new simulation on new DTS basecase

The process includes:

- Load DBXZ DTSPSM NETMOM into "New" Clone, and load DBxy DTSPSM NETMOM into "Old" Clone.
- Run MigrateDTS tool.
- Revalidate migrated NETMOM and put into DBxz basecase.
- Clean up the errors and alarms.
- Create new simulation savecase for DBxz.
- Fine-tune the migrated DTS case with automated scripts.

Modifications for Specific Simulations

The DTS engineer used to modify the new DBxx case in study clone to meet specific simulations with a work process described below (Fig. 5.10).

It includes five steps:

- Load DBXX Basecase DTSPSM NETMOM into STNET.
- Modify NETMOM in STNET to meet simulation needs.
- Revalidate modified NETMOM and put into DBxx basecase. Add scenario events.
- Clean up the errors and alarms.
- Create new simulation savecase for DBxz.

5.3.2.5 DTS Relay Models and Setup

The DTS relay models allow the simulation to have some of the automation of response that occurs in the real-time operations.

5.3.2.5.1   Generic Under-Frequency Load Shed

The Generic Frequency Relays monitor frequency for an entity area within an island. When the frequency goes below a set threshold, the load in that island for the entity is multiplied by the scale factor for that threshold.

- Values are per unit (p.u.)

  .9830 = 58.98 Hz
  .9670 = 58.02 Hz
  .9500 = 57.00 Hz
  .95 = 5% shed
  .90 = 10% shed
  .85 = 15% shed

5.3.2.5.2   Frequency for Generation Tripping

The Frequency Relays are used for tripping units when the frequency excursion is beyond the set thresholds for a time period

$$0.9137 = 54.822\,Hz \qquad 1.0830 = 64.98\,Hz. \qquad 90\,seconds$$

5.3.2.5.3   Under-Voltage Load Shed

The Voltage Relays are used for tripping lines when the voltage excursion is beyond the set thresholds for a time period.

0.6 p.u. low voltage, 2.0 p.u. high voltage, 10 s

5.3.2.5.4   Synchronizing

The Synchronizing Relays are used to keep a switching element from operating when such operation would not be possible with the operational conditions.

5.3.2.5.5   Impedance

While Peak RC hasn't used Impedance Relays, the capability exists.

5.3.2.5.6   Remedial Action Scheme (RAS)

RAS schemes can be operative in the DTS if they are modeled in RTCA properly. There are status points which indicate whether a RAS is armed or in service. These points are Non-Simulated Data (NSD) to keep the status at the appropriate level and with good data quality.

5.3.2.5.7   Non-Simulated Data (NSD)

This is data that is not being driven by the simulator but was being received by the online system. This data is normally static for the purpose of the simulation and defined in the NSD to keep it showing with a good data quality.

The RAS arming points are examples of NSD to keep the status at the appropriate indication and with good data quality.

Analog point examples include an entity's L_{10} value and are NSD to keep the value at the appropriate level and with good data quality.

5.4   System Restoration Drills Hosted by Peak RC

5.4.1   Introduction

The system restoration drill is a primary DTS use case. There were quite a few success stories published in IEEE Journals [15–17] to demonstrate DTS blackstart training experience.

System-wide restoration training in the Western Interconnection was a response to the need to address system restoration issues that were identified in the findings reports for the blackout events.

The practice of starting a restoration from a degraded power grid and the worst-case situation of a complete blackout can be hard for a system operator to realize all the complication that can happen without having gone through it. Peak RC uses an advanced simulator to put participants through many of the complications that are experienced during an actual restoration event. These complications can include picking up unexpected load through failure to properly clear system devices; over- and under-frequency conditions that impact generator operation; managing reactive resources and cold load pickup; waiting for a "hot line" to provide start-up power for beginning to restore a system; and coordinating the additional load and generation in a "dance" to restore a system while maintaining stability of the voltages and frequency.

Restoration drill participation was required for the Peak RC area RCSOs, BAs, and TOPs. GOPs, LSEs, and Alberta (AESO and the associated TOPs) were invited to be part of the drills.

5.4.1.1 Full Restoration

A Full Restoration is a complete blackout of the power grid as far as the trainees are concerned. There is a small island that remains which is a requirement for the simulator to run. This can leave some participants waiting for a "hot line" to be able to start their restoration if no blackstart resources are available.

5.4.1.2 Partial Restoration

A Partial Restoration has several key islands energized for trainees to start from. Hot lines are available to all that don't have a blackstart capability. This requires the trainees to evaluate the grid more closely to understand boundaries between utilities.

5.4.2 Western System Restoration Drills

5.4.2.1 Overview

Since 2011, Peak RC (formerly WECC RC) hosted annual west-wide system restoration drills. The drills were performed at various system levels:

1. A full WECC system.
2. Three regional systems, i.e., Area A, Northwest; Area B, Rocky Mountains and South Deserts; Area C, California and North Baja, Mexico.
3. Combination of two area systems. For the full system restoration drills, the total number of participants in a single drill was more than 250. It was recorded as the largest blackstart training activity across North America.

5.4.2.2 Main Topics of System Restoration Drills

- Frequency ranges on restoration (59.95–60.05 HZ and 58.5–61.5 HZ)
- Load pickup (% of gen capacity)
- Ferranti effects (MVAR injections)
- Shunt devices (light and heavy loading)
- Modeling deficiencies
- Cold load pickup and zero-droop unit
- AGC and generation ramping
- Interconnection restoration
- *Element decisions for realism*
- Full/partial restoration
- Relaying
- Unit frequency response

- Conditional events in INSTRUCT EVENTS
- Blackstart unit setup
- DC blackstart capability
- Adding SBO to DTSPSM displays for CB actions
- Transaction schedules for multi-BA islands

5.4.2.3   Challenges

The Western System Restoration Drills were participated by over a thousand operation personnel annually from the utilities in the West. The drills were delivered in the whole system or individual region or combined with two of three regions.

There were a number of issues found from WECC's blackstart drills between 2011 and 2013:

- The DTS interface used by participants was based on a third party web client that was more challenging than the current cloud interface.
- A manual DTS model build process was tedious, time-consuming, and deficient of consistency of quality.
- Unit governor response and islanding frequency control performance were unrealistic.
- A portion of blackstart units were modeled incorrectly in the WSM (e.g., wrong AVR and ramp rate settings).
- DTS power flow engine became vulnerable under many simultaneous network switching activities, and there was no good way to pinpoint which user actions caused the DTS simulation to stop or loss of an island.
- Islanding AGC control was not effective.
- No good indication on available spinning reserve on each island.
- Over-/under-frequency and sync check relays were missing or not modeled properly.
- Coordination and communication between trainers and DTS engineers were not ideal.
- No scorecards or performance metrics developed to evaluate training effectiveness.

Since 2014, Peak RC has enhanced the DTS support team and the DTS software for development of advanced DTS features in collaboration with the vendor to meet the blackstart drill and other RC training program objectives.

5.4.2.4   Special Configuration in Modeling for Restorations

There are special configurations needed to the DTS model when doing a full or partial restoration. This is due to the fact that the power grid in the normal state of operation has many adjustments that have been made without consideration of there being nothing else energized around any blackout area. Several of these adjustments require modification for the model to run effectively. These modifications include:

- **Unit participation factor** - Generating units may be modeled with a zero (0) participation factor in simulator network model as well as the State Estimator network model. While this isn't an issue for the State Estimator, it can cause initialization problems for the simulator. To correct this a HDBRIO script is run to change any units in the simulator network model with a participation factor value of 0 to have a small value. Peak RC has used 0.001 for this value.
- **Units with negative output for MW** - Generators which can also act as pumps need to be set offline if it is operating with a negative mw flow.
- **Island frequency controller changes** - For proper operation of the island frequency controller, the DTSMOM UNIT record needs the/DISFREQ flag set to true to keep from having the frequency jump when controlled.
- **Other tuning scripts (automated)**

  - Add aux loads to units using a script to make them blackstart capable.
  - Modify unit parameters including inertia, damping, and participation factors.
  - Modify regulation node of all units to local node to reflect blackstart capability.
  - Model protective relays to the DTS model and validate model.
  - Perform manual tuning of the case and get rid of voltage and thermal violations.
  - Activate cold load pickup option.

5.4.2.5   Blackstart Drill Modeling Improvements

To improve blackstart training performance, the DTS engineers made a number of modifications on the WSM by scripts to better suit restoration drill needs and to improve the operators' experiences with DTS. There was a complaint that the simulator is designed only to train the transmission operators and does not provide a realistic opportunity to train the generation operators. This was mainly due to the fact that the operators do not see a bus voltage on the generator bus until some load is picked up. This was the main requirement to form an island and only when an island is formed will the voltages be calculated and displayed. Until then, the bus showed up as dead bus. This prevented the generation operators from energizing a unit on no load and then pick up prescribed load block. In order to overcome this, the minimum number of buses needed to form an island was reduced to two from ten in default WSM case. Also, a fictitious aux load of 0.1 MW was added to every generator bus. It was the best way to mimic no-load start-up for generators by just closing the unit breaker.

5.4.2.5.1 Cold Load Pickup

One of the restoration principles was not modeled in the WSM-DTS until 2018. In 2018 restoration drills, cold load pickup was implemented successfully and received excellent feedback from the operators. As a result, operators had to be cautious with the load pickup amount in order to maintain the island frequency and reserves.

5.4.2.5.2 Protective Relay Modeling

One of the outstanding issues from restoration drills until 2017 was that the islands did not collapse when the frequency was unrealistic (e.g., 120 Hz). The ultimate resolution was to model Frequency Relays and Under-Frequency Load Shed (UFLS) relays in the WSM-DTS. Frequency relays were modeled for every unit to trip if the unit frequency stays below 55 Hz or above 65 Hz for more than 30 s. The reasoning behind the relaxed threshold was to give the operator an opportunity to recover from any unintentional operation or a simulator tool issue. This also prevented the islands from being energized with unrealistic frequency.

Peak RC DTS engineers implemented UFLS relays and synch check relays starting from 2017 drills. UFLS relays helped arresting frequency decline when excess load was picked up unintentionally. Modeling UFLS relays reduced DTS simulation interruptions during the restoration drill. Synch check relays were used to avoid unintentional synchronization between islands. The islands could be synchronized only when the parameter thresholds were met.

Peak RC DTS engineers added Under-Voltage Load Shedding (UVLS) relays to the model in 2018 blackstart drills. UVLS was intended to isolate a part of an island where the voltage decline would have eventually collapsed the whole island. After the relays were introduced, the number of DTS interruptions due to unintentional load pickup was further reduced. The threshold for voltage excursion was set to 0.7 p.u. below which the load breaker would be opened. Next step would be to extract individual UVLS settings from WECC planning dynamic data file and model them in the WSM-DTS.

The addition of UFLS, synch check, and UVLS relays in the WSM-DTS helped create a more realistic training environment for the operators and engineers.

5.4.2.5.3 Island Frequency Control

Since the 2011 blackstart drills, the most concerning issue for the participating operators was that frequency response in smaller islands was unrealistic. The frequency change for relatively small load pickup was so drastic that the operators were not even able to pick up recommended load blocks. To overcome this challenging issue, Peak RC DTS engineers collaborated with GE developers in two areas: (1) implementing "island frequency controller" software enhancement and (2) performing unit dynamic model fine-tuning.

5.4.2.5.4   Island Frequency Controller

From previous blackstart drills, there was an inherent issue with the simulator using units on zero droop. Another limitation was when there were multiple islands within a single balancing authority, there can be only one frequency source for AGC to control in the constant frequency control mode of operation. There was no automatic way to control frequencies in multiple islands within the same balancing authority. Hence, the frequency could realistically be controlled in only one island for a single balancing authority.

To resolve the issues, GE developed a software enhancement called "Island Frequency Controller," which was the alternative method to put unit(s) on frequency control. This feature enables the operators to control the island frequency more accurately during the restoration drill. With this new feature, every island has its own frequency control setting, frequency target, tolerance/dead band, and the ability to choose frequency regulating unit(s).

Peak RC implemented this enhancement in the WSM-DTS and exercised the feature in the 2016 blackstart drills for the first time. In spite of the positive feedback received on the new Island Frequency Controller enhancement, there were still issues with frequency control until 2018. Unit dynamic model parameters need be well tuned to resolve those issues.

5.4.2.5.5   Unit Dynamic Model Parameters Tuning

With the support of GE developers, Peak RC DTS engineers narrowed down to two important modeling parameters impacting frequency control the most: unit inertia and damping coefficient. The default GE recommended value of 1 p.u. MW/p.u. Hz has been used since 2011, but an analysis of the frequency response by modifying the modeling parameter was needed to achieve realistic results. The default value works fine for the non-blackstart simulations as the inherent frequency response is enabled. But in case of blackstart drill cases, due to DTS software deficiency, the prime mover frequency response is disabled to make it compatible with Island Frequency Controller. As a result, the damping coefficient needs to be modified. After a few iterations, it was identified that the damping coefficient should be updated to 4 p.u. MW/p.u. Hz. It is important note that the, WSM-DTS used an unique inertia value for each unit which was extracted from the WECC dynamic data file. The only modification done to the inertia values in the simulator model was that there should not be any zero inertia units. Zero inertia units tend to cause one of the simulator processes to crash, and hence a very small intertie value of 0.01 s is added to those units with zero inertia. After changing the damping coefficient, the frequency response was more realistic as expected. One of the trainers commented – "… I believe our operators will enjoy it at 2018 Peak RC training. Also, new frequency control feature having the island isochronous unit is good. We all know that it took about 10 years to make a good frequency control in DTS."

	CO	STATION	CURRENT MW ON FC		REMAINING MW ON FC	FC ON/OFF	FREQUENCY
ISLAND 1	PGAE	DELTA	63		187	ON	60.01
ISLAND 2	PGAE	BALCHS	57		51	ON	59.99
ISLAND 3	SDGE	ENCINA	143		202	ON	60.02
ISLAND 4	PGAE	COLGAT	10		165	ON	60.00
ISLAND 5	BANC	CAMINO	77		73	ON	60.01
ISLAND 6	PGAE	REDWDSUB	16		27	ON	60.02
ISLAND 7	SCE	MNDALY	66		149	ON	60.02
ISLAND 8	SCE	ALAMIT	99		383	ON	60.02
ISLAND 9	CEN	PJZ230	81		79	ON	60.02
ISLAND 10	BANC	FOLSOM	8		136	ON	59.98
ISLAND 11	LADWP	HARBOR	77		91	ON	60.02
ISLAND 12	TID	DPEDRO	47		63	ON	60.01
ISLAND 13	PGAE	BLACK	6		78	ON	60.03
ISLAND 14	PGAE	GRNLF2	26		38	ON	60.02
ISLAND 15	BANC	DPEDRM	33		22	ON	60.02
ISLAND 16	LADWP	HAYNES	124		106	ON	60.02
ISLAND 17	PGAE	FRIANT	15		12	ON	60.01

Fig. 5.11 DTS island reserve status display

5.4.2.5.6   Island Reserve Monitoring

When the Island Frequency Controller tool was introduced, the operators had difficulty in monitoring the reserves on these zero droop unit(s). This resulted in abnormal frequencies frequently because of no up-/downregulation reserves. GE DTS RTGEN Reserve Monitoring function did not prove useful because the reserves needed to be calculated on island level and not balancing authority level. Peak RC DTS engineer developed a custom SCADA display built in with extensive scripting which provided real-time reserve status for each island. The display helped Peak Reliability's System Operators (RCSO) and transmission operators or generator operators from entities to obtain reserve level awareness of each island during the drills.

It played a pivotal role in maintaining the island frequency within thresholds and the reserves at adequate levels. This display had a number of features including flashing red and green bar charts when the reserves were less than 5% which immediately caught operators' attention. An example of the DTS island reserve monitoring display is shown in Fig. 5.11.

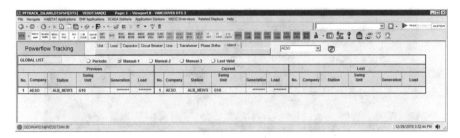

Fig. 5.12 Main display of DTS power flow tracking function

5.4.2.6   Power Flow Tracking Capability

During restoration training, there are large numbers of switching, load/generation adjustments, and other system activities conducted by multiple trainees. These activities sometimes cause loss of islands or DTS power flow solution issues with no obvious reasons. These types of issues are very hard to debug and track the cause, especially in a training environment where there is no time to run study application to analyze the cause. To address this issue, the Peak RC DTS team internally developed a tool that logs all topology changes to narrow down potential causes for unsolved power flow. Since it was not very user-friendly, Peak RC requested GE to develop a tool to continuously track and report user actions that cause loss of islands or power flow to go unsolved in user-friendly manner to the instructor. The main display of the tool is shown on Fig. 5.12.

The **e-terra***simulator* has a feature of creating system snapshots when the simulation is running. The snapshots are saved in memory. The instructor can request the retrieval of the snapshot to reconstruct the system to a prior time when power flow fails to converge to a solution.

To track the system changes, the memory residing snapshots can be leveraged to accomplish this. The simulation snapshot can be compared with the power system states in current simulation time to capture both manual input and calculated solution changes. The compared results will help to identify the changes made by the trainees potentially causing island loss or other power flow issues. The tool display has a number of tabs to separate various equipment categories like units, loads, circuit breakers, lines, transformers, shunt reactive devices, and phase shifters. There is a log that maintains the list of all the islands that disappeared from the list. There are three possible reasons when an island doesn't appear from the current island list:

- When the power flow cannot converge for the island and the island was lost.
- When the island merges with another island, the island name changes.
- When the swing bus in an island changes due to any reason, the tool identifies the island as possibly lost because the swing bus for the island is one of the unique identifiers for an island.

Fig. 5.13 Synchroscope display

5.4.2.7   Miscellaneous Display Enhancements

5.4.2.7.1   Synchroscope

In order to assist the transmission operators with synchronizing of islands, a synchroscope display (shown in Fig. 5.13) were developed in DTS. These displays are associated with all breakers across which synchronizations can potentially occur. The display provides the basic information about the two islands including frequencies, phase angle deltas, voltages, and breaker information.

5.4.2.7.2   SBO Display Change

Since no equipment was owned or operated by Peak RC, the SCADA displays were not set up to control/switch equipment in DTSPSM. As a result, all switching was done from DTSPSM displays. In default DTSPSM displays, there was no secondary confirmation needed to operate any equipment. One of the drill issues identified was that some operators were switching equipment much faster than they could do in a real-time environment. This caused some unintentional circuit breaker operations and typographical errors resulting in simulator interruptions. There was also simulator performance degradation with more than 100 operators making too many switching operations in every scan cycle. The Peak RC team modified the DTSPSM display set to add a Select Before Operate (SBO) capability to all switchable devices that would pop-up an execute selection display (Fig. 5.14). This resulted in a slowdown in the number of operations per cycle for each operator and an overall decrease in the operation per cycle.

Fig. 5.14 Peak RC custom
SBO display

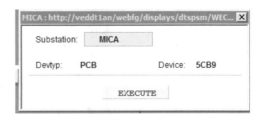

Table 5.2 Statistics of 2016 area C blackstart drills

Items\timeline	Week 1	Week 2	Week 3	Week 4	Week 5
Number of islands	10	12	11	11	7
Total load (MW)	10,919	11,181	9,454	12,100	14,078
Net load restored	**5,187**	**5,449**	**3,722**	**6,368**	**8,346**
Synchronization #	2	1	1	1	3
Total interruptions	**10**	**9**	**12**	**5**	**4**

5.4.3   Results

Peak RC fielded a mature WSM-based DTS tool for the restoration training process.
The DTS tool was able to support large-scale system restoration drills with high
quality. The Peak RC EMS application was fully replicated. A Cloud service was
used that allowed up to 300 people participating in the drills either locally or
remotely. The DTS handled island frequency controlling properly and simulate
complicated system conditions for up to 50 islands. It has DTS software features for
power flow tracking and Peak RC custom displays for Island Reserve Status and
SBO control functionality.

5.4.3.1   Achievements of 2018 Drills

Main achievements are seen from the following statistics:

- Increase in simulator availability from 81.4% in 2017 to 90.5% in 2018.
- Reduction in operator errors that caused simulator pauses has been reduced by
  27% from 2017 to 2018. Use of SBO displays alone resulted in a 24% reduction
  in simulation pauses related to operator mistakes. Total simulation run time
  increased by 8%.
- The addition of the island reserve monitoring screen was a great participant aid.
  The participants now had a visualization of reserve left on a frequency control
  unit. This addition, combined with tuning of frequency control parameters,
  resulted in fewer relay firing, i.e., a reduction of 23%.
- Tables 5.2 and 5.3 show the statistics of 2016 and 2018 Area C blackstart drills.
  To compare Net Load Restored and Total Interruptions between the two tables,
  one can see significant improvements achieved in 2018.

Table 5.3 Statistics of 2018 area C blackstart drills

Items\timeline	Week 1	Week 2	Week 3	Week 4	Week 5
Number of islands	10	9	9	6	6
Total load (MW)	12,998	12,248	9,614	23,041	15,050
Net load restored	**8,191**	**7,442**	**4,808**	**18,235**	**10,244**
Synchronization #	2	5	1	5	3
Total interruptions	5	4	6	6	2

Table 5.4 2019 peak reliability restoration drill statistics

2019 RESTORATION DRILLS		AREA	Drill 1 Quantity	Drill 1 Time Paused	Drill 2 Quantity	Drill 2 Time Paused	Drill 3 Quantity	Drill 3 Time Paused	Drill 4 Quantity	Drill 4 Time Paused	Drill 5 Quantity	Drill 5 Time Paused	AREA	Avg Quantity	Avg Time Paused
Type of Pause	Modelling	A	0	0 Min	0	0 Min	0	0 Min	0	0 Min	0	0 Min	A	0	0 Min
		B	0	0 Min	0	0 Min	0	0 Min	0	0 Min	0	0 Min	B	0	0 Min
		C	0	0 Min	0	0 Min	0	0 Min	0	0 Min	0	0 Min	C	0	0 Min
	Operator	A	4	16 Min	7	22 Min	9	27 Min	7	23 Min	11	35 Min	A	7.6	24.6 Min
		B	9	35 Min	10	33 Min	14	45 Min	11	54 Min	9	33 Min	B	10.6	40.0 Min
		C	9	34 Min	5	14 Min	6	27 Min	1	1 min	2	3 Min	C	1.8	6.8 Min
	System	A	1	15 Min	0	0 Min	0	0 Min	0	0 Min	0	0 Min	A	0.2	3 Min
		B	0	0 Min	0	0 Min	0	0 Min	1	28 Min	0	0 Min	B	0.2	5.6 Min
		C	0	0 Min	1	9 Min	0	0 Min	0	0 Min	0	0 Min	C	0.2	1.8 Min
Entity Synchronization		A	3		8		8		5		7		A	6.2	
		B	4		6		5		6		4		B	5.0	
		C	1		1		5		4		2		C	2.6	
Protection Schemes	Tripped Undervoltage Relays	A	14		13		42		17		13		A	17.8	
		B	20		6		6		12		14		B	11.6	
		C	12		4		16		0		14		C	9.2	
	Tripped Underfrequency Relays	A	9		7		8		9		7		A	8.0	
		B	10		8		8		12		9		B	9.4	
		C	4		6		7		8		4		C	5.8	
General Metrics	Simulation Run Time (Hour:Min) out of Possible 05:15 available	A	04:44 (90.2%)		04:53 (93.0%)		04:48 (91.4%)		04:52 (92.7%)		04:40 (88.9%)		A	04:47 (91.2%)	
		B	04:40 (88.9%)		04:42 (89.5%)		04:30 (85.7%)		03:53 (74.0%)		04:42 (89.5%)		B	04:29 (85.5%)	
		C	04:41 (89.2%)		04:52 (92.7%)		04:48 (91.4%)		05:14 (99.7%)		05:12 (99.0%)		C	04:57 (94.4%)	
	Total Load Restored (MW)	A	6,992		7,712		5,913		7,924		11,617		A	8,032	
		B	10,283		7,446		6,979		10,812		11,318		B	9,368	
		C	4,013		3,344		10,421		8,657		9,341		C	7,424	
	Max Number of Logins(Participants)	A	86 (97)		80 (108)		68 (70)		81 (95)		86 (95)		A	80.2 (93.0)	
		B	105 (150)		120 (153)		127 (152)		126 (140)		112 (135)		B	118.0 (146.0)	
		C	65 (96)		70 (90)		65 (95)		69 (92)		65 (84)		C	66.8 (91.4)	
	Number of Entities (BAs, TOPs, GOPs, & LSEs)	A	17		17		15		16		20		A	17.0	
		B	22		22		21		22		18		B	21.0	
		C	14		13		14		17		14		C	14.4	

AREAs > A is the Pacific Northwest, B is the Rocky Mountain/Desert Southwest, and C is California/Nevada/Mexico

- A total of 1,330 participants received over 9,700 NERC CEHs from the Peak RC system restoration drills of 2018. Participants simulated restoration of over 136 GW of system load during 2018 drills.

5.4.3.2   Comparison of 2019 Drills vs. 2018 Drills

The results from the last 2 years of the drills are shown in Tables 5.4 and 5.5.

A total of 1,212 participants received over 9,500 NERC CEHs from Peak RC system restoration drills of 2019. Participants simulated restoration of over 124 GW of system load during 2019.

5.4.4   Lessons Learned

There were several important lessons learned from these blackstart drill sessions:

Table 5.5 2018 peak reliability restoration drill statistics

2019 RESTORATION DRILLS		AREA	Drill 1		Drill 2		Drill 3		Drill 4		Drill 5		AREA	Drill Averages	
			Quantity	Time Paused	Quantity	Time Paused	Quantity	Time Paused	Quantity	Time Paused	Quantity	Time Paused		Quantity	Time Paused
Type of Pause	Modelling	A	0	0 Min	0	0 Min	0	0 Min	0	0 Min	0	0 Min	A	0	0 Min
		B	0	0 Min	0	0 Min	0	0 Min	0	0 Min	0	0 Min	B	0	0 Min
		C	0	0 Min	0	0 Min	0	0 Min	0	0 Min	0	0 Min	C	0	0 Min
	Operator	A	4	16 Min	7	22 Min	9	27 Min	7	23 Min	11	35 Min	A	7.6	24.6 Min
		B	9	35 Min	10	33 Min	14	45 Min	11	54 Min	9	33 Min	B	10.6	40.0 Min
		C	9	34 Min	5	14 Min	6	27 Min	1	1 min	2	3 Min	C	1.8	6.8 Min
	System	A	1	15 Min	0	0 Min	0	0 Min	0	0 Min	0	0 Min	A	0.2	3 Min
		B	0	0 Min	0	0 Min	0	0 Min	1	28 Min	0	0 Min	B	0.2	5.6 Min
		C	0	0 Min	1	9 Min	0	0 Min	0	0 Min	0	0 Min	C	0.2	1.8 Min
Entity Synchronization		A	3		8		8		5		7		A	6.2	
		B	4		6		5		6		4		B	5.0	
		C	1		1		5		4		2		C	2.6	
			Quantity		Quantity		Quantity		Quantity		Quantity			Quantity	
Protection Schemes	Tripped Undervoltage Relays	A	14		13		42		17		13		A	17.8	
		B	20		6		6		12		14		B	11.6	
		C	12		4		16		0		14		C	9.2	
	Tripped Underfrequency Relays	A	9		7		8		9		7		A	8.0	
		B	10		8		8		12		9		B	9.4	
		C	4		6		7		8		4		C	5.8	
General Metrics	Simulation Run Time (Hour:Min) out of Possible 05:15 available	A	04:44 (90.2%)		04:53 (93.0%)		04:48 (91.4%)		04:52 (92.7%)		04:40 (88.9%)		A	04:47 (91.2%)	
		B	04:40 (88.9%)		04:42 (89.5%)		04:30 (85.7%)		03:53 (74.0%)		04:42 (89.5%)		B	04:29 (85.5%)	
		C	04:41 (89.2%)		04:52 (92.7%)		04:48 (91.4%)		05:14 (99.7%)		05:12 (99.0%)		C	04:57 (94.4%)	
	Total Load Restored (MW)	A	6,992		7,712		5,913		7,924		11,617		A	8,032	
		B	10,283		7,446		6,979		10,812		11,318		B	9,368	
		C	4,013		3,344		10,421		8,657		9,341		C	7,424	
	Max Number of Logins(Participants)	A	86 (97)		80 (108)		68 (70)		81 (95)		86 (95)		A	80.2 (93.0)	
		B	105 (150)		120 (153)		127 (152)		126 (140)		112 (135)		B	118.0 (146.0)	
		C	65 (96)		70 (90)		65 (95)		69 (92)		65 (84)		C	66.8 (91.4)	
	Number of Entities (BAs, TOPs, GOPs, & LSEs)	A	17		17		15		16		20		A	17.0	
		B	22		22		21		22		18		B	21.0	
		C	14		13		14		17		14		C	14.4	

AREAs > A is the Pacific Northwest, B is the Rocky Mountain/Desert Southwest, and C is California/Nevada/Mexico

First, good preparation can make a restoration exercise go smoother. Peak RC offered "Train the Trainer" (TtT) sessions to familiarize entity trainers with the simulator so they could handle basic questions.

Second, with so many internal and external users, there was always technical issues that arise. There was usually connection, system security, or other technical issues that occur when an entity logs in for the first time. Peak RC provided each entity the opportunity to test their connections to Peak RC Cloud Services (PCS) in advance. This allowed time to correct any problems, so they didn't affect participants on the day of the exercise. Multiple DTS engineers provided support for each exercise. Each engineer was responsible for supporting a number of companies for technical issues during the restoration drill. These assignments were made in advance based on the complexity of the companies to prevent any confusion during the drill. The individual entities were provided with contact information for the engineers, so they could be contacted directly.

Third, maintaining availability of the simulator during the exercises was very important. Serious operator errors, such as failure to clear loads off of a line being energized or excessive load pickup, can cause the simulator to pause. This was done to prevent voltage collapse in the affected island. The engineers investigate and correct the issue and restart the simulator. Knowledge of the variety of logs and visualization tools the DTS offers has been helpful in maximizing the simulator availability. Also, each engineer has a different approach to identifying the problem. Some are very skillful in data analysis and iteration summaries to find problems, while others can review the operator logs and effectively locate the problem. A diversity of approaches has been helpful in maintaining a high level of simulator availability. As experience and expertise grow, simulator availability has gone up steadily.

The improvements to the simulator have been noticed by the entities. Many comment on the increased availability of the simulator during the restoration exercises. The improvements to generator frequency control and island reserve status have been appreciated, as well.

Changes in tools have contributed to improved performance by the participants of the restoration drill. One problem noted with previous years' drills was that operators treated the blackstart simulation as a race or challenge to restore as much load as possible in a short period of time. Numerous simulator pauses resulted from reduced attention to what was going on.

Frequency control had been a difficult problem for operators during previous blackstart exercises. Island frequency control was not smooth in previous drills. Operators also noted difficulty with monitoring their reserve levels in an island, which contributed to too much load being picked up in an island. Under-frequency relays would trip and shed load accordingly.

5.5   July 2, 1996, Blackout Simulation

5.5.1   Training Objectives

The July 2, 1996, blackout broke up the WECC system into five islands and resulted in loss of 11,850 MW of load in total to over two million customers who were left out of electricity service from 30 min to 6 h [15].

To help operators learn from the event, in 2019 Peak RC created a training activity to mimic the July 2, 1996, blackout in the WSM-DTS. The simulation began from a system intact condition and finished with the interconnect split into four islands. The goals of the training exercise were to cover the sequence of events for that day, to practice restoration skills, and to gain an appreciation of how quickly a major system disturbance can occur. This training was made available to all entities within the Peak RC footprint. Access was granted for them to participate remotely using PCS.

5.5.2   DTS Simulation Case Creation

Creation of the simulation began with establishing the initial operating conditions, which presented an initial challenge since the WSM was not in existence in 1996. The decision was made to start with a current system model and make modifications to it that would reflect 1996 conditions. Lines and generation that did not exist were simply removed from service. Exact flow and loading conditions were not known for all facilities at the time of the disturbance, so area loads and generation were adjusted to reflect heavy system loading in the Pacific Northwest and other areas.

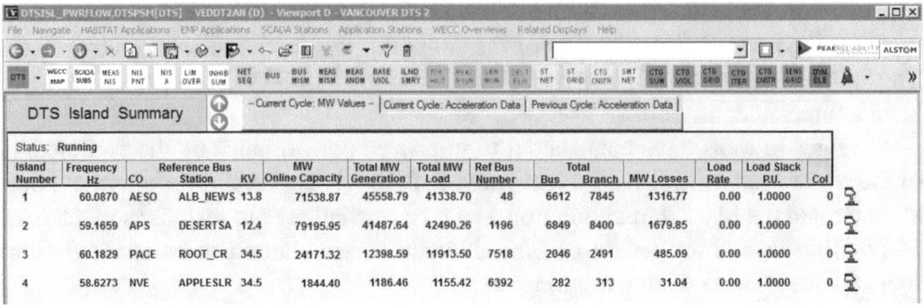

Island Number	Frequency Hz	CO	Reference Bus Station	KV	MW Online Capacity	Total MW Generation	Total MW Load	Ref Bus Number	Total Bus	Total Branch	MW Losses	Load Rate	Load Slack P.U.	Col
1	60.0870	AESO	ALB_NEWS	13.8	71538.87	45558.79	41338.70	48	6612	7845	1316.77	0.00	1.0000	0
2	59.1659	APS	DESERTSA	12.4	79195.95	41487.64	42490.26	1196	6849	8400	1679.85	0.00	1.0000	0
3	60.1829	PACE	ROOT_CR	34.5	24171.32	12398.59	11913.50	7518	2046	2491	485.09	0.00	1.0000	0
4	58.6273	NVE	APPLESLR	34.5	1844.40	1186.46	1155.42	6392	282	313	31.04	0.00	1.0000	0

Fig 5.15 DTS island summary on July 2, 1996, event

The WECC disturbance report was used to create the events for the simulation. Initial attempts were made to reproduce them on a step by step basis, which worked until the Northeast (NE)/Southeast (SE) separation scheme would have activated. The scheme had been disabled in the field prior to the event. The simulator initially would not solve after this point without the scheme in place. The remainder of the steps all occurred within 4.5 s following the point where the separation scheme would have activated. Since they occurred in such a short time frame with respect to the 4 s simulator refresh rate, the decision was made to group the remaining switching steps into a single event. This final event would include the remaining switching steps plus any small adjustments needed to stabilize voltages. The simulator then solved successfully and split the interconnection into nine unstable islands, five of which tripped immediately.

The remaining four islands were shown on the following figures, which required operator interaction to stabilize and restore. The entire event occurred in approximately 1 min. The decision was made to group the initial steps (prior to the separation scheme firing) into a single event, as well as to make the timing more realistic (Figs. 5.15, 5.16, and 5.17).

5.5.3 *Challenges and Resolutions*

Running complex simulations like this one required two people. The first initiated the Quick Fire event steps, monitored the overall simulation progress and stability, and interacted with the training group for unscripted changes to the simulation. The second also monitored the simulation stability and progress. This person also assisted in making changes to the simulation to mimic the desired actions of the participants and to ensure that the simulation ran properly. Changes made during the simulation for stability were recorded so that they could be used to update the simulation for a better result during the next week. Simulations were run for 6 weeks.

Peak RC typically ran its simulator in an interrupted mode. The simulator will pause when faced with a condition that would cause loss of an island or voltage collapse. This allows engineers to correct and explain the error to the trainee without

Fig 5.16 WECC system map – north island on the July 2, 1996, event simulation

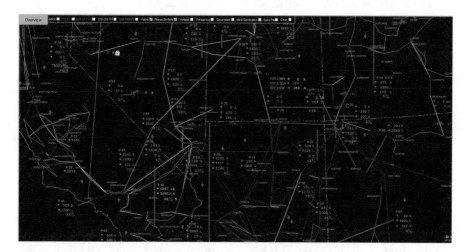

Fig 5.17 WECC system map – south island on the July 2, 1996, event simulation

them actually losing an island that they were working with. When initiating these large-scale simulations, it was better to operate in a continuous mode. The simulator would not pause when faced with a voltage collapse situation. The affected area or island would be lost. This made the system breakup and voltage collapse in Idaho much easier to replicate. Rather than creating individual actions to remove generators from service, we just let them be part of an unstable island that just blacked out.

Once the simulation was initiated, the simulator was switched back to interrupted operation.

In summary, the July 2, 1996, blackout and restoration training successfully provided the RCSO and the entities impacted by the event of how the event occurred and the speed at which it happened. At the end of one training session, one system operator gave her feedback – "I enjoyed the July 02 1996 blackout training, which looks realistic and beneficial to me." Note that this operator is the one who made a right decision to save the system on the day of July 3, 1996.

5.6 September 8, 2011, Blackout Event Replay

5.6.1 Replay of Historical Data

The DTS has an ability to use historical data stored on the EMS to review the system events. This ability is through the REPLAY application. The uses for REPLAY include:

- Real-time event analysis on SCADA and Network Apps
- Allow to initialize DTS back the history at any time point.
- Network Applications are able to run when replaying SCADA.
- Minimize the trainer's efforts to deal with training logistics.
- Training opportunity to support staff.

It can replay an event recorded in real time or a DTS simulation while keeping DTS in synch with EMS including database and displays.

Under a replay session, all EMS Application, i.e., SCADA, ALARMS, SE, RTCA, Dynamic Ratings, Sensitivity Analysis, and Study features, are all available, except QKNET (Fast Topology Processor) because it doesn't run in replay mode.

5.6.2 Setup for REPLAY

The setup for REPLAY requires a number of things, but first is the identification of the time frame to be replayed. This may seem obvious, but often the need to review a historical time frame comes much later than the present.

The sequence of tasks shall be done via the DTS Replay Control display shown as follows:

1. Retrieve the proper database and displays.
2. Switch DTS mode from simulation to replay.
3. Load HDR files from EMS to DTS.
4. Set up the starting and ending time of replay
5. Initialize the replay.

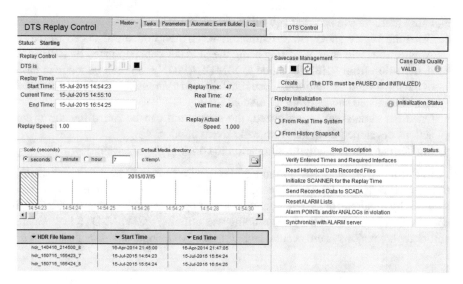

Fig. 5.18 DTS Replay Control Menu

6. Continue replay and monitor the SCADA and Network Applications.
7. Pause the replay.
8. Switch DTS mode from replay back simulation.
9. Initialize the simulation from HDR.
10. Continue the simulation (Fig. 5.18).

5.6.3   Automatic Event Builder

The DTS Replay Automatic Scenario Builder task is a utility that allows the instructor to automatically create an EVENTS scenario based on the changes recorded within the available Historical Data Recording files on the DTS Replay directory.

The **e-terra**_simulator_ can be used for large event analysis by replaying recorded SCADA data files (i.e., HDR files) to show what the system saw as the event was happening. The **e-terra**_simulator_ also provides the capability to filter out the unneeded data from the HDR files from substations that aren't important to the analysis. This can be done through the Automatic Event Builder tool in REPLAY.

5.6.4   Transition to Simulation

Peak RC DTS engineers found the Replay function very powerful in multiple ways:

• REPLAY can be used to step through an event on the system.
• The replay speed is adjustable where the replay can be slowed down or sped up.

- Pause markers can be set at key times during the replay.
- Once the replay starts, State Estimator, RTCA, and other Advanced Network Applications will start executing using the SCADA data populated by REPLAY.
- A transition from REPLAY mode to SIMULATION mode can be made in case of having a valid SE solution. Once transitioned to simulation, the DTS will need to be initialized successfully by manual tuning.
- Once the simulation is initialized, the simulation can be run using the event actions developed from the Automatic Event builder.
- Studies can be run using the simulation to check "What if" scenarios.

In Spring 2017, Peak RC hosted a series of "Dynamic of Disturbances" (DoD) training class taught by an independent instructor. The DOD training was well received by hundreds of system operators and support engineers from Peak RC and entities. One highlight of the DoD training was utilizing the WSM-DTS to simulate and/or replay major system events, including the 2011 Southwest Blackout.

5.6.5   The 2011 Southwest Blackout Event Replay

The 2011 Southwest Blackout event directly impacted five TOPs across three BAs and indirectly impacted all of WECC with some generation being tripped off as far away as northern Canada. While the event lasted 11 h before all customers were returned to power, key events occurred within 11 min. Peak RC DTS engineers were able to retrieve the HDR files and the WSM databases and EMS displays matching the event time and replay the SCADA data of the event while keeping SE and RTCA up running. Replay data could provide the best in-depth analysis for the cause and also for understanding the overall effect of the event. Moreover, the WSM-DTS provides the wide-area view of how an event affected the system around it.

The DTS September 8, 2011, blackout event training replay session was initialized from the HDR file captured before the North Gila-Hassaympa 500 kV line trip occurred at 15:23:53, Mountain Standard Time. The major outages are listed in the table below for reference (Table 5.6).

From SCADA Overviews and Alarms and PI displays, participants attending the training could gain a big picture of system stress development on a sequence of cascading outage events. In addition, from RTCA summary results in Fig. 5.19, one may notice the number of unsolved contingencies jumps from 4 (pre-disturbance) to 11 after outages at step 2 happened from the replay session. Newly unsolved contingencies occur in SDG&E and SCE footprint, which indicates the areas are at high risk of affecting system reliability.

In summary, DTS replay can be a useful tool in analyzing and training for large event lessons learned. Large events can require resizing of the EVENTS database from the current sizing. Transitioning from REPLAY to simulation for a large event can provide opportunity for "What if" scenario execution and get feedback based upon the operational data at the time.

Table 5.6 Study outages by steps

Study step	Event time	Outage equipment IDs
1	15:27:43	North Gila-Hassaympa 500KV line
2	15:28:24	Two 230/92 transformers and one 161/92 KV transformer at Coachella Valley
		Coachella Valley-Ramonz 230 KV line
3	15:32:25	Blythe-Niland 161 KV line
		Coachella Valley-Niland 161 KV line
		Unit-GT2 at Niland
		Unit-Comic at Coachella Valley
4	15:36:59 ~ 15:38:09	Pilot Knob-Yucca 161 KV line
		Pilot Knob 161/92 KV transformer
		Cenroa-CCM230 230 KV line
		El Centro-Imperial Valley 230 KV line
		El Centro-Pilot Knob 161 KV line

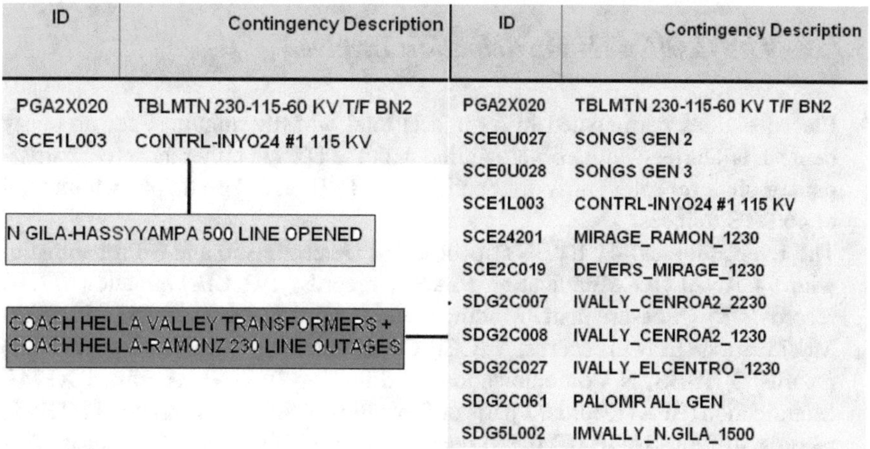

ID	Contingency Description	ID	Contingency Description
PGA2X020	TBLMTN 230-115-60 KV T/F BN2	PGA2X020	TBLMTN 230-115-60 KV T/F BN2
SCE1L003	CONTRL-INYO24 #1 115 KV	SCE0U027	SONGS GEN 2
		SCE0U028	SONGS GEN 3
		SCE1L003	CONTRL-INYO24 #1 115 KV
N GILA-HASSYYAMPA 500 LINE OPENED		SCE24201	MIRAGE_RAMON_1230
		SCE2C019	DEVERS_MIRAGE_1230
		SDG2C007	IVALLY_CENROA2_2230
COACH HELLA VALLEY TRANSFORMERS + COACH HELLA-RAMONZ 230 LINE OUTAGES		SDG2C008	IVALLY_CENROA2_1230
		SDG2C027	IVALLY_ELCENTRO_1230
		SDG2C061	PALOMR ALL GEN
		SDG5L002	IMVALLY_N.GILA_1500

Fig. 5.19 Unsolved contingencies after step 1 and 2 outages

5.7 GridEx DTS Simulation

NERC's Grid Security Exercise (GridEx) is an opportunity for utilities to demonstrate how they would respond to and recover from simulated coordinated cyber and physical security threats and incidents, strengthen their crisis communications relationships, and provide input for lessons learned. The exercise is conducted every 2 years. GridEx aims to:

- Exercise incident response plans
- Expand local and regional response
- Engage interdependent sectors

- Increase supply chain participation
- Improve communication
- Gather lessons learned
- Engage senior leadership [3]

Peak RC participated in the GridEx exercise for GridEx II (2013), GridEx III (2015), and GridEx IV (2017).

5.7.1   The GridEx II Participation Involved

- Internal participation only
- The Event File had three EVENT records (three Deterministic) with four CB Operation (ECB) records, six Load Modification (ELD) records, and four Unit Modification (EUN) records. This is a total of 21 records.

5.7.2   The GridEx III Participation Involved

- Eleven entities participated in 63 injects total with the min/max for an entity being 1/1. Thirteen entities were online with the DTS in either an active participant or observer role. Thirty-four entities and 12 Peak RC participants for a total of 46 DTS trainees.
- The Event File had 41 EVENT records (38 Deterministic and 3 Probabilistic) with 94 ANALOG Modification (EANL) records, 192 CB Operation (ECB) records, 35 Measurement Modification (EMEAS) records, 57 STATUS Modification (EPNT) records, 3 ICCP Communication Modification (ESITEs) records, 3 SUBSTN Communication Modification (ESTN) records, 4 XFMR Modification (EFX) records, 3 pairs of Probabilistic Setup (COND and PRTEST) records, 48 MESSAGE (EMESS) records, and 76 TEXT (EDOC) records. This is a total of 607 records.

5.7.3   The GridEx IV Participation Involved

- Nineteen entities participated in 165 injects total with the min/max for an entity being 3/19. Thirteen entities were online with the DTS in either an active participant or observer role. Forty-five entities and ten Peak RC participants for a total of 55 DTS trainees.
- The Event File had 245 EVENT records (all Deterministic) with 57 ANALOG Modification (EANL) records, 564 CB Operation (ECB) records, 94 Measurement Modification (EMEAS) records, 58 STATUS Modification (EPNT) records, 16

ICCP Communication Modification (ESITE) records, 60 SUBSTN Communication Modification (ESTN) records, 45 UNIT Modification (EUN) records, 2 LINE Modification (ELN) records, 0 XFMR Modification (EFX) records, 0 MESSAGE (EMESS) records, and 124 TEXT (EDOC) records. This is a total of 1255 records.

5.7.4   GridEx IV Lessons Learned

From implementing and running the GridEx drill, there were a few lessons learned:

- To handle an EVENTS database of this size, a better means was needed for working with the database than just entering the items separately into the database. This was especially true with the modifications that kept coming in. Peak RC obtained the EVENTIO tool from GE to facilitate this work. It exports and imports records from the EVENTS database as CSV files just like the other RIO tools. However due to some contract delays, it wasn't received until late in the GridEx model development process. The entity inject pieces were tested individually but didn't have a chance to get everything together until the final week before GridEx. There was only limited time to get familiar with the tool and the associated strengths and weaknesses. The tool also clears out the database before adding the import file.
- An Excel spreadsheet was used to manage all the injects from the entities and developed individual sheets for each Event Action type (EANL, ECB, EPNT, etc.). The event actions were organized under the EVENT records for each inject and scheduled according to the event time provide by the entities. Many large injects were broken into separate EVENT records with slight time delay (seconds) between each to ensure that there wouldn't be an execution problem.
- The INSTRUCT Message Log doesn't show that ESITE or ESTN events were executed by INSTRUCT. The execution occurs, but there's no indication in the log or in ALARM when the points go suspect. The SCADA Substation display and Tabular display show the points as SUSPECT. The loss of communications by ESITE or ESTN event action needs to cause an ICCP_DOWN alarm for ESITE (an existing alarm) or a new alarm for ESTN. In the EMS, the ICCP_DOWN alarms come from the OAG servers which are not part of the DTS environment. The DTS utilizes DTSCOMM of the DTS Common Utilities to simulate our communications structure. Peak RC has no RTUs as all points come into the system via ICCP. The STATION structure of DTSCOMM will be used to approximate a RTU structure. (See Design note below.)
- A BPA Inject involved multiple Line Open End tripping at CAPTJACK, CUSTER, BOUNDARY, and RAVER. The inject did execute in that order, and it was at this point that a DTS pause occurred. This pause occurred due to the inject opening CUSTER-INGLEDOW before BOUNDARY-NELWAY was opened in the inject. When this inject was tested initially, the order was reversed

and didn't have a problem. This was an inject that should have been broken into sub-injects and the BOUNDARY-NELWAY go first with a delay to the CUSTER-INGLEDOW executing. An overcurrent relay on the BOUNDARY-NELWAY line would also solve the problem.

5.7.5   Design Considerations for Future Injects

- As Peak RC got all its data from ICCP, there is no individual RTU modeling that matches the communications structure of an entity. If an event needs to fail a RTU in a substation, an ESTN record is used to do that. This usually works fine for substation with a single RTU. However larger substations typically have multiple RTUs, and this presents a problem when only one RTU fails. Each individual status point and analog for that part of the substation will need to be failed individually unless the station can be taken out as a whole. (This happened in 2015 with the TESLA 230 kv yard RTU failure, but the remainder of the yard had communications.)
- Any large number of analogs that need to be shown as changing randomly or with a bias need to be identified early in the process to be sure sufficient EMEAS records will be available to accommodate the modeling for them.
- Future submissions for GE e-terra platform using entities can include an EVENTS database savecase that has what an entity wants for the draft injects. The file would be exported and brought into spreadsheet and adjusted for naming differences between the WSM-DTS Model and the entity. Also, any limitations due to the model design differences will be reconciled (e.g., sub 100 kV or RTU failure issued discussed above).
- Large injects would be broken into sub-injects to be sure there isn't an execution issue and that all are started at close to the same time versus waiting for each to execute. If there is a dependency upon an earlier sub-inject for a subsequent sub-inject, please note the dependency to be sure the dependency is captured when defining the sub-injects.
- Relays need to be implemented on the BOUNDARY-NELWAY Line terminals to be sure that separation occurs for an overload condition on that line.
- Be Sure to have sufficient time for debugging large simulations. Set a cutoff date for changes and keep to it.
- Need to make sure that the Message Log for INSTRUCT options is set to FILE with the Device field left blank so that a file with the INSTRUCT Messages is saved in the HABITAT_LOGDIR directory once the queue is full. Without message dump to a file, the message queue is overwritten. The message queue has 200 entries, and with the VSA_ROSE values being brought in, it can fill up quickly.
- Having HDR recording turned on during the large simulation allows for replaying the simulation events after the fact. This can help in evaluating performance and trouble areas.

5.8   Peak RC-CAISO Joint Training on RT-VSA Tools in 2017

5.8.1   Objectives

- Correctly determine which IROL to use in the operating scenario.
- Evaluate correct functionality of tools used to determine IROL value.
- Take appropriate actions when tools used to determine IROL value are not operating correctly due to tool failure.
- Utilize the online and offline VSA tools to calculate SDG&E/CENACE IROL value.
- Utilize STNET to validate the SDG&E/CENACE IROL value.
- Demonstrate ability to troubleshoot incorrect IROL value.
- Take appropriate actions when the SDG&E/CENACE IROL margin value changes.

5.8.2   SDGE System Conditions:

- Moderately warm summer day for SDG&E with area peak load expected to reach around 4000 MW.
- Hxxx Beach unit-2 and Synchronous condenser-4 are on planned outage.
- Sxxx-Serxxx 220 KV Line on planned outage.
- Palxx and Otayxxx in service with a limitation of 300 MW max on the whole plant.
- Encxx-2 online with a max of 50 MW. Rest of the Encxxx units not available.
- IV Thermals (Txx and LAxxx) online.
- IV Renewables online.
- SDG&E is having GAS shortage and most of the internal generation is limited. Only 200 MW of generation (Pxxxers, QFs, spin, etc.) is available beyond what is online already.
- If Path-45 flow is higher than normal rating, ignore it. There is some issue with meter reading. Assume flow is following the schedule.

5.8.3   Script for CAISO Joint Training

5.8.3.1   Preparation

1. Savecase DB86_CAISO_SDGE_IROL_TRNG_3 will be loaded prior to beginning simulation.
2. Quick Fire scenario name DB86_CAISO_SDGE_IROL will be retrieved into DTS

3. Simulator time will be reset to PST time before starting DTS
4. PI will be started in DTS to allow for baseline data to populate
5. All alarms will be cleared from the alarm browser
6. Export NETMOM will be aligned to send data to the online VSA tool from DTS running simulation
7. The online VSA tool will be started and operating prior to running exercises
8. Communication will be established between Peak RC DTS engineers and CAISO trainers to coordinate the events for the training scenario

5.8.3.2 Simulation

- ROE/RCSO will have opportunity to log in and bring up tools and to become familiar with simulator and loaded case
- This basecase will be starting with the SDGE/CENACE IROL being monitored

5.8.3.3 Evaluation

1. For each exercise, the ROE/RCSO will be expected to exhibit good communication techniques:

   (a) Three-way communication will be evaluated, if applicable.
   (b) Messages sent will be evaluated on clarity.
   (c) Good listening skills will be tested, as well.

2. For each exercise, the ROE/RCSO will be expected to correctly determine the IROL to be used for the particular situation.
3. For each exercise, the ROE/RCSO will be expected to perform studies using various tools to correctly validate IROL.
4. The ROE/RCSO will be expected to perform studies to calculate the correct limit at appropriate times.
5. The ROE/RCSO will be expected to take actions to identify cause of incorrect IROL values, when appropriate.
6. The ROE/RCSO will be expected to take appropriate actions in response to changing IROL margins.

5.8.3.4 Simulation

- This exercise will be initiated by DTS engineer initiating a trip of the Impxxx Vaxxx-Ocoxxx 500 kv line from the Quick Fire Toolbox.

   – This action will trigger a reduction of the SDG&E/CENACE IROL of approximately xxx MW.
   – The RCSO/ROE will notice reduction in IROL and take steps to validate the limit. This limit should be able to be validated using the offline VSA tool.

- When limit has been validated, the RCSO will take actions appropriate for the new margin value.
- Load in the SDG&E area will be increased by 700 MW using the "INCR LOAD SDG&E – SCHEDULE CHNGS FOR xx MW" button in the Quick Fire Toolbox.
- When an operating instruction is issued to reduce load, that load will be shed using the "TAKE SIM DOWN xxx MW AND BRING BACK IN 30 MIN" button in the Quick Fire Toolbox.
- Evaluation

   The ROE/RCSO identifies the reduction of the operating limit as Pass/Fail. The ROE initiates actions to validate the new limit:

   - Identification of limiting contingency as pass/fail.
   - Perform validation of limit using Study VSA tool, if applicable, as pass/fail.
   - Perform validation of limit using offline VSA study tool.

      - Correctly identify most limiting contingency as pass/fail.
      - Verify that most limiting contingency is logical for operating condition as pass/fail.
      - Correctly identify critical bus for the contingency as pass/fail.
      - Validate that the critical bus is logical for operating condition as pass/fail.
      - Determine which RAS are being triggered in the calculation of the new limit as pass/fail.
      - Validate that the triggered RAS are correct for the operating scenario as pass/fail.
      - Initiate a conference call with appropriate entities to compare calculated limit and to resolve any discrepancies as pass/fail.
      - Establish new operating limit and verify operating margin as pass/fail.
      - Take actions appropriate to the correct import procedure as pass/fail.

   The RCSO will take appropriate actions for reducing SDG&E/CENACE import margin. The actions taken will be in accordance with the "Monitoring of SDG&E/CENACE Import" procedure

   - At the 800 MW margin value, the RCSO initiates conference call to discuss margin and to evaluate possible steps to increase as pass/fail.
   - At the 500 MW margin value, the RCSO initiates conference call to discuss margin and to implement steps to increase as pass/fail.
   - At the 200 MW margin value, the RCSO initiates conference call to discuss margin and to verify steps are being taken to increase as pass/fail.
   - When imports exceed the import limit or if the limit has been exceeded, the RCSO will verify that all steps to increase margin have

been exhausted. RCSO will then issue operating instruction to shed load in the San Diego or CENACE area to mitigate the exceedance. Pass/Fail

The ROE/RCSO exhibits good communication techniques throughout the exercise

- Three-way communication will be evaluated, if applicable, as pass/fail.
- Messages sent will be evaluated on clarity as pass/fail.
- Good listening skills will be tested, as well, as pass/fail.

This joint RT-VSA Tool training was completed successfully. Operation personnel in both Peak RC and CAISO sides were satisfactory with the outcome of the DTS training simulation and overall settings.

5.9   Online Transient Stability Analysis Tool (TSAT) and Use Case Training

5.9.1   Operation Conditions That TSAT Looks at but RTCA Can't

- Loss of units out of step due to fault: While loss of a few small units may not be a concern for reliability of the interconnection, cascading loss of units or entire plants might be. Currently 1000 MW.
- Frequency response: Large penetration of renewable sources can significantly affect frequency response in the future. Dynamic simulation provides a method to predict the frequency response measure in real time and provide situational awareness to operators when there is a high risk of triggering Under-Frequency Load Shedding or when the interconnection may be deficient of frequency response to meet the Frequency Response Obligation requirements.
- Damping: Post-contingency damping for the credible single and multiple contingencies can also be evaluated with real-time dynamic simulation. Negative damping would cause uncontrolled increase in amplitude of MW flows and angle oscillations of units. This can lead to triggering of protection on overloaded lines due to large flows or generation tripping for out of step that might lead to cascading. SOL methodology requires a minimum of 3% damping
- RAS interactions: Given the ability to model RAS actions in real-time tools, the effect of RAS actions on the transient stability assessment can be evaluated. Also, the modeling of RAS actions and their activation in the simulations will ensure that their effect on neighboring systems is evaluated.
- Fast voltage collapse: Since part of the load is modeled as induction motor, TSAT can provide an answer if the system is in danger of fast voltage collapse due to induction motors accelerating following fault clearing.

- N-1/N-k: When running RTCA occasionally, some contingencies cannot be solved due to impossibility to accurately represent RAS model in power flow (transient RAS). For these contingencies TSAT can be used with accurate RAS model to evaluate actual effect of the contingencies correctly.

5.9.2 TSAT Training Topics

- Contingency analysis

  - ATR or Colstrip RAS Operation awareness
  - Assessment of current system for a handful of N-1 and some credible N-2
  - Evaluates system for Margin, Voltage Drop/Rise, and Frequency Drop/Rise
  - Frequency Dip Monitoring against contingencies, i.e., FTL violation event forecast
  - System and individual BA Inertial Calculations
  - PV Solar Generation Momentary Cessation Impact Risk Assessment

- Transfer analysis

  - Analyzes flow boundaries against a specific set of contingencies
  - Source and Sink concept to determine import limit
  - There will be no work performed using transfer analysis except backup to the RT-VSA tool for real-time assessment of the Western IROLs
  - Source/Sink defined in the scenario, along with the contingencies
  - These five transfer analysis scenarios can be modified/improved as needed

5.10 Integrating the RT-VSA Tool into DTS Simulation

5.10.1 Purpose

To provide instruction and serve as reference source to DTS engineers about different scripts that are being used for running POM Voltage Stability Assessment in DTS environment.

5.10.2 Introduction

The Voltage Stability Assessment (VSA) application runs every 5 min in real time on a stand-alone server. It uses a network model export from RTNET. For the EMS environment the ExportNetmom process creates export file (WSMExport.csv) from RTNET.EMS clone and send it to V&R server. But in DTS, the RTNET.EMS clone is not available. The ExportNetmom process needs to be modified for DTS to export the data from RTNET.DTS clone. As a work around, a batch script is developed to create NETMOM export file from RTNET.DTS clone and transfer it to V&R server.

Once V&R server receives WSMExport.csv file, it runs the voltage stability assessment and stores the stability limits in a text file (EMSAlarm.csv). In EMS, FileLink process reads the alarm files, converts them to OAG format, and sends them to EMS through OAG server. As there is no OAG server in DTS, a python script is created to read the data from FileLink EMSAlarm.csv file and send the data to SCADA points through DTS/SCADA non-simulated data function. So, the corresponding SCADA points need to be added to non-simulated SCADA database (NSSCDB).

5.10.3   Description of Tools

5.10.3.1   Export NETMOM for DTS – ExportNetmomDTS.bat

ExportNetmomDTS.bat script is executed every 5 min on all DTS servers by windows scheduler. In DTS environment, only one VSA server is available, and it should receive WSMExport.csv file from only one active DTS server. User can configure the active DTS server by setting "habdata90\ExportNetmom\ExportNetmomSetting.txt" to **True**.

Note It should be set to TRUE only on one server. If this is set to TRUE on multiple servers, multiple servers will send WSMExport.csv file to V&R server and might overwrite each other. Edit the file and set False on all remaining servers (Fig. 5.20).

For each execution, ExportNetmomDTS.bat script performs the following actions:

• Verifies that the D:\Eterra\habdata90\ExportNetmom\ExportNetmomSetting.txt file was set to TRUE and continues to next step only if it is set to TRUE

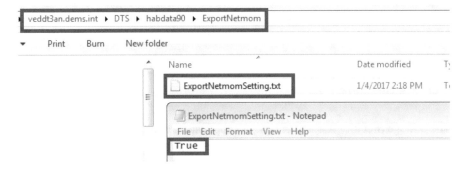

Fig. 5.20 DTS server location for the scripts

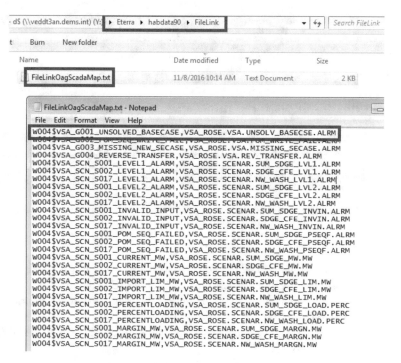

Fig. 5.21 Example of FileLink OAG/ICCP SCADA mapping file

- Checks the DTS state and continues only if it is running or in replay mode
- Checks RTNET solution status and continues to next step only if it is either "Valid Solution" or "Solved W. Mismatch"
- Exports the data and copies WSMExport.csv and SPMEAS.csv to V&R server

5.10.3.2   FileLink for DTS – FileLinkDTS.py

Once the V&R server receives WSMExport.csv file, it performs voltage stability assessment and stores the results in EMSAlarm.csv file. The FileLink process watches for any new EMSAlarm.csv file, creates a new Alrm.csv file with ICCP object IDs, and copies the Alrm.csv file to "D:\Eterra\habdata90\FileLink" folder on all DTS servers (Fig. 5.21).

FileLink ICCP Object ID and SCADA Point mapping are maintained in "D:\Eterra\ habdata90\FileLink\FileLinkOagScadaMap.txt" file.

FileLinkDTS.py script is executed every minute by windows scheduler. For each execution, the script performs the following actions:

- Checks if there is any new Alarm.csv file is available in D:\Eterra\habdata90\ FileLink folder and continues to next steps only if there is new Alrm.csv file.

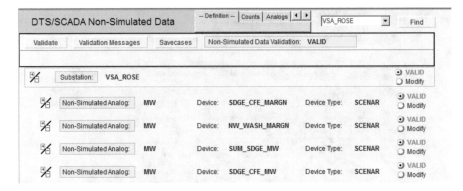

Fig. 5.22 DTS non-simulated data display

- Checks if DTS is running and continues to next steps only if it is running.
- Process Alarm.csv file and updates the points in NSSC database.
- Once the data is updated in NSSC database, DTS process reads the data and sends it to SCADA points.
- All the SCADA data processing, alarms/calculations are done as if the SCADA point is received from ICCP.

5.10.3.3   Creating Non-simulated Points in DTS

These non-simulated data points can be viewed from DTS-> Related displays ->Local Control and Measurements->DTS/SCADA Non-simulated Data (Fig. 5.22).

SCADA analogs and points associated with VSA process can be created/updated in this display.

Also, a NSSCDB savecase should be saved and available in all DTS servers which has all necessary points and analogs added after each modification.

5.11   Summary

Peak RCSO built and maintained a full mature western system model in its DTS tool for RC training programs. Among various DTS training simulation scenarios we created over years, system restoration or blackstart is the one experiencing the most challenges and impacting West Interconnection the most. Peak RC's WSM-DTS tool successfully supported large scale system restoration drills and other DTS training programs with high quality as follows:

- It is able to fully replicate Peak RC's EMS Application suite and mimic historical system events in WECC footprint accurately.

- It's fully integrated with Cloud service that allows up to 300 people participating the drills either locally or remotely.
- It's capable for both blackout event simulation and blackout event replay modes.
- It is able to handle island frequency controlling properly and simulate complicated system conditions including up to 50 islands.
- It implemented many new DTS software features, e.g., power flow tracking, and Peak RC custom displays, e.g., Island Reserve Status, SBO, etc.
- It has modeled various replays required for the blackstart simulation.
- The Peak RC DTS system over the period of 2014 to 2019 provided restoration training to over 8,600 participants (certified and non-certified) earning almost 58,000 NERC CEHs. Over the period of 2016 to 2019, the participants restored over 412 GW of load in these drills.
- A productive DTS team, automated model build and migration processes, efficient workflow, procedures and performance tracking metrics, effective communication channels, close collaboration with internal and external stakeholders, and many years hands-on training experience.

As Peak RC winded down, some practice and lessons learned could be useful for the Western RCs. The new Western RCs may develop a more interesting and complete western system restoration scenario including a few key measures for success:

- Start with normal operation stressed by heavy load due to hot weather.
- Enable modeling all necessary RAS and relays and mimicking practical implementation in the field.
- Introduce cascading outage events and blackout a large portion of WECC system, i.e., break the system into three areas and each area consists of over ten islands.
- Start three independent system restoration drills in parallel on three different DTS servers: one drill for each area.
- Merge three restoration drill simulation cases into one, and make it ready for a single large blackstart drill including all participants from three areas.
- Continue the drills for island synchronization to restore the whole system loads in 2 days.
- Review performance metrics and scorecard.

Such system restoration drill will ultimately enable operators to be familiar with the whole system event process and exercise their skills and knowledge in a high-fidelity simulation environment.

References

1. Federal Energy Regulatory Commission, "*Arizona-Southern California outages on September 8, 2011: Causes and recommendations,*" FERC and NERC Staff, Apr 2012. [Online]. Available: https://www.ferc.gov/legal/staff-reports/04-27-2012-ferc-nerc-report.pdf
2. Zhang, H., Wangen, B.. (2012, July 22–26). Implementation of a full western bulk system operational model for reliability monitoring. *Proceedings of IEEE PES General Meeting.*

3. Taylor, C. W., & Erickson, D. C. (1997, January). Recording and analyzing the July 2 cascading outage. *IEEE Computer Applications in Power, 10*(1), 26–30.

4. Kosterev, D. N., Taylor, C. W., & Mittelstadt, W. A. (1999, August). Model validation for the August 10, 1996 WSCC system outage. *IEEE Transactions on Power Systems, 14*(3), 967–979.

5. NERC Report, *"Review of Selected 1996 Electric System Disturbances in North America"*, August 2002. [Online]. Available: https://www.nerc.com/pa/rrm/ea/System%20 Disturbance%20Reports%20DL/1996SystemDisturbance.pdf

6. Ross, H. B., Zhu, N., Giri, J., & Kindel, B. (1994, August). An AGC implementation for system islanding and restoration conditions. *IEEE Transactions on Power Systems, 9*(3), 1399–1410.

7. Podmore, R., Giri, J. C., Gorenberg, M. P., Britton, J. P., & Peterson, N. N. (1982, January). *An advanced dispatcher training simulator. IEEE Transactions on Power Apparatus and Systems.*

8. DyLiacco, T. E., Enns, M. K., Schoeffler, J. D., Quada, J. J., Rona, D. L., Jurkoshek, C. W., & Anderyon, M. D. (1983, November). Considerations in developing and utilizing operator training simulators. *IEEE Transactions on Power Apparatus and Systems.*

9. Prais, M., Johnson, C., Bose, A., & Curtice, D. (1989, August). Operator training simulator component models. *IEEE Transactions on Power Systems, 4*(3), 1160–1166.

10. Prais, M., Zhang, G., Chen, Y., Bose, A., & Curtice, D. (1989, August). Operator training simulator: Algorithms and test results. *IEEE Transactions n Power Systems, 4*(3), 1154–1159.

11. Waight, J. G. et al. (1991, May) *An advanced transportable operator training simulator.* IEEE Proceedings of the PICA conference, pp. 164–170.

12. Vadari, S. V., Montstream, M. J., & Ross, H. B., Jr. (1995, November). An online dispatcher training simulator function for Real-time analysis and training. *IEEE Transactions on Power Systems, 10*(4), 1798–1804.

13. Wangen, B., Zhang, H. (2012, July 22–26). Monitoring for post-contingency system operating limit exceedance in the Western Interconnection. *Proceedings of IEEE PES General Meeting.*

14. Shin, B., Gibson, P. F., Wangen, B., & Perez, L. A. (2011, July 24–29). Wide-area DTS implementation in the Western Electricity. *IEEE PES General Meeting.*

15. U.S. – Canada Power System Outage Task Force, *Final Report on the August 14, 2003 Blackout in the United States and Canada: Causes and Recommendations*, Apr 2004, Chapter 7, p.105; Available: https://ferc.gov/industries/electric/indus-act/reliability/blackout/ch7-10.pdf

Chapter 6
Implementing Synchrophasor Applications for Grid Monitoring

Hongming Zhang

Organization Acronyms

AEP	American Electric Power
AESO	Alberta Electric System Operator
BC Hydro	British Columbia Hydro
BPA	Bonneville Power Administration
CAISO	California Independent System Operator
ComEd	Commonwealth Edison
ENTSOE	European Network of Transmission System Operators
EPG	Electric Power Group
ERCOT	Electricity Reliability Council of Texas
FERC	Federal Energy Regulatory Commission
FPL	Florida Power & Lighting
GE	General Electric
GPA	Grid Protection Alliance
ISO-NE	New England Independent System Operator
MISO	Middlewest Independent System Operator
NERC	North American Electric Reliability Corporation
NYISO	New York Independent System Operator
PAC	PacifiCorp
PG&E	Pacific Gas and Electric
PJM	PJM Interconnection
PowerTech	PowerTech Research Labs Inc.
RTE	French Transmission System Operator
SCE	Southern California Edison
SDG&E	San Diego Gas & Electric
SPP	Southwest Power Pool
TVA	Tennessee Valley Authority
UTK	The University of Tennessee, Knoxville

© Springer Nature Switzerland AG 2021
H. Zhang et al., *Advanced Power Applications for System Reliability Monitoring*, Power Systems, https://doi.org/10.1007/978-3-030-44544-7_6

WECC   Western Electricity Coordinating Council
WSU   Washington State University

NERC/WECC Acronyms

AGC   Automatic Generation Control
BA   Balancing authority
BES   Bulk Electric System
COI   California Oregon Intertie
EI   Eastern Interconnection
EMS   Energy Management System
GOP/GO   Generator Operator
HTTP   Hyper Text Transfer Protocol
HVDC   High Voltage Direct Current
ICCP   Inter-Control Center Communications Protocol
JSIS   WECC Joint Synchronized Information Subcommittee
MVWG   WECC Model Validation Work Group
PDC   Phasor Data Concentrator
PDCI   Pacific DC Intertie
PMU   Phasor Measurement Unit
PRSP   Peak Reliability Synchrophasor
RAS   Remedial Action Scheme
RC   Reliability Coordinator
RCO   Reliability Coordinator Office
RCSO   Reliability Coordinator System Operator
SCADA   Supervisory Control and Data Acquisition
SMS   NERC Synchronized Measurements Subcommittee
WI   Western Interconnection
WISP   Western Interconnection Synchrophasor Program
WSM   Wide Area System Model

Other Acronyms

DMO   Damping Monitor Offline
DSA   Dynamic Stability Assessment
eLSE   (EPG's) enhanced Linear State Estimator
ePMU   (PowerTech's) emulated PMU
FFDD   Fast Frequency-Domain Decomposition
FODSL   Forced Oscillation Detection and Source Location
FSSI   Fast Sub Space Identification
GEP   Gateway Exchange Protocol
GUID   Globally Unique ID
MAS   Modal Analysis Software
MVRA   Maximal Variance Ratio Algorithm
ODM   Oscillation Detection Module
OMS   Oscillation Monitor System
PCS   Peak RC Cloud Services

PI   (OSIsoft's) Plant Information
PI AF   (OSIsoft's) PI Asset Framework
PLI   Power-Load Imbalance
PMA   Pattern Mining Algorithm
PP   (GE's) PhasorPoint
RTDMS   (EPG's) Real-Time Dynamics Monitoring System
SSAT   (PowerTech's) Small Signal Analysis Tool
SMART   Synchronized Measurements and Advanced Real-Time Tools
SDK   System Development Kit
TSAT   (PowerTech's) Transient Stability Assessment Tool
WAMS   (GE's) Wide Area Management System

6.1   Historical Oscillation Events

During the last decades, a number of large-scale inter-area oscillation events occurred in major interconnections of North America and other continents:

6.1.1   WSCC August 10, 1996, Disturbance

On August 10 1996, at 15:48 PDT, a sequence of major transmission outages and bulk generation tripping eventually resulted in a poorly damped inter-area oscillation developed across the Western Electricity Coordinating Council (WECC) system. Generation in the Upper and Lower Columbia River hydro projects picked up much of the lost generation, further stressing transfers across the system. Eventually the oscillation became negatively damped, and the PDCI began oscillating in response to the poor AC voltage. As the oscillations reached about 1000 MW and 60 kV peak-to-peak at Malin 500 kV substation, the voltage collapsed (see Fig. 6.1).

This severed the ties between the Pacific Northwest and California and eventually caused the other ties between California and its neighbors to also open due to low voltage and out-of-step protection. This disturbance broke up the WECC system into four islands and resulted in loss of ~30 GW load to 7.5 million customers, who lost power for periods ranging from several minutes to 6 h.

6.1.2   WECC November 29, 2005, Oscillation Event

On November 29, 2005, a failed control valve at Nova Joffre cogeneration facility in Alberta, Canada, injected a 20 MW peak-to-peak forced oscillation occur at 0.27 Hz into the system. This forced oscillation due to the failure at the plant excited

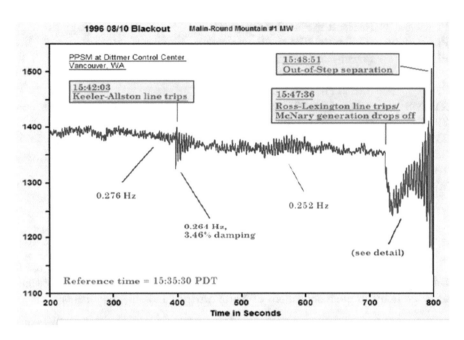

Fig. 6.1 August 10, 1996, WECC system oscillation. (Source: WECC)

the North-South system mode, resulting in 200 MW peak-to-peak oscillations on the COI (see Fig. 6.2).

Fortunately, the oscillation persisted for only about 5 min until the steam supply diminished. There was no severe consequence for loss of generations and loads, but this event is useful in illustrating the importance of identifying system modal characteristics to understand how unexpected forced oscillations can interact with the system modes.

6.1.3 January 11 2019, EI Oscillation Event

This oscillation was observed across entire Eastern Interconnection from 08:44:41 UTC (03:44:41 EDT) to 09:02:23 UTC (04:02:23 EDT), January 11, 2019. The oscillation frequency of 0.25 Hz aligns with inter-area mode frequency across EI. Power swings around Florida of 200 MW, around ISO-NE of 50 MW for about 15 min (see Fig. 6.3). FPL, TVA, AEP, Duke Energy, NE-ISO, ATC, ComEd, etc. observed significant power swing on their EMS systems during the event. RCs identified oscillation on PMU data and notified RC Hotline. A repowered combined cycle plant in FPL footprint was identified as the oscillation source. The oscillation was removed after the plant was manually tripped by operator. UTK provided videos of oscillation event and GOs noticed oscillation on power plants across EI. Per the investigation by NERC and relevant entities, the plant injected 0.25 Hz forced oscillation because of failure of Power-Load Imbalance (PLI) controls [1].

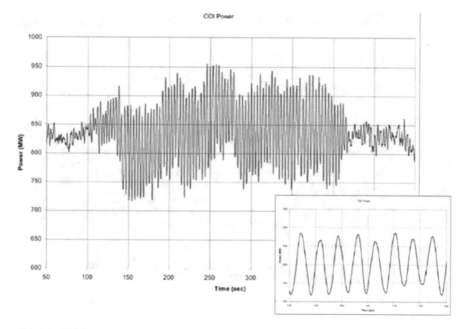

Fig. 6.2 COI flow during November 2005 oscillation. (Source: BPA)

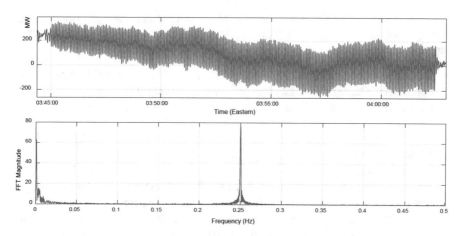

Fig. 6.3 SCADA data – line flow oscillation and dominant F. (Source: AEP)

6.1.4   Inter-area Oscillations in Continental Europe

On December 1, 2016, at 11:18 AM, a nearly non-damped oscillation at 0.22 Hz was developed when unexpected opening of the circuit breaker (without fault) at Cantegrit substation => Argia-Cantegrit tripping. Many Transmission System Operators (TSOs) within ENTSOE: European Electricity Network across 34 inter-connected countries observed this oscillation. Prior to the incidence, there were

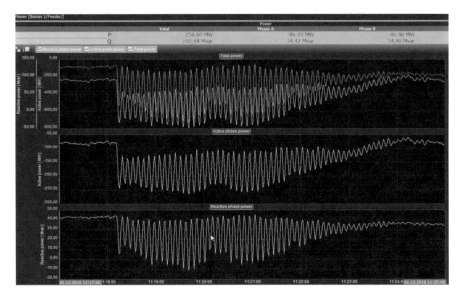

Fig. 6.4 France-Spain HVDC flows where the oscillations were observed. (Source: RTE)

high imports from Spain to France at ~2300 MW and power exchange from Portugal to Spain at 3000 MW. The following event analysis was presented by French TSO (RTE) [2]:

- The event started at 11:18 am, unexpected opening of the circuit breaker (without fault) at Cantegrit substation causing Argia-Cantegrit to be tripped. The system appeared normal in static situation, but low-frequency power oscillation with near zero damping started (see Fig. 6.4).
- At 11:21 am, the Spain to France schedule was reduced from 2250 MW to 1000 MW to restore [N-1] security successfully.
- A similar oscillation event occurred on March 10, 2017, at 11:17 am (see Fig. 6.5). Both oscillation events indicate the high import flow from Spain to France affecting the damping of inter-area mode at 0.22 Hz.
- Note that prompt coordination between the TSOs played a vital role in the mitigation of the transient faults which identify the root cause of the oscillation and mitigate the issue timely.
- The observed inter-area mode (East-Center-West mode) was accurately reproduced by the nonlinear time domain simulation due the large-scale ENTSOE model.

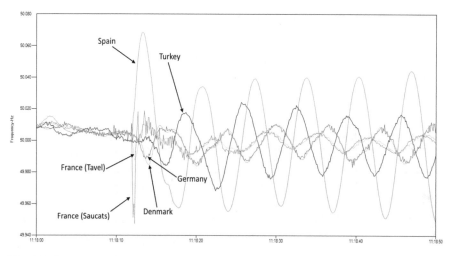

Fig. 6.5 Frequency variations during April 10, 2017 oscillation. (Source: RTE)

6.1.5   Observations and Lessons Learned

A number of large-scale outages and near miss events have included oscillations as either a cause or effect of the event. These events demonstrate that coincidence and combination of different factors can affect the system stability. Each factor may not normally be critical itself, but in particular scenarios the combined effect decreased the general system damping.

RTE's study shows the damping of the West-Center-East mode at 0.22 Hz decreases, while active power import from Spain to France increases (see the table below). The scenario becomes worse under higher penetration of renewable energy across ENTSOE.

Active power exchanges France to Spain	Damping of the West-Center-East mode (~0.22 Hz)
2800 MW	17,09%
1400 MW	15,64%
−850 MW	8,47%
−2800 MW	5,68%

In the WECC system, it's observed that the phenomena of the damping of North-South inter-area modes decline, while the system inertia (real time) decreases [see the table below]. Instantaneous renewable generation penetration has increased steadily over the last decade. From below tables, one can see CAISO system exceeded over 50% renewable penetration continuously for more than 20 min on October 8, 2017. Peak RC recorded a 23.66% of instantaneous renewable penetration in April 17, 2018. It's projected that a combination of lower system inertia and

Table 6.1 WECC high renewable penetration case

Date/time	Total (MW) renewable	Solar	Wind	Total gen	Renewable % of WECC generation
4/17/18 14:05	22,055	11598.81	10456.2	93226.27	23.66%

Source: Peak Reliability

Table 6.2 CAISO high renewable penetration cases

Date/time	Total renewable generation	Solar gen	Wind gen	Total gen	Renewable % of CAISO total gen
08-Oct-17 10:25:00	10632.80	8125.51	2507.29	20244.92	52.52%
08-Oct-17 10:30:00	10732.46	8162.15	2570.32	20251.69	53.00%
08-Oct-17 10:35:00	10672.15	8098.73	2573.43	20133.65	53.01%
08-Oct-17 10:40:00	10566.27	7974.30	2591.97	20204.22	52.30%
08-Oct-17 10:45:00	10596.76	7974.67	2622.10	20276.08	52.26%

Source: Peak Reliability

higher power transfer across COI and PDCI will worsen damping of western system inter-area modes (Tables 6.1 and 6.2).

Some factors such as HVDC control mode, unit control setting, and PSS settings across the impacted power systems for oscillation modes are significant.

A large-scale oscillation could disturb multiple RCs' footprints under an interconnection such as EI, WI, ENTSOE, etc. Prompt coordination between the TSOs or RCSOs plays a vital role in the mitigation of the inter-areas' oscillation issues.

Dynamic evaluations on system behavior are becoming more and more necessary. A full system model, initialized from pre-disturbance SE basecase solution, is essential for system model validation and post-event analysis.

6.2   Western Interconnection Synchrophasor Projects

6.2.1   Introduction

The oscillatory characteristics of power systems due to cyclic load variations were first uncovered in the mid-1960s [1]. After the August 10, 1996, blackout which involved a large oscillation, Bonneville Power Administration (BPA) endeavored to develop Phasor Measurement Unit (PMU)-based software to detect and monitor system oscillation events in real time [3–5]. Starting from 2010, WECC initiated the Western Interconnection Synchrophasor Program (WISP), co-funded by US Department of Energy (DOE) and 19 industry partners, to deploy Synchrophasor technology throughout the US portion of the Western Interconnection. The project completed in 2014 through installing or upgrading 584 PMUs and 77 phasor data concentrators (PDCs), among which 393 PMUs and 57 PDCs used WISP funding, respectively. WISP developed a wide-area communication network in Peak RC's control rooms (formerly WECC RC before the bifurcation in February 2014), to support PMU data transfer, information technology infrastructure, and advanced grid analytics software.

While the WISP project was completed in 2014, most of new synchrophasor applications were deployed in Peak RC's control rooms yet could not realize operational benefits, primarily due to poor PMU data quality and lack of oscillation detection tools baselining and operational use cases. In October 2014, Peak Reliability Synchrophasor Program (PRSP, and widely viewed as WISP 2.0) was funded by DOE and nine cost sharing entities. Subsequently, Peak RC collaborated with the partners to deliver PMU Registry, validate and improve PMU data quality, implement new oscillation detection and source locating tools, and develop operational use cases.

After the PRSP project ended in 2016, Peak RC strived to develop a framework of online oscillation detection, baselining and source locating suitable for control room solution. Together, the tools in the framework increase RCSO's visibilities into bulk power system conditions in near-real time, enable earlier detection of problems that threaten grid stability or even cause cascading outages, and facilitate sharing of information with neighboring control areas.

6.2.2   PMU Registration Function

6.2.2.1   Introduction

Success of the WISP project is based on accurate, reliable, and efficient collection of synchrophasor data measurements from geographically diverse stations own or operated by WECC's partner entities. As the WISP synchrophasor system is built and partner's employ varying equipment types, topologies, standards, and conventions within their own infrastructure, providing a central repository where knowledge of each partner's unique capabilities and configuration can be maintained in a common form will be critical to the project's success. Synchrophasor measurement data can then be used as an integrated set of information with all parties knowing from, where, and how the measurements are taken.

In its general form, the WISP PMU Registry, or just Registry, is an Asset Management repository for information about synchrophasor measurements and the associated assets and entities used to deliver and manage those signals. The PMU Registry does not contain the synchrophasor signal or measurement data itself just the metadata.

The types of metadata provided by the PMU Registry include device installation status, equipment type (manufacturer and model), physical location, signal ranges, units, and the point on the electrical system where measurements are made. The

Registry also contains additional information such as device IDs and tags that would more specifically identify or relate a one asset to another.

A primary function of the Registry is to provide a catalog of available synchrophasor measurements within the Western Interconnection and act as a type of Name Server. A basis for this functionality will be the assignment of globally unique key to each asset and measurement. Additionally, the Registry will also support the navigation and selection and requests for data from the synchrophasor archive.

6.2.2.2   System Architecture and Basic Components

Implementation of the WISP PMU Registry begins from a perspective that any of the items managed within the registry can be generalized as an asset. Each asset then has common attributes and relationships as well as attributes and relationships specific to the asset type. Using this generalized approach will allow the registry to grow and adapt with the needs of the organization while also providing a simpler more maintainable system.

Peak RC designed and developed the PMU Registry function based on the system architecture shown in Fig. 6.6. The tool became functional for the WISP member entities to use in 2015.

Initially the PMU Registry will support only Measurement Device Assets and basic Electrical Assets but will be able to include other asset types as needed. The primary method of interacting with the registry will be with a self-service web application, but it will also provide system interfaces through a service-oriented architecture or Web services.

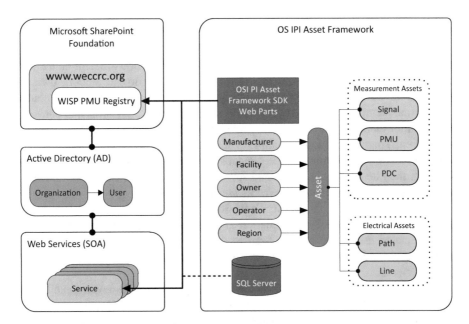

Fig. 6.6 PMU registry architecture with OSI PI asset framework. (Source: Peak Reliability)

With the heavy work of template configuration and data storage handled by PI AF Repository and its System Development Kit (SDK), the bulk of development for the Registry will involve creating the user interface and Web service end points to interact with that repository.

6.2.2.2.1 OSI PI Asset Framework

OSIsoft's PI Asset Framework (PI AF) is a flexible product that will enable WISP to define a consistent representation of Registry Assets. PI AF has a built-in user interface and tools allowing Asset Templates to be created and edited efficiently based on the specific attributes of the asset type. AF also allows creating relationships or organizational structure such as categories, hierarchical or connected models, and associate specific measurement data with the elements created from the templates. AF also supports calculations, rules, and formatting to assist in the presentation of the associated data.

Once defined, individual assets can be organized grouped and regrouped, without limitation, creating meaningful and effective representations of the assets and their relationships. AF can also provide direct links to the real time and time series historical data maintained in the historical trending archives.

6.2.2.2.2 Microsoft SharePoint

Peak RC used Microsoft SharePoint services as the hosting platform for their partner facing presence, and the Registry application user interface will be hosted as a sub-site. Users can access the Registry through the familiar and intuitive look of SharePoint, and it can be deployed rapidly with less disruption to current work and with less demand for training and support.

6.2.2.2.3 Active Directory: Security and Data Access Control

PMU Registry will be deployed as part of PeakRC.ORG and leverage AD and SharePoint security. Specific users, groups, and roles will be defined that will ensure simple integration with the existing host website.

6.2.2.2.4 Web Services

Web services will provide interfaces designed to support interoperable machine-to-machine interaction over the local or wide area network. The interface specifications describe, in a machine interpretable format (specifically WSDL), the available functionality and specific methods available from that service or end point. Web services typically utilize an HTTP transport with XML serialization.

6.2.2.2.5 Registry Metadata Architecture

Not specifically shown in the diagram, but a key component of the Registry is the metadata that is maintained within and alongside PI AF.

6.2.2.2.6 Asset Tag Naming Conventions

Stations, equipment, signals, and measurements will be visibly identified by what is referred to as a "tag." The origin of the tag comes from a physical metal tag or placard fixed to equipment and stations used to uniquely identify them within the station or entity in which it resided.

Bringing organizational entities together with disparate naming conventions and opportunities for name or "tag" duplication across the entities, makes the ability for the Registry to function as Name Server a bit more complex, but also increases its overall value. The Registry will accommodate whatever naming convention or duplicate tags that may exist across the partner entities by assigning a specific and unique tag to each asset created. The naming convention used by Peak RC to assign these tags is described below. Additionally, each asset will also maintain Globally Unique ID (GUID) that will insure the asset global uniqueness.

6.2.3 Asset Tag Management

The Registry Metadata Architecture defines various tags for assets, which are associated with one or more partner entities. Based on the specific asset tag type, certain rules for "uniqueness" must be enforced. The following table defines these rules [6] (Table 6.3).

Table 6.3 Peak RC Office (RCO) tag types

Tag/ID attribute	Element usage	Uniqueness
RCO tag	All	Unique within the **region**
RCO signal tag	Signal	Unique within the **region**
ICCP tag	Signal	Unique within the **region**
Device tag	Signal	Unique within the owning **PMU**
Device tag	All	Unique within the owning **entity**
Signal tag	Signal	Unique within the owning **entity**
Owner tag	All	Unique within the owning **entity**
Other tag	All	Not enforced
Protocol ID	PDC	*NA*
	PMU	

6.2.3.1 PMU-Related Tags

The phasor measurement or "signal" provide by a PMU is a complex number whose components are magnitude and phase angle.

In normal use, this measurement is managed and viewed in its compound form. Other systems such as SCADA or historical archiving may not support the complex type and may require the components be mathematically extracted and represented individually. In these cases, individual components are represented with an additional tag segment, such as

- PSM – Positive sequence magnitude
- PSA – Positive sequence phase angle

6.2.3.2 RCO Tag Naming Standards [6]

1. The elements shown in the table below are combined to form a specific tag.
2. A decimal point or "dot" is used to separate the elements.
3. Only upper-case alphanumeric characters (0–9, A–Z) and underscore "_."
4. There are inherent system character length limitations based on the tag's usage and the number of elements being used within a specific system.
5. Duplicate station IDs have either the owner entity abbreviation included or a numerator added as a suffix.
6. Consistency and structure are the goal; however readability within a limited number of display characters may impact the actual tag name (Table 6.4).

6.2.3.3 Tag Naming Examples

- **VINCENT.PMU.PMU123**

  - VINCENT.LN.VINC_MIDW_1500.SP.F
  - VINCENT.LN.VINC_MIDW_1500.SP.DF
  - VINCENT.BUS.VINC_MIDW_1500.SP.PSEQ.V
  - VINCENT.LN.VINC_MIDW_1500.SP.PSEQ.V
  - VINCENT.LN.VINC_MIDW_2500.SP.PSEQ.A
  - VINCENT.LN.VINC_MIDW_3500.SP.PSEQ.V
  - VINCENT.LN.VINC_MIDW_1500.SP.PSEQ.A
  - VINCENT.LN.VINC_MIDW_2500.SP.PSEQ.V
  - VINCENT.LN.VINC_MIDW_3500.SP.PSEQ.A

- **MIDWAY1.PMU.PMUXYZ**

  - MIDWAY1.LN.VINC_MIDW_1500.F
  - MIDWAY1.LN.VINC_MIDW_1500.DF
  - MIDWAY1.LN.VINC_MIDW_1500.SP.PSEQ.V
  - MIDWAY1.LN.VINC_MIDW_2500.SP.PSEQ.A

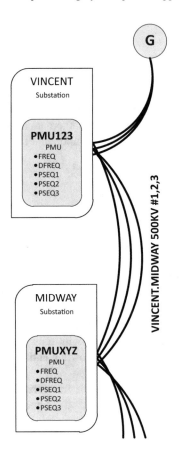

Table 6.4 Tag naming convention definition

Element	Length	Description
Station ID	8	Physical location of the device, typically a substation
Device type	4	Type of device being referenced.
		Examples:
		LN – line
		BUS – system bus
		CP – reactive inductor or capacitor
		BRKR – breaker
		XFMR – transformer
		RLAY – relays
		PMU – Phasor Measurement Units
Device Id	14	The format and actual content vary depending on the device type but often imply multiple properties within the ID
		Example:
		VINC_MIDW_2500 implies the second 500 kV line between the Vincent and Midway substations
Measurement type	4	**Example:**
		MEA – analog
		MEP – digital status point
		F – frequency
		DF – delta frequency or rate of change ($\Delta f/ \Delta t$)
		PSP – PMU status point
		SP – phasor s
		PSEQ – positive sequence
		NSEQ negative sequence
		ZSEQ – zero sequence
		SPA, SPB, SPC – phase A, B, or C
		Phasor components
		SPM -magnitude (component of phasor)
		SPA – angle (component of phasor)
Measurement	6	The measurement tag often implies unit of measurement (UOM)
		Example:
		V – voltage (kV)
		A – amps (I or current)
		W – watts (kW, MW)
		Hz – hertz (frequency)

6.2.4 System Design (Fig. 6.7)

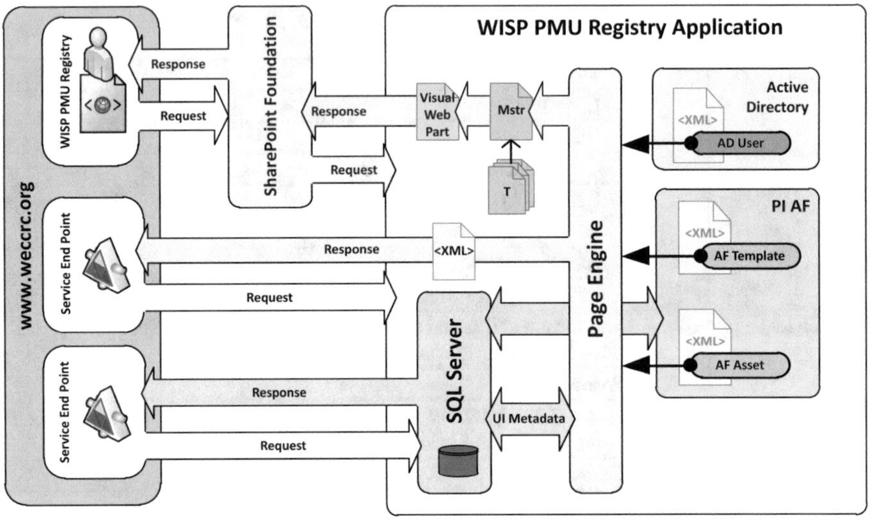

Fig. 6.7 WISP PMU Registry system design. (Source: Peak Reliability)

6.2.5 Web User Interface

The Registry was deployed as a SharePoint Line of business application. Leveraging its capabilities as much as possible will lead to more efficient, stable, and supportable application. The Registry Client UI Navigation is illustrated in Fig. 6.8.

There are emerging technologies that may be employed in building the Page Engine such as Microsoft's ASP.NET Dynamic Data, a tool which dynamically creates a simple but fairly complete website to maintain any database using an Entity Framework Data Model. ASP.NET Dynamic Data lets you create extensible data-driven Web applications by inferring at run time the appearance and behavior of data fields and by deriving UI behavior from it.

The Registry Application Page Engine will follow those same patterns for viewing and editing data contained in the PI-AF Repository, the difference being the data or objects will come from the PI-AF Model rather than an Entity Framework Data Model. To accomplish this, wrapper library will be created to expose the PI-AF PISystem.AFDatabase object as a DynamicData.ModelProvider.

Fig. 6.8 WISP PMU Registry UI navigation. (Source: Peak Reliability)

6.2.6 UI Meta Data

Because the PI-AF data repository does not provide the granularity needed to define the characteristics and roles of each asset attribute within the user interface, a separate UI data model will be built to provide those details. The details of the UI data model can be found in User Interface Metadata section of this document.

6.2.7 System Interfaces

Web service interfaces and end points will be defined and built as needed by the WASA Applications. The actual WASA vendor was not selected and no current requirements exist.

6.2.8 Security

User access authentication and control to the registry will be provided by the hosting SharePoint Site Services which is fully integrated with active directory. Additional details of the can be found in the WECC Application Technology Standards. Once a user is authenticated and associated with active directory, the security context of the user will be maintained throughout the session to determine access rights and capabilities. PI-AF is fully integrated with active directory and will be used to enforce asset access.

6.2.9   PMU Registry Use Cases

Since 2015, Peak RC Registry function has been extensively used by internal customers and WISP participating entities to manage PMU asset information. There were a number of use cases developed for the Registry system, including, but not limited to:

- Create a Region Asset
- Create an Entity Asset
- Create a Path Asset
- Create a Manufacturer Asset
- Create a Manufacturer Model Asset
- Create a Station Asset
- Create a PDC Asset
- Create/update/remove a PMU Asset (associated with an existing PDC and/or station)
- Create a Signal Asset (associate with an existing PMU and/or Path)
- Add/disable/delete Entity User Account
- Clone a PDC Asset with or without PMUs
- Clone a PMU Asset with Signals

6.2.10   WISP PMU Registry Record Export

As part of Peak RC wind down process, the WISP PMU Registry retired by end of 2019. Therefore, Peak RC developed a PMU and Signal Template to export the PMU records from the Registry to a Master Excel spreadsheet. The Template mimics PMU Registry hierarchy and data structure. It also retains the TREE structure as one can see in Fig. 6.9. The Template will be used WECC Entities to manage PMU Registry information in the Excel spreadsheet.

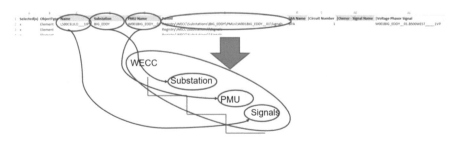

Fig. 6.9 Example of WSIP PMU and signal template. (Source: Peak Reliability)

6.3 Real-Time Oscillation Detection and Monitoring Frameworks

6.3.1 A Framework of Online Modal Analysis Software (MAS)/Mode Meters

6.3.1.1 Introduction

Online estimation on system oscillation modes has been of high interest to the industry [3–5, 7–9]. The MAS tool deployed in BPA's control center was the first implementation of an online oscillation detection software for operational use [10]. As a primary component of the MAS, Mode Meter is designed to continuously track known western system modes. Mode Meter delivers a "mode estimate" as its result, including estimates of the frequency, damping, energy, and shape of the system mode. Another key component of MAS is Oscillation Detection Module (ODM), which is designed to detect various oscillations classified in four bands bounded by 0.01 Hz, 0.15 Hz, 1.0 Hz, 5.0 Hz, and 14 Hz [10]. Recent research work to improve the MAS algorithms were published to detect periodic oscillations and estimate electromechanical modes and forced oscillation simultaneously [11–13].

6.3.1.2 Integrating Montana Tech MAS in GE Software Platform

Through the WISP and PRSP grants, Peak RC successfully integrated MAS engine into GE's Wide Area Management System (WAMS)-PhasorPoint and were able to pass real-time solution results of the MAS and intrinsic WAMS applications to WSM-EMS system in early 2016. The integration framework and application architecture are illustrated on Fig. 6.10 [14].

Below are main components of the architecture:

- Data acquisition. By July 2019, Peak RC was able to receive live data stream of 435 PMUs from individual PDC of 16 entities via Harris network and manages incoming raw PMU data in GPA's openPDC.
- Measurement based applications. openPDC sends synchronized PMU at 30 samples per second to PhasorPoint to perform system dynamics and stability monitoring in several areas: system oscillation monitoring and detection by MAS engine:

  - Islanding detection and system re-synchronization/black start
  - Composite event analysis
  - Wide angle separation, WECC paths flow calculation

- MAS solution along with other PhasorPoint application results are updated every 10 sec and are transferred to GE/EMS GSA via Web service. GSA has an interface with SCADA for alarming of violations.

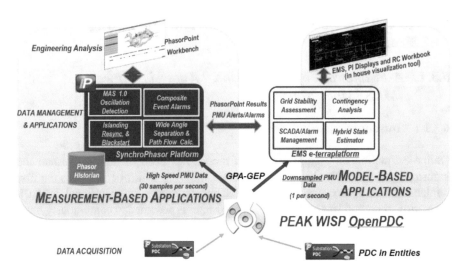

Fig. 6.10 System architecture of GE-EMS and WAMS integration. (Source: Peak Reliability)

- Model-based applications. Peak RC uses GPA's Gateway Exchange Protocol (GEP) application to send downsampled PMU (one sample per sec.) to GE EMS/SCADA system at first and then to get into a hybrid SE for solving at 1-minute interval.
- The PMU application results available in GSA, SCADA, and SE are stored in EMS archive cases, PI historian, and other in-house data repository tools, respectively.
- By implementing the above framework, Peak RC integrated MAS's both ODM and Mode Meter in PhasorPoint, as shown in Fig. 6.11. PMU data are archived in both PhasorPoint Archive and OSI PI Phasor Historian.

Peak RC was able to transfer MAS results to the GSA application for centralized visualization and alarming in EMS. Peak RC also enhanced GSA in 2017 to enable storing all MAS results in PI for event analysis and baselining.

Figures 6.12 and 6.13 show how Mode Meter results are separately displayed in the GSA and PI Processbook displays.

6.3.2   Framework of Forced Oscillation Detection and Source Locating

As of September 2019, nearly 9000 substations were modeled in the WSM. Peak RC received live data streams of 440 PMUs located at nearly 200 substations, but only a dozen of the stations are generation power plants. Limited PMU visibility on power plants is one of primary challenges for the MAS/ODM to locate source units initiating a forced oscillation.

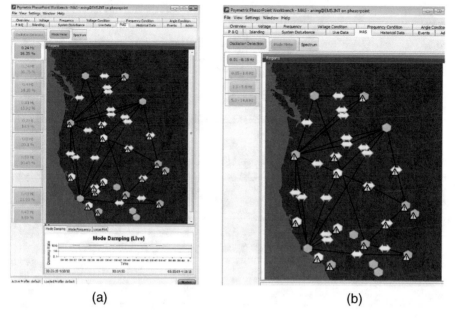

Fig. 6.11 GE PhasorPoint integrated with the MAS engine. (**a**) ODM results. (**b**) Mode meter results. (Source: Peak Reliability)

MODE ID	MODE FREQUENCY (Hz)	DAMPING RATIO (%)	TIME
0			14-Jun-2018 13:53:59
1			14-Jun-2018 13:53:59
2	0.45 Hz	2.4 %	14-Jun-2018 13:53:59
3	0.45 Hz	4.0 %	14-Jun-2018 13:53:59

Fig. 6.12 MAS-Mode Meter results displayed in EMS/GSA. (Source: Peak Reliability)

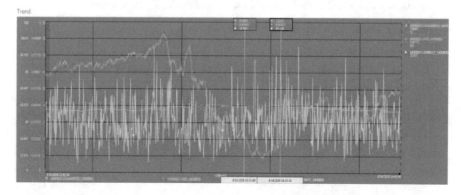

Fig. 6.13 PI trend on MAS-Mode Meter results in EMS/GSA. (Source: Peak Reliability)

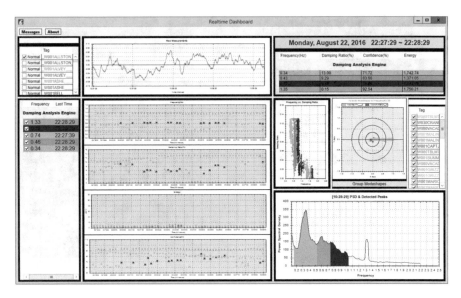

Fig. 6.14 WSU OMS tool dashboard view. (Source: Peak Reliability and Washington State University)

On the other hand, the ICCP measurements received in the WSM cover more than 2800 generators, including both real and reactive power outputs on each measured generator. The question is whether we can leverage both PMU data and ICCP measurements from generating units for the FODSL [15].

Since 2015 Peak RC has worked with Washington State University (WSU) for development and implementation of a new of FODSL framework using both PMU and ICCP measurements. The framework embraces three main components:

1. Enhanced version of WSU's Fast Frequency-Domain Decomposition (FFDD) online engine to estimate dominant oscillations
2. The FODSL to check FFDD output results for forced oscillation detection and source locating if a forced oscillation is detected
3. UI display to visualize and replay FODSL results in PI Processbook

The FFDD engine has several merits:

1. Use all available PMU signals.
2. Automatically detected dominant modes (frequency, damping ratio, mode shape, energy, confidence level).
3. 1-minute moving window which yields a shorter reaction time to system changes.
4. It is a universal oscillation monitoring tool which could also catch forced oscillation modes.

The tool is updated every 10 s and provides user a dashboard shown in Fig. 6.14 to interpret the results quickly.

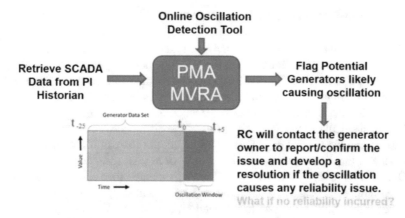

Fig. 6.15 Conceptual workflow of the FODSL. (Source: Peak Reliability)

The FODSL starts with post-processing FFDD estimation results. It runs every 1 min, uses 5–10 min buffer of FFDD results, and groups the oscillation results by oscillation frequency. FFDD by itself does not indicate whether the oscillation modes it detected are forced oscillation or not. Hereafter, FODSL will have to bifurcate the results by applying multiple cross-check rules, such as:

- Sustained oscillations until source mechanism mitigated
- Near zero damping ratio
- Mostly fixed oscillation frequency
- High oscillation energy on some measurements
- High estimation confidence if applicable

The source locating process is triggered upon detection of a new forced oscillation or updated for an ongoing forced oscillation. The only information needing to be passed from the detection module to the source locating module is oscillation start and end time. The end time would be the current time if the oscillation continues. It retrieves all available generator MW and MVAR SCADA values based on the start time less than the ambient window length and end time of a forced oscillation event. A conceptual workflow of FODSL is given on Fig. 6.15, where PMA and MVRA are two original algorithms jointly developed by Peak RC and WSU to locate source units [13].

6.3.3   The FODSL Algorithms and Implementation

Peak RC started to implement FODSL in its development and testing environments since 2016. FODSL embraces three main modules as follows:

6.3.3.1 Pattern Mining Algorithm (PMA)

Once a new oscillation event is detected by FFDD or MAS with specific event start time and end time, one should input the SCADA data of generators. One would apply a 25-point median filter and subtract the median filtered data from the raw data for detrending. The threshold (denoted the 3σ) is set to be three times the standard deviation of the detrended data in the ambient window [15].

One then calculates the absolute values of the differences between the raw measurements and the filtered data. One would reject the channel if the maximal absolute value of the differences is less than 1 MW or MVAR (user configurable).

One counts NUM_{osc_i} and NUM_{amb_i}. The PMA method then counts the number of the high-amplitude peaks in the raw measurements whose detrended values are outside the 3σ threshold in the oscillation window, denoted NUM_{osc}, and in the ambient window, denoted NUM_{amb}. In this context, the amplitude of the peaks is ignored per se. The ranking index of each channel is then formulated as

$$K_{PMA\_i} = \frac{NUM_{osc\_i}}{Length_{osc}} - \frac{NUM_{amb\_i}}{Length_{amb}}, \quad i = 1, 2, \ldots, n, \tag{6.1}$$

where $Length_{osc}$ and $Length_{amb}$ represent the lengths (total number of samples) of the oscillation and the ambient windows, respectively.

In order to determine relative ranking between the generators, the main steps of the pattern-mining algorithm are summarized as below:

1. Input SCADA data of generators and the oscillation event time as detected by FFDD using PMU data.
2. Data sanity check.
3. Apply the median filter, and subtract the median filtered data from the raw data for detrending.
4. Calculate the absolute values of the differences between the raw measurements and the filtered data.
5. Reject the channel if the maximal absolute value of the differences is less than 1 MW or MVAR.
6. Count NUM_{osc_i} and NUM_{amb_i}.
7. Compute the ranking index K_{PMA_i} based on (1).
8. Apply steps 2–7 for the rest of channels.
9. Select top 3 channels based on the ranking index.
10. Inspect the MW outputs of the possible oscillation sources for manual verification.

6.3.3.2 Maximal Variance Ratio Algorithm (MVRA)

This method is similar to the PMA in terms of workflow. But the MVRA uses a third order bandpass filter to detrend SCADA data at the 0.1 Hz sampling rate (i.e., 10 s SCADA update rate), and the corner frequencies are set to be 0.005 Hz and 0.035 Hz for the bandpass filter.

In addition, two key factors are considered when calculating the new ranking index K_{MVRA} in this approach. One is the number of times the data values cross their mean value within the oscillation window N_{osc}, which indicates how much the MW data is showing sustained oscillations. The other one is the average standard deviation of the SCADA signal, which is a measure of the oscillation amplitude [15].

In order to accommodate the slow sampling rate of the SCADA data, MVRA first computes the standard deviation σ_1 of a defined analysis window, say 30 samples (5 min). MVRA then moves the analysis window along the time axis with a fix step, say 6 samples (1 min). Next, one would calculate the standard deviation σ_2 of the new window. One should keep moving the analysis window and computing the standard deviation σ_i ($i = 3, 4, \ldots$) until the end of the data. The moving standard deviations are calculated over both the initial ambient window and the oscillation window, and the averages of the moving standard deviations for the two windows are denoted as STD_{amb} and STD_{osc}, respectively.

MVRA computes the ranking index for each signal as

$$K_{MVRA\_i} = N_{osc\_i} \frac{STD_{osc\_i}}{STD_{amb\_i}}, \quad i = 1, 2, \ldots, n. \tag{6.2}$$

The main steps of the maximal variance ratio algorithm are summarized below.

1. One inputs the SCADA data of generators and the oscillation event time from FFDD analysis of PMU data.
2. Data sanity check.
3. Calculate the average of the moving standard deviations for the initial ambient window. Reject the channel if the maximal difference of the data during the window is less than a preset multiple of the average standard deviation (Filter 1).
4. Calculate the average of the moving standard deviations for the oscillation window. Reject the channel if this average standard deviation is less than the minimum oscillation threshold (Filter 2).
5. Detrend using the bandpass filter.
6. Reject the channel if the number of spikes inside the oscillation window is no less than the spike count threshold (Filter 3).
7. Apply the clipping limiter for the oscillation window.
8. Count the number of times the data values cross their mean value within the oscillation window.
9. Reject the channel if $Nosc\_i$ is less than a preset factor of the number of samples inside the oscillation window (Filter 4).
10. Recalculate the moving standard deviations over both the initial ambient window and the oscillation window for the filtered data, and compute the average $STDamb\_i$ and $STDosc\_i$ for the two windows, respectively.
11. Compute the ranking index $KMVRA\_i$ according to (2).
12. Apply step 2–11 for the rest of channels.
13. Select top 3 channels based on the ranking index.

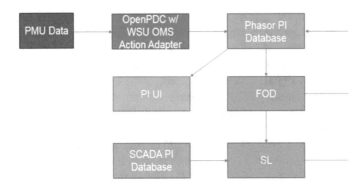

Fig. 6.16 Data flow between FODSL modules

14. Inspect the MW outputs of the possible oscillation sources for manual verification.

Before sharing source locating results of PMA and MVRA with RCSO and the owners of the units, the Peak RC Network Applications engineers are required to inspect the MW outputs of the possible oscillation sources for manual verification. Typically, the source location results are most credible, while both algorithms report the same candidate units.

6.3.3.3   The FODSL Work Flow

Overall data flow of FDOSL modules is given in Fig. 6.16. PMU data from all entities are streamed into the PEAR RC's openPDC, which also has FFDD running as a customized adapter within it. FFDD modal estimate results are saved into PI database. The oscillation detection module, FOD, grabs the results from PI and issues a triggering signal with start time to the source locating module, SL. Upon receiving the signal, SL grabs relevant SCADA data and runs the analysis locating the likely oscillation sources. Both FOD and SL write their results back into PI database. Finally, PI user interface retrieves FODSL results from PI database for visualization.

6.3.3.4   RT-FODSL UI Display

The FODSL UI display is built in PI Processbook to visualize up to four oscillation mode estimates. For example, a few most suspected units identified by PMA and MVAR algorithms are highlighted by red and white solid circles on Fig. 6.17, respectively.

For a given forced oscillation, the size of circles is normalized in proportion to the ranking index of each suspected unit by PMA and MVAR. Once a new forced

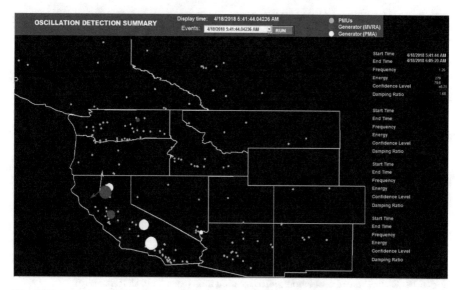

Fig. 6.17   Real-time FODSL tool UI display. (Source: Peak Reliability)

oscillation notification is received, the Peak RC Network Application engineers will review the source locations calculated by the online tool.

It must be emphasized before contacting RCSO and entities for an oscillation event, the engineers need to validate online source location results by running offline source locating tool and trending historical generations of the suspected units for an extended time window. The offline study tool menu is shown as Fig. 6.18.

6.3.4   Framework of North-South Modes Monitoring

Among five inter-area modes, monitoring North-South (N-S) modes "A" and "B" are primary concerns from a reliability perspective. These two modes influenced multiple system disturbances, including the August 10, 1996, blackout. Major hurdles for monitoring N-S modes by RCSOs are:

1. Mode Meter generates false alarms when any input PMU channel goes wrong.
2. Mode Meter results could be biased by a forced oscillation that has a similar oscillation frequency to one of N-S modes.
3. Lack effective means for RCSOs to evaluate low damping risk and develop mitigation plan for N-S Modes on particular contingencies in real time.

Peak RC implemented an N-S mode monitoring and alarming diagram in PI Processbook, as shown in Fig. 6.19.

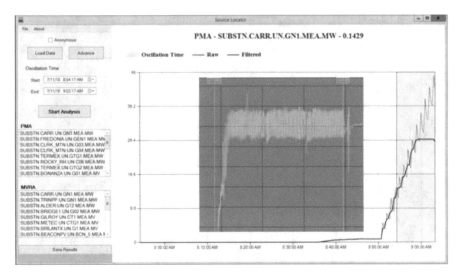

Fig. 6.18 Offline forced oscillation source locating tool. (Source: Peak Reliability)

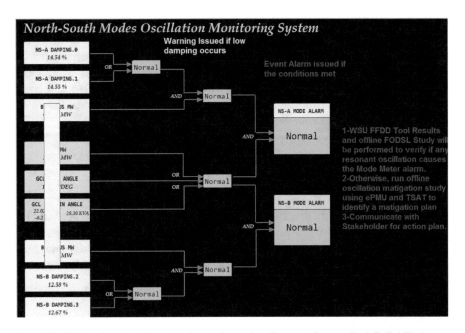

Fig. 6.19 N-S modes A and B monitoring and alarming diagram. (Source: Peak Reliability)

It consists of the following execution steps:

1. Retrieve latest Mode Meter results on N-S modes A and B from EMS PI. Each mode uses two separate channels to calculate primary and backup mode results. If the damping ratio of N-S "A" or N-S "B" mode is greater than x% or y%, respectively, both modes are within normal, and no further action is needed. Otherwise, proceed onto step 2.
2. Retrieve DS-PMU based MW flows on WECC Path 66, i.e., COI, Path 3 (BC Hydro-US interface), and BPA internal Path-NOH from EMS PI. Retrieve wide angle separations, i.e., GCL-JD and GCL-ML (500 kV bus angle pairs). Compare three path MW flows and two angle pair differences with their respective limits or thresholds.
3. For N-S "A" mode, if the damping ratio %, Path 3 flow, NOH path flows, both angle pair differences are exceeded concurrently, issue a low damping alarm on N-S "A" mode. Or if the damping ratio %, Path 66 flow, NOH path flows, both angle pair differences are exceeded concurrently, issue a low damping alarm on N-S "B" mode and proceed onto step 4. Otherwise, only issue low damping alert on either mode. No immediate action is needed for an alert notification.
4. Upon receipt of low damping alarms on either mode, Peak RC engineer retrieves WSU FFDD dashboard view and real-time FODSL UI display to verify if a forced oscillation appears and meets the following criteria:

    (a) Oscillation frequency is mostly fixed and close to mode "A" or "B"'s dominant frequency, i.e., $|diff| < 0.02$ Hz.
    (b) Near zero damping ratio.
    (c) High oscillation energy band.
    (d) The oscillation sustains over 10 min.
    (e) A few candidate units for the oscillation source are reported.

    If all criteria are met, the engineer needs to run the offline FODSL tool to confirm the correctness of oscillation source units and oscillation start time. Then contact the owner of the likely oscillation source units to report the issue and request for necessary mitigation actions.
5. If none of forced oscillation source units is identified by FFDD and FODSL tools for the low damping alarm, immediately contact BPA to cross-check their MAS tool results, and discuss if there are other causes such as PDCI controller failure and if BPA is to mitigate the oscillation risk by following operating procedure.
6. The N-S modes monitoring logic was put in effect on the Peak RC Test environment in early 2018. Since then a few N-S mode low damping alerts were received, but none of them were ultimately escalated to the level of alarm. Peak RC investigated each oscillation alert and tracked down the root cause. Case studies are reviewed in the next sections. To perform baselining study on N-S modes, Peak RC leveraged its online TSAT to simulate the 1996 blackout events and generate simulated PMU signals using TSAT intrinsic ePMU module and then ran studies with offline oscillation analytical tools.

6.4 Implementation Experience and Lessons Learned at Peak RC

6.4.1 WISP Network Statistics and PMU Data Utilization

The WISP Harris network hosted by Peak RC started to receive stable live PMU data streams (at a rate of 30 samples per second) from participating entities in 2015. The WISP was feeding the PMU data into the Peak RC's synchrophasor applications and downsampled into the EMS back. Since 2015, the WISP performance was improved steadily, and the number of live streaming PMUs was increased significantly [16]. The WISP Harris network stopped running on July 18, 2019, when the WISP network replacement project by CAISO took in place as follows.

Harris WISP Network Replacement Timeline

As of July 18, 2019, there were 791 PMUs in the WISP Registry, while 437 PMUs were streaming data to the WISP network.

Figure 6.20 shows the geographic map of individual PMUs in the WECC system. The PMUs in the WISP network were applied in multiple applications such as:

- Hybrid State Estimator (using downsampled PMU data) in Production EMS
- Wide Area Angular Separation Monitoring in Production EMS (Monitoring)
- Corridor Flow Calculation in EMS (Monitoring only)
- PhasorPoint (PP)/MAS/Mode Meter in EMS (Monitoring)
- WSU Oscillation Monitoring System (OMS) and Forced Oscillation Detection Source Locator (FODSL) in Dev and Test EMS environment for engineering validation
- EPG enhanced Linear State Estimation (eLSE) deployed in Dev and Test servers for validation (Fig. 6.21)

The statistics of the WISP PMUs for each participating entity is summarized in Fig. 6.22.

On May 30, 2019, Peak RC provided the last update on the WISP PMU data quality metrics as shown in Fig. 6.23.

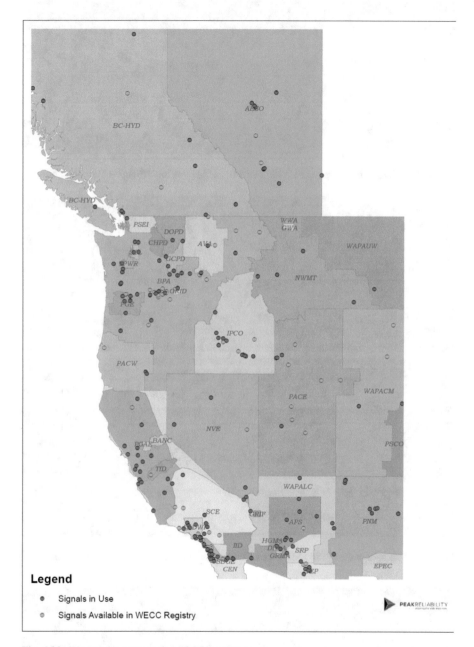

Fig. 6.20   Western interconnection PMU location map

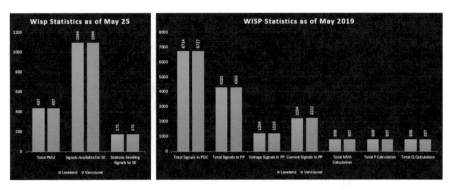

Fig. 6.21 PMU signal utilization in WISP network and applications. (Source: Peak Reliability)

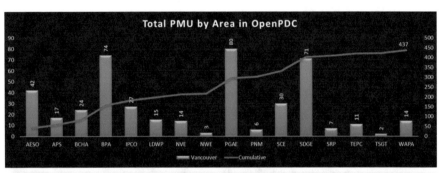

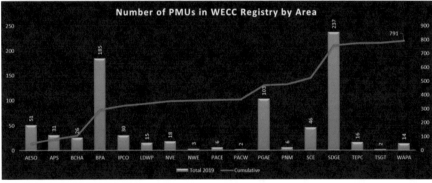

Fig. 6.22 Statistics of the WISP PMUs for each participating entities. (Source: Peak Reliability)

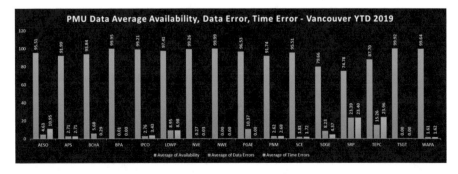

Fig. 6.23 PMU signal utilization in WISP network and applications. (Source: Peak Reliability)

6.4.2   Oscillation Detection and Source Locating Case Studies

Since 2015 Peak RC has used both MAS and OMS software (in both offline and real-time modes) to identify inter-area mode low damping events and impactful forced oscillations. On July 18, 2019, the entities switched for sending live PMU data streams to CAISO as replacement of the WISP network. In the past years, the Peak RC's online Mode Meters and FODSL tool were proved by a number of success stories in identifying specific root cause of low damping events on the dominant modes. A few examples are provided as follows.

6.4.2.1   September 5, 2015, N-S Mode B Resonant Oscillation

On September 5, 2015, Peak RC online MAS engine detected very low damping on the N-S "B" mode. Later we confirmed the low damping oscillation event by offline study with DMO. MAS Mode Meter and DMO study results are compared in Fig. 6.24. Neither major topology changes (quick) nor gen/load/tie line flows drops (slow) happened concurrently.

By running offline DMO and FODSL study on the event, Peak RC engineer detected a hydro power plant that is connected to a 500 kV line. The PMU data on Fig. 6.25 confirms the line flow oscillated at 6 MW during the event time. This forced oscillation caused a resonant effect on North-South mode B, which has excited COI flow by a power oscillation of 40 MW on Fig. 6.26.

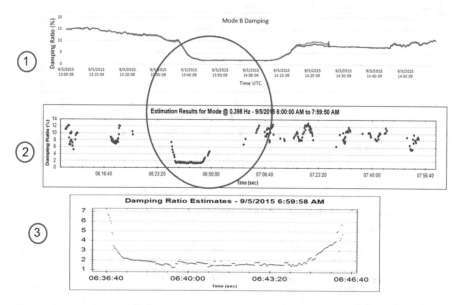

Fig. 6.24 September 5, 2015, N-S mode B low damping detected by MAS/Mode Meters and WSU OMS/FFDD. (Source: Peak Reliability)

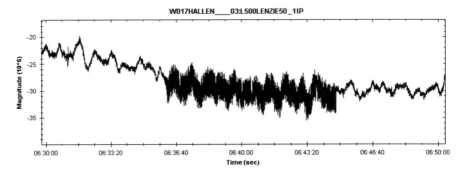

Fig. 6.25 A 500 kV line flow oscillation. (Source: Peak Reliability)

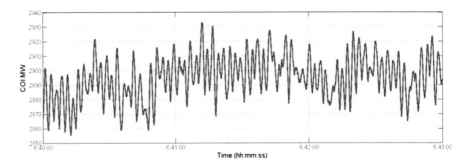

Fig. 6.26 COI flow oscillation. (Source: Peak Reliability)

In this case a "false alarm" could be issued by MAS/Mode Meters due to ambient engines: the actual system mode damping is high, but another forced oscillation at the close frequency is causing negative interaction. To distinguish between two frequencies that are very close, it becomes mathematically difficult for frequency domain modal analysis approaches. Instead the engineer performed further analysis using WSU Fast Sub Space Identification (FSSI) tool, which is a time domain simulation algorithm. The two separate oscillations are distinguishable clearly as shown in Fig. 6.27.

The engineer further contacted the operator of the suspected unit and confirmed the source location result by PMA is correct. The investigation results are summarized as follows:

The root cause actually stems from a faulty combustion turbine MW transducer. There are two MW transducers for this unit and the control logic uses a select logic of the two values to control the output to a MW set point. Of course, if the measured output is incorrect or oscillating, then the control system attempts to correct it to the set point, and there will be real output swings.

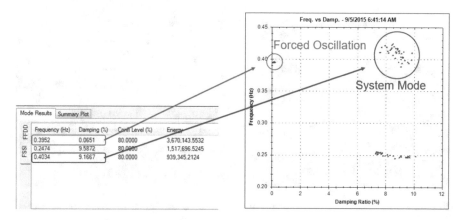

Fig. 6.27 FSSI analysis results on September 5, 2015, N-S mode B low damping resonant event. (Source: Peak Reliability)

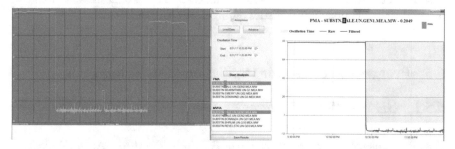

Fig. 6.28 August 31, 2017, 1.23 Hz FO event source locating results and PI trend. (Source: Peak Reliability)

6.4.2.2 August 31, 2017, 1.23 Hz Forced Oscillation Event

This 1.23 Hz forced oscillation (FO) event sustained for 6 h. The BPA MAS/ODM tool issued alarms for the event correctly but didn't locate the source unit. Peak RC was requested to run offline source locating tools: PMA and MVRA. The engineer quickly identified one hydro unit as the oscillation source in a PacifiCorp (PAC) power plant. PAC was contacted with the source locating results shown in Fig. 6.28.

After the investigation, PAC confirmed the oscillations by the comments, "It's due to the extreme turbulence and cavitation occurring during the flooded motoring combined with the PSS operation at the same time ..."

6.4.2.3 05/23/2018 PDCI Probe Test

Peak RC MAS/Mode Meters, integrated in GE EMS and interfacing with PI, correctly detected low damping on North-South mode B during BPA PDCI probe test on May 23, 2018. As expected, the FODSL was run to find no forced oscillation

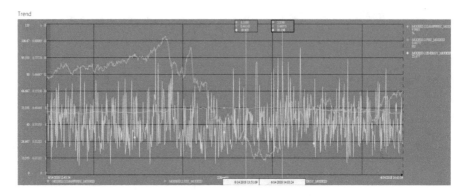

Fig. 6.29 May 23, 2018, BPA probe test: N-S mode B oscillation PI trend. (Source: Peak Reliability)

source associated with the low damping in this case. Figure 6.29 shows PI trend on the mode B oscillation characteristics during the PDCI probe test.

6.5 Progression of Vendor Oscillation Monitoring Tools

6.5.1 GE WAMS Applications

Peak RC implemented the early version of GE (formerly Alstom/Areva) Wide Area Measurement System (WAMS), including PhasorPoint integrated with the custom Montana Tech MAS engine and the Grid Stability Assessment (GSA) application. GE WAMS production is currently recognized as one of the leading synchrophasor platforms deployed in many custom sites, including Peak RC WISP, Middle West ISO (MISO), ISO-New England (ISO-NE), New York ISO (NYISO), India National Grid and Brazil Grid, etc.

A newer version of WAMS 3.2 was released in 2017. It introduced a few new enhancements:

- Seamless Integration with EMS and Dynamic DTS shown in Fig. 6.30. The Dynamic DTS is intended for operation training for transient and dynamic simulation of the system, including oscillation phenomena.
- Angle-based grid management, i.e., predictive angle separation analysis and mitigation
- Sub-synchronous oscillation analytic
- Enhanced oscillatory stability monitoring
- Enhanced island management
- Enhanced disturbance management

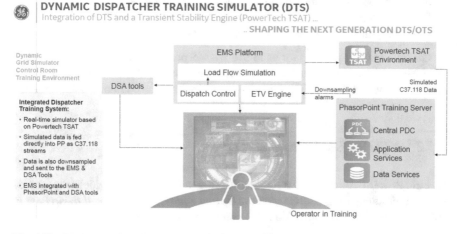

Fig. 6.30 GE dynamic DTS integration architecture and key components. (Source: GE)

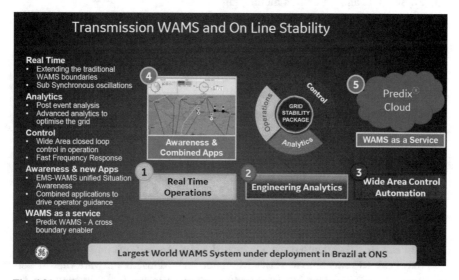

Fig. 6.31 GE transmission WAMS and online stability deployment in Brazil. (Source: GE)

GE deployed the world largest system WAMS system in Brazil, with key components/features summarized on Fig. 6.31.

GE's new Dynamic DTS is the integration of DTS and a transient stability engine (PowerTech TSAT) with the following highlights:

- Integrated Dispatcher Training System
- Real-time simulator based on PowerTech TSAT

- Simulated data is fed directly into PP as C37.118 streams
- Data is also downsampled and sent to the EMS and DSA tools
- EMS integrated with PhasorPoint and DSA tools

GE rolled out a road map for its future WAMS products in several focused areas: (1) edge to Cloud technology; (2) automated advisory mode for engineering analysis; (3) autonomous mode for wide area control with reduced human intervention.

6.5.2 EPG RTDMS Product

Electric Power Group (EPG) is another leading vendor offering widely recognized RTDMS products for *real-time monitoring and wide-area situational awareness* in industry. The RTDMS applications for control room use cover

1. Wide-area situational awareness – dashboard
2. Oscillation detection and monitoring
3. Phase angle and grid stress monitoring
4. Automated event analyzer
5. Voltage sensitivity monitoring
6. Frequency stability monitoring
7. Inter-area power transfer
8. Generation trip detection
9. Islanding detection
10. Intelligent alarms with composite logic

    - Example: contingency alarm such as for N-1, N-2 conditions, etc.

11. GridSmart – Reports event list for the last 24 h (i.e., intelligent alarms) and comparison of intertie flows between yesterday and today

In 2017, Peak RC worked with EPG for a pilot and demo project to validate RTDMS Oscillation Monitoring Applications using WISP PMU data stream and WSM-based enhanced Linear State Estimation (eLSE) information.

The Peak RC EPG demo project achieved a few positive outcomes or progress.

First, Peak RC was able to verify Montana Tech MAS2.0 software (newer version than Peak RC's) was integrated and configured in the RTDMS platform for both Mode Meters and Oscillation Detection Module (ODM) properly, as shown in Figs. 6.32 and 6.33, respectively.

Second, EPG provided Peak RC a feasible plan shown in Fig. 6.34 to integrate WSU's OMS software tools into the RTDMS. Per the plan, users will probably have an option to include both the latest Montana Tech MAS 2.0 and WSU comprehensive OMS tools in a single RTDMS platform for oscillation detection and mode meters monitoring in real time.

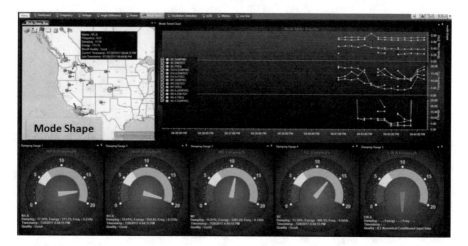

Fig. 6.32 Peak RC Mode Meter display built in EPG demo project. (Source: EPG)

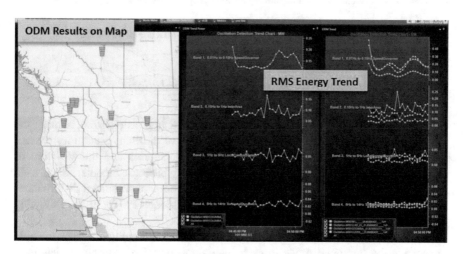

Fig. 6.33 Peak RC ODM displays built in EPG demo project. (Source: EPG)

After the demo project was completed, Peak RC didn't move forward due to non-technical concerns, including the RC's wind-down path. To date, EPG's RTDMS applications have been deployed already or are under deployment to CAISO, Southwest Power Pool (SPP), ERCOT, and PJM, leading utilities.

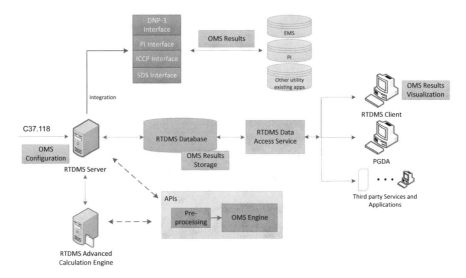

Fig. 6.34 WSU oscillation tool integration plan EPG proposed for Peak RC. (Source: EPG)

6.6   New Attempts on Model-Based Oscillation Modal Analysis

6.6.1   Gap in Planning Model-Based Simulations

The three major interconnections in the North American power system, namely, the Eastern, Western, and ERCOT Interconnections, have different oscillatory characteristics because of differences in generation-load patterns and the structure of transmission lines. To benchmark EI and WI planning basecase models, NERC used PSLF and PSSE simulation software platforms to compare simulated disturbances with actual disturbance data captured from the wide-area PMUs in the following tables [1] (Tables 6.5 and 6.6).

Overall, benchmarking results demonstrated that the planning models were able to recreate the dominant mode in each interconnection. The current models look sufficient to capture the dominant mode shapes. The value for the oscillation modal frequencies was within 12% of the PMU data.

However, the damping ratios simulated from the planning models were between half and double values for the primary dominant mode. This raises a concern over whether a system damping control and mitigation plan developed from planning basecases is adequately applicable for real-time operation monitoring and deployment. Based on prior operation experience in Western Interconnection, the low damping alert levels for North-South mode A and B are set around 7% and 4%, respectively. A gap of 4–8% damping estimations calculated from actual PMU and planning basecase simulation separately will result in false alarms of low damping or underestimate real low damping issues, while damping control and mitigation

Table 6.5   EI dominant mode comparison

	Dominant mode 1 simulated	Dominant mode 2 simulated	Dominant mode 3 simulated	Dominant mode 1 actual	Dominant mode 2 actual
Frequency (Hz)	0.32	0.71	0.53	0.32	0.67
Damping ratio (%)	17.8	6.4	6.7	20	13
Relative energy (%)	44	37	12	79	16

Source: NERC

Table 6.6   WI dominant mode comparison

	Dominant mode 1 simulated	Dominant mode 2 simulated	Dominant mode 1 actual	Dominant mode 2 actual
Frequency (Hz)	0.37	0.25	0.42	0.29
Damping ratio (%)	8	16	12	8
Relative energy (%)	72	27	51	31

Source: NERC

schemes are developed and validated by pre-selected planning basecases. It's not trivial to align planning basecase system condition settings with the actual system operation on a given system event. Many changes, such as system topology, generation output, and load variations, can make basecase simulation outcome off of characteristics in a system event.

6.6.2   Effect of Forced Oscillations to Inter-area Mode Monitoring

Forced oscillation is one kind of disturbance, which has large-area influence on the system. It may lead to cascading outages, sustained mechanical vibrations, etc. The causes of forced oscillation can be various. Most of them are caused by equipment failures, malfunctioning control, or abnormal operations. It lasts long until the source of the disturbance is cleared. Therefore, a quick and accurate detection and localization of forced oscillations are critical for the system security.

Two main concerns for forced oscillation are as follows:

1. When the oscillation frequency happens to be close to a dominant inter-area mode, it causes a large power swing on the grids and jeopardizes the system reliability.

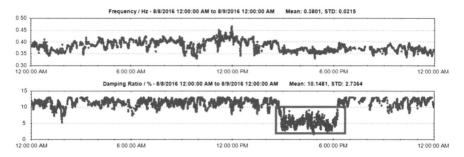

Fig. 6.35 N-S mode B low damping alarm on August 8, 2016. (Source: Peak Reliability)

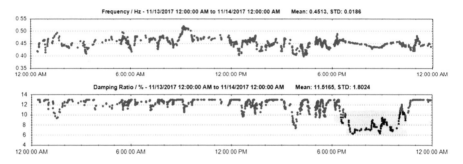

Fig. 6.36 N-S mode B low damping alarm on November 13, 2016. (Source: Peak Reliability)

2. Real-time oscillation detection engines, if using the PMU signals from a trouble-some power plant for oscillation mode estimation, could be biased to generate false alarm of low damping on a dominant system mode.

The EI oscillation on January 11, 2019, and other historical events clearly show negative impact of forced oscillation to the grid operation stability if causing reso-nant effect to a dominant inter-area mode.

Peak RC had been running real-time modal analysis engine (MAS)-Mode Meters (developed by Montana Tech) in Production and Oscillation Monitoring System (developed by Washington State University) in a Test environment. As shown on Figs. 6.35 and 6.36, North-South mode B reported that the false damping drop alarms by the Mode Meter and OMS tools several times.

On the August 9, 2016 incidence, a forced oscillation at around 0.39 Hz was caused by a hydro unit while it's operating in a "rough zone" (i.e., 25–60% rating). On the November 13, 2017 incidence, a forced oscillation around 0.40 Hz was injected by another hydro unit that was under manual control due to maintenance.

To validate and remove the impact of forced oscillation to inter-area mode damp-ing estimation, model-based approaches are needed to complement the inadequacy of synchrophasor applications. For example, BPA monitored wide-area angle separation and selected WECC path flows in addition to Mode Meter's calculated damping ratios before issuing relevant operator alarms on low damping violation on a dominant inter-area mode. Peak RC attempted to use online TSAT simulation cases for analysis of system oscillation modes by SSAT and ePMU data streams.

6.6.3   Real-Time WSM-Based Simulations

Before the West-wide System Model (WSM) was built for real-time operation in 2009, WECC Model Validation Work Group (MVWG) used to take 6 months or even 1 year to develop a WECC dynamic simulation basecase to validate a single system event. Since 2013, Peak RC worked with GE and WECC MVWG to develop a new framework that enables GE/PSLF importing a pre-disturbance WSM-SE solution savecases to create a new system event validation basecase in several weeks. Besides, after Peak RC cut over its online TSAT tool for operation decision in October 2018, the TSAT basecase archive zip file (created for every 2–5 min) can be readily used for MOD-033 study with minimal case fine-tuning, such as actual fault definition. Peak RC has sent out a bunch of the TSAT archive cases for simulation and validation of selected system events to NERC, WECC, and operating entities upon data request. Values of the real-time WSM SE and TSAT autosave or archive cases for model validation and event analysis are widely recognized by industry.

The WSM-TSAT simulation on Fig. 6.37 calculates bus frequency on a system event including multiple 500 kV line outages and RAS gen drop actions and compares the simulations with PMU recording. By integrating online TSAT tool with EMS State Estimator to import real-time SE snapshot basecase for transient simulation, Peak RC engineers were able to run small signal stability analysis and generate ePMU data streams from online TSAT archive cases for offline study on system oscillation behaviors.

To start with an offline oscillation mode analysis by SSAT, engineer retrieved a historical WSM-TSAT archive case. In this system condition denoted by "basecase[0]," the system is initially lightly loaded, especially the COI is carrying less than 1900 MW. Then the engineer increased the loads in SDG&E by 2500 MW while raising the generation in AESO and BC-HYD accordingly. The engineer ran another SSAT study for the stressed condition denoted by "stressing[1]." Next, the engineer increased the load in SCE and PGAE areas each by 500 MW and raised the BPA generations accordingly. The engineer then ran the SSAT study again for the more stressful operation condition denoted by "stressing[2]." The modes frequency and damping ratios for three operation conditions are presented as follows (Table 6.7).

From the simulation results in the table above, one can clearly observe damping ratios of North-South modes decrease when COI loading increases. By contrast, East-West mode is not sensitive to COI flow variation. SSAT study is performed on a given system snapshot. To evaluate continuous impact of system changes to a dominant mode, Peak RC generated ePMU data stream for a period of interest from

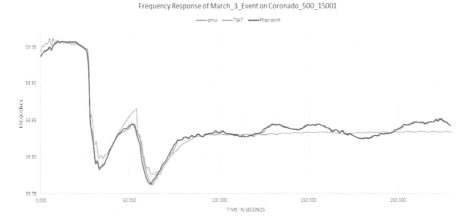

Fig. 6.37 WSM-TSAT simulation benchmarking with PMU on a system event. (Source: Peak Reliability)

Table 6.7 WI dominant modes: oscillation frequencies and damping ratios (%) simulated in SSAT

Dominant system modes	Simulated frequency (basecase0)	Simulated damping% (basecase0)	Simulated frequency (stressing[1])	Simulated damping% (stressing[1])	Simulated frequency (stressing[2])	Simulated damping% (stressing[2])
E-W mode	0.4141	8.38	0.3995	8.33	0.3947	8.33
N-S mode B	0.4058	8.95	0.3933	4.56	0.3763	3.46
N-S mode A	0.2988	11.82	0.278	6.38	0.2661	4.53

Source: Peak Reliability

the TSAT engine and imported the ePMU into WSU's FFDD tool for mode estimate. Figures 6.38 and 6.39 show an example on how to leverage real-time model-based simulation results from TSAT & SSAT into modal analysis tools via ePMU data to analyze the inter-area mode damping change behaviors and validate whether a system change leads to a damping drop or not.

6.7 Summary

As stated in NERC SMS guidelines, "Oscillations across all interconnections in North America have been observed over the years, with different phenomena causing those oscillations. Some are localized to one or a group of power plants while others are experienced across a wide area."

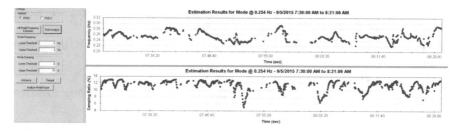

Fig. 6.38 N-S mode A oscillation analysis by WSU-FFDD using ePMU data. (Source: Peak Reliability)

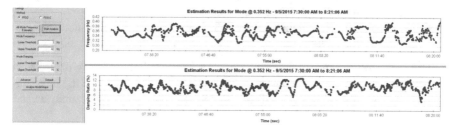

Fig. 6.39 N-S mode B oscillation analysis by WSU-FFDD using ePMU data. (Source: Peak Reliability)

The Western Interconnection, for example, has spent significant effort understanding the oscillatory behavior of interconnection due the small signal stability risks experienced in the 1996 blackout[1] and continued oscillatory risks during highly stressed operating conditions (although rare). Through the Synchronized Measurements and Advanced Real-Time Tools (SMART) work group, Peak RC worked with Western operating entities to make solid progress in monitoring dominant western system modes in real time and development of forced oscillation source locating tools. From Peak RC's practice, we can conclude a few lessons learned:

- A work group like SMART is an efficient mechanism to build strong collaboration among operating entities within one or more interconnections. In observance of Peak RC winds down, SMART WG was transitioned to WECC Joint Synchronized Information Subcommittee (JSIS) in June 2019 to play a leadership role in industry continuously.
- Leading vendor software products such as GE WAMS and EPG-RTDMS were well established to support entities' objectives on oscillation monitoring. Montana Tech MAS engine and WSU OMS software are promising for system oscillation mode analysis and real-time monitoring.

---

[1] 1996 System Disturbances – NERC, https://www.nerc.com/pa/rrm/ea/System%20Disturbance%20Reports%20DL/1996SystemDisturbance.pdf

- Forced oscillations have negative impact to the oscillation events: resonant effect on inter-area modes and "bias" signals input to the analytical tools. To overcome the impact, effective forced oscillation source locating tools need be developed and deployed in control rooms.
- To clearly understand resonant oscillation phenomena and estimate mode damping accurately for development of correct mitigation plans, it's essential to do the following:
  - Develop a full Interconnection-wide operational model for real-time screening of oscillation changes under contingencies to facilitate online TSAT and SSAT simulation.
  - Have an effective mechanism to coordinate inter-area mode monitoring and action-taking under certain circumstance.
  - Develop high-performance online FSSI tools for deployment in control room to decompose an inter-area mode and a forced oscillation automatically.

- Use synchronized measurements across the interconnection during grid disturbances or abnormalities to baseline the oscillatory performance of the interconnection. The following data will be required for oscillation event analysis and simulation:
  - Synchronized PMU measurements collected for multiple system events
  - Online TSAT archive cases captured before the events
  - System Topology Changes stored in EMS SE auto savecases and/or PI
  - AGC control measurements
  - WECC Path/Cutplane Flow and SOL changes
  - Unit dynamic models and RAS
  - System load variations and composite load impact study

References

1. NERC SMS Report-Interconnection Oscillation Analysis [Online]. Available: https://www.nerc.com/comm/PC/SMSResourcesDocuments/Interconnection_Oscillation_Analysis.pdf
2. Florent, X. (2019). *Interarea oscillations in Continental Europe: Analysis and impact of HVDC links*. Presented in May 2019 WECC JSIS meeting [Online] Available: https://www.wecc.org/Administrative/14_RTE-Interarea%20oscillations.pdf
3. Hauer, J. F., & Vakili, F. (1990). An oscillation detector used in BPA power system disturbance monitor. *IEEE Transactions on Power Systems, 5*(1), 74–79.
4. Pierre, J. W., et al. (1997). Initial results in electromechanical mode identification from ambient data. *IEEE Transactions on Power Apparatus and Systems, 12*(3), 1245–1251.
5. Trudnowski, D. J., et al. (2008). Performance of three mode-meter block-processing algorithms for automated dynamic stability assessment. *IEEE Transactions on Power Apparatus and Systems, 23*(2), 680–690.
6. WECC WISP Technical Notes PMU Identification Numbers Unique 16-Bit Identifiers Assigned to the PMU Registry Technical Notes by Dan Brancaccio [Online]. Available: https://www.wecc.org/Reliability/WISP%20PMU%20Identification%20and%20%20Signal%20Naming.pdf

7. Zhou, N., et al. (2007). Robust RLS methods for online estimation of power system electrome-chanical modes. *IEEE Transactions on Power Apparatus and Systems, 22*(3), 1240–1249.
8. Zhou, N., et al. (2008). Electromechanical mode online estimation using regularized robust RLS methods. *IEEE Transactions on Power Apparatus and Systems, 23*(4), 1670–1680.
9. Donnelly, M., Trudnowski, D., Colwell, J., Pierre, J., & Dosiek, L. (2015). *RMS-energy filter design for real-time oscillation detection.* Proceedings of the 2015 IEEE PES general meeting, July 2015.
10. Kosterev, D., Burns, J., Leitschuh, N., Anasis, J., Donahoo, A., Trudnowski, D., & Donnelly, M. (2016). *Implementation and operating experience with oscillation detection application at Bonneville power administration.* CIGRE US National Committee 2016 grid of the future symposium.
11. Follum, J., & Pierre, J. W. (2016). Detection of periodic forced oscillations in power systems. *IEEE Transactions on Power Systems, 31*(3), 2423–2433.
12. Follum, J., Pierre, J. W., & Martin, R. (2017). Simultaneous estimation of electromechanical modes and forced oscillations. *IEEE Transactions on Power Systems, 32*(5), 3958–3967.
13. Nezam Sarmadi, S. A., & Venkatasubramanian, V. (2016). Inter-area resonance in power systems from forced oscillations. *IEEE Transactions on Power Systems, 31*(1), 378–386.
14. Zhang, H., Ning, J. A., Yuan, H., & Venkatasubramanian, V. (2019). *Implementing online oscillation monitoring and forced oscillation source locating at peak reliability.* In: Proceedings of IEEE NAPS meeting, October 14–15, 2019.
15. O'Brien, J., Wu, T., et al. (2017). *Source location of forced oscillations using synchrophasor and SCADA data.* In: Proceedings of the 50th Hawaii international conference on system sciences.
16. WISP PMU Data Statistics presented in WECC JSIS May 2019 meeting [Online]. Available: https://www.wecc.org/Administrative/11a_JSIS_PMU_Data_Statistics_Peak_2019.pdf

Chapter 7
Bringing EMS Node-Breaker Model into Offline Planning Tools

Slaven Kincic

Acronyms

AGC	Automatic generation control
AVC	Automatic voltage control
AVR	Automatic voltage regulation
BA	Balancing authority
BCS	Basecase
BES	Bulk electric system
BPA	Bonneville Power Administration
COI	California-Oregon Intertie
CIM	Common Information Model
EMS	Energy management system
FERC	Federal Energy Regulatory Commission
HVDC	High voltage direct current
ICCP	Inter-Control Center Communications Protocol
ISO-NE	ISO–New England
LTC	Load tap-changing transformers
MDF	Master dynamic file
MMWG	Multiregional Modeling Working Group
MVWG	Modeling and Validation Work Group
NERC	North American Electric Reliability Corporation
PDCI	Pacific Direct Current Intertie
PST	Phase shifting transformer
RAS	Remedial Action Scheme
RC	Reliability coordinator
RCSO	Reliability coordinator system operator
ROSE	Region operation security existence
RTCA	Real-Time Contingency Analysis
RT-VSA	Real-Time Voltage Stability Analysis
SCADA	Supervisory Control and Data Acquisition

© Springer Nature Switzerland AG 2021
H. Zhang et al., *Advanced Power Applications for System Reliability Monitoring*, Power Systems, https://doi.org/10.1007/978-3-030-44544-7_7

SMVTF	System Model Validation Task Force
SPS	Special Protection Scheme
SSR	Sub-synchronous resonance
SVC	Static VAR compensator
SVD	Static var devices
TSAT	Transient Stability Analysis Tool
TVA	Tennessee Valley Authority
UFLS	Under frequency load shedding
ULTC	Under loading tap changer
UVLS	Under voltage load shedding
V&R	V&R Energy Research Laboratory
WBRTF	West-wide Model and Basecase Reconciliation Task Force
WECC	Western Electricity Coordinating Council
WI	Western Interconnection
WSM	West-wide System Model
ZBR	Zero-impedance branches

7.1 Introduction

The power system model used in EMS systems for state estimation and other near real-time applications (also referred to as full topology model) are based upon detailed node-breaker representation. The node-breaker model preserves the details of substations such as nodes and switching devices. Real-time operations model needs to keep track of changes in network topology and follow system operating conditions, which requires flexibility provided by inserting switching devices into the model. The network topology and system conditions are automatically updated in real time based upon changes in status of switching devices and available field measurements. Nodes are point of connection in between two devices, and each node can become a separate bus or multiple nodes are collapsed into one single bus depending on status of switching devices. On the other hand, power systems planning models are based upon bus-branch representation where nodes connected by closed switches or breakers are consolidated to form buses and details of switches and breakers are not preserved, thereby simplifying the model and reducing data and memory requirements. Such model cannot be used in real-time operation due to lack of flexibility. In addition, it is very difficult and time-consuming to adjust such a model on desired system study conditions.

The West-wide System Model (WSM) is the state estimator operation model based upon detail node-breaker representation of the WECC system. The WSM was used by Peak Reliability (formerly WECC Reliability Coordinator Office). As described in the previous chapters, WSM is used for state estimation (SE), Real-Time Contingency Analysis (RTCA), Real-Time Voltage Stability Analysis (RT-VSA) and online dynamic stability assessment/transient stability analysis tool (DSA/TSAT) and for near real-time and next-day studies. WSM is continuously updated and benchmarked against real-time measurements and currently contains approximately 180,000 real-time mapped measurements allowing the model to follow real-time operating condition. Peak RC ran state estimation every minute, RTCA, RT-VSA, and DSA runs every 5 min, and a snapshot of the system representing the given operating conditions from real time was archived every 5 min. These real-time cases can be promptly used to perform system studies. On the other hand, WECC's planning models, also known as basecases, are based upon bus-branch representation of the WECC system. The basecase is assembled by WECC staff and Systems Review Work Group. The planning models typically represents heavily stressed or lightly loaded conditions and used by planning and operations community to perform short-term and long-term system studies. There are typically dozen of such cases prepared annually for operation and planning studies, and this system model is often referred as planning model.

The difference in modeling philosophies between operations and planning can be primarily attributed to historical limitations on memory, computational power, and different time frames of studies to be performed. Advances in computational power and reduction in the cost of memory have eliminated the first two barriers. However, there are fundamental differences that are difficult to overcome. Widespread use of a bus-branch model in the past several decades for power systems planning has led to the use of bus identifier (bus number or name and base voltage) as the primary key to identify or specify any power system equipment with the assumption that bus identifiers would typically not change. Widely used industry formats like EPC and DYD in GE PSLF or RAW and DYR in PSS/E are defined based upon this assumption and use bus identifiers. Data submission guidelines for planning models like WECC DPM [1] or Multiregional Modeling Working Group (MMWG) guidelines [2] require use of bus identifiers when specifying power system equipment data. An operations model on the other hand uses EMS labels as the primary key to identify or specify any power system equipment. Bus numbers change in the operations model unlike planning models, based upon status of switching devices, so these numbers are not suitable as unique identifiers for real-time and near real-time studies.

WECC is a stability-limited region, and several major disturbances have occurred in WECC in past which makes it more important to validate the planning models against real-time measurements [3, 4]. The new NERC MOD 033 [5] standard requires validating planning models every 24 months using actual disturbance data. One of the reasons for the deployment of large number of PMUs across the WECC is to allow for model validation [6, 7]. The use of two different models in operations and planning for the same power system makes it extremely difficult and time-consuming to use real-time operations data to validate planning models or to use if for disturbance analysis. The use of two different models for the same power system model also leads to discrepancies between model parameters and consequently to discrepancies in results that need to be overcame as recommended in FERC/NERC outage report [4].

Typically a EMS model is used only in the control centers for real time operation. The first step in approaching operation and planning model is to migrate full topology model from EMS system into the planning tools and to make it available to system planners in full topology as well as in collapsed bus-branch format and to match it to dynamic database. Peak Reliability worked with major software vendors (GE PSLF, V&R Energy, and PowerWorld) to complete development of such a tool. This chapter shows how WSM is brought from Peak Reliability State Estimator into the GE Positive Sequence Load Flow (PSLF), PowerWorld, and V&R Energy software in full topology format and validated by means of dynamic simulations and to benchmark simulation results from both bus-branch and node-breaker models against each other and against synchrophasor measurements (PMUs) data. Bringing node-breaker model into the planning software allows system planners to use operation cases from real time and to get used to switching device in the model, EMS names instead of bus numbers, and eventually lead to single interconnection model. It will also allow performing more frequent model validation using cases from real-time and disturbance data.

7.2   Advantages of Node-Breaker Model Over the Bus-Branch Model

There are multiple advantages of representing a power system using node-breaker model. One of the primary benefits is that the power system is represented more accurately with detailed substation configuration. Presence of switching devices allows for more accurate modeling of contingencies and Remedial Action Scheme and Special Protection Scheme devices (RASs and SPSs).

Presence of switching devices eliminates the need to manually split buses as currently done in planning models to simulate bus splitting actions. Node-breaker model also facilitates simulations for stuck breaker conditions that are common occurrence in power system operation. Moreover, planning models are rarely validated against measurement due to difficulty to adjust model at desired operating conditions.

Additional advantage of having modeled switching devices in system model is that reconfiguration of substation can be used as a remedial action. Currently many planning software use only line switching to mitigate violation, but reconfiguration of substation is not attempted since in bus-branch model, there is no knowledge of substation configuration.

On the other hand, the size of the system represented with node-breaker model can be typically 4–8 times greater than the corresponding bus-branch model due to the presence of switching devices and nodes. Engineers would require new tools and techniques in existing software products to analyze and visualize such large amount of data. Utilities would have to train engineers in use of such new tools and features, if they are not intuitively designed to minimize learning to use them.

Although node-breaker model represents the system in more detail, there are still several advantages of representing power systems or a portion of the system with bus-branch model for planning studies. Planning studies are primarily focused on analyzing future systems where it is difficult to estimate breaker configuration in new or planned substations. In addition, it is often not possible to represent the entire system in operations model that is to be used in real time due to lack of complete visibility and network observability. Modeling non-observable part of the system in real time can lead to convergence problem since measurements are used to adjust model to operating conditions. Such parts of the system can still be represented in planning models with bus-branch model since planning model does not need to change in real time.

7.3   Planning Software Requirements to Enable Use of Node-Breaker Model

To help planning community process and analyze the large volume of data associated with node-breaker or hybrid model, some new features are required. This section describes the features developed in PSLF to help WECC and others in the industry to use node-breaker or hybrid model.

First, it is important to group switches based upon their capability to clear the fault (breakers vs. disconnect). This is important in order to add automated functionalities in contingency processors or other tools for switching schemes, fault actions, and breaker failure schemes.

Unlike planning models, bus numbers change in the operations model based upon status of switching devices during topology processing, and this makes it impractical to define equipment based upon their bus identifiers. Every piece of the power system equipment in the operations model is uniquely defined and identified based upon the EMS equipment labels. In WECC, BPA and Peak Reliability agreed to use same labeling structure to facilitate standardization [9]. WECC members have agreed to store the EMS equipment labels for units in the planning model as long identifiers field creating a permanent link in between planning and operation model. PSLF enables WECC users to define any power system equipment based upon either traditional bus identifier or EMS labels within the software as well as in the ASCII file formats (power flow, short circuits, or dynamics) used for data sharing.

To facilitate the use of EMS labels for equipment which exist in the planning and operations model, a one-time effort was required to map all the generators in the WSM to the corresponding generators in WECC planning basecase models. In addition, incremental updates to the mapping files based any future changes in operations and/or planning models that need to be performed over the time.

Planning models typically define control areas based upon geographical coordinates to study load growth, generation capacity and technology mix forecasts, as well as planned projects. On the other hand, an operation model is focused upon balancing authority (BA) areas, which are based upon the metered boundaries of the balancing authorities. The control areas defined in the planning basecases doesn't always correspond to the operations balancing authority area which introduces another difference between these models. PSLF enables WECC users to model balancing authority areas in addition to control areas to bridge this gap between the planning and operations model. Availability of balancing authority area in PSLF would facilitate use of detail and accurate AGC models in WECC planning studies and post-transient simulations. The traditional area interchange solution option available in planning software also needs to be extended to balancing areas.

New visualization tools are needed to view full topology model. SCADA displays can be used for individual substation. However, it is often impractical to see breakers. For this reason, PSLF developed different tiers of visualization capabilities so that system can be seen in detail with breakers and switches or view can be converted to bus-branch representation or substation representation.

A state estimator snapshot contains bus mismatches. The estimation errors are modeled as pseudo loads – artificial positive and negative loads. Although the magnitude of individual pseudo loads are very small, the presence of a large number of such pseudo loads can have significant impact on major transmission paths and line flows. PSLF allows the users to import such pseudo load(s) as an option to study impact of such estimation errors on the system including path and interface flows. This will also facilitate EMS modelers to better understand and consequently improve the EMS models. These type of loads are grouped separately from other type of loads, so that they are ignored when running different types of analysis such as PV or scaling loads.

Since planning community is new to node-breaker model, it is very important to provide a suite of debugging tools and assist users to detect suspect data or issues with imported operations data. Features like a bay table showing mismatches at primary nodes after topology processing or different tables showing difference between State estimation flows and calculated flows would help identify suspect data like incorrect transformer taps, impedance table errors, or other such data errors which might lead to differences between measured and calculated flows.

Finally it is very important to design features in the software packages to make it seamless for the users to use bus-branch or node-breaker. The users should not be required to run different commands or actions when using the bus-branch versus the node-breaker model. The software features should be the same regardless of which type of underlying data model is being used. For example, to solve a power flow case, the PSLF users only need to load the case and solve it without having to worry about any intermediate steps when using node-breaker or hybrid model or bus-branch model. Similarly, running a dynamic simulation in PSLF is a simple four-step process (load the power flow, load dynamics data, initialize the case, and run the case) regardless of whether the user is using the bus-branch or node-breaker model.

7.4 Consolidation of the Models

There are several challenges with the task of aligning the operations and planning models. It is important to reconcile these modeling differences between the operation and planning models in addition to updating planning software packages to be able to support these models.

Often, a group of generators are modeled as a lumped unit in operations case, while they are modeled as individual generators in planning models or vice versa. A unit needs to match one to one in both models. Not matching units one to one prevents matching dynamic model to power flow. To ensure consistency of results and to reconcile differences between the two major WECC models, different WECC stakeholders have joined forces to accomplish the following tasks [15]:

- Assess differences between the two models.
- Provide guidelines on how to overcome differences.
- Provide modeling recommendations for the node-breaker model.
- Conduct inspection of models and correlate equipment and equipment parameters.
- Ensure proper mapping between WSM and basecases.
- Evaluate and find out areas where additional work needs to be performed to synchronize the two models.
- Coordinate between the Reliability Coordinator and the planning community.
- Assess limitations in both modeling environments and work with vendors to fill the gaps.
- Standardize the process of developing power flow case from EMS SE snapshot cases.
- Benchmark models against each other and against PMU measurements through means of dynamic simulation of past system disturbances.

7.4.1 Common Differences in Between Operation and Planning Models

This section discusses some of the most common differences between two models.

Generators on different shafts in a cross-compound unit are often modeled as one aggregate generator in planning models instead of modeling the generators on each shafts separately. The dynamics data is correctly specified for units on different shafts separately in the planning models. Modeling cross-compound generators as an aggregate unit eliminates the chances of potential errors when defining generator contingencies and also eliminates errors with inconsistent generator dispatch in planning studies. Contrary to planning models, a generator on each shaft of cross-compound generator is modeled individually in an operation model like WSM, which leads to reconciliation challenges. To reconcile this difference, generators on

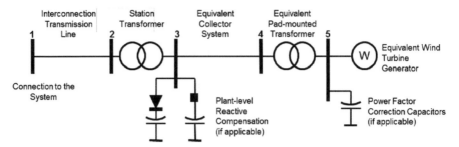

Fig. 7.1 Steady state model of wind farm as recommended per WECC MVWG

each shaft of a cross-compound machine should be modeled individually in planning model as seen in operations model.

Historically, station service loads are used to comprise 5–10% of gross output of generators in fossil or nuclear power plants. Presently, station service load can range from 7 to 15% of the gross output of generators in large fossil power plants, and their percentage continues to increase due to stricter environment regulations. WECC and NERC require explicit representation of station service load for large generators in the planning models. WECC planning models illustrates this modeling good practice. However, operations model often does not represent station service loads due to lack of detail measurements from within a power plant. Station service loads are often netted from the gross generation in such cases. This leads to model validation challenges using simulation tools and leads to optimistic results.

Another common issue is that wind farm and other renewable do not match one to one in two models. Figure 7.1 show a typical wind farm model in steady state as recommended per WECC Modeling and Validation Work Group (MVWG).

In the real-time model, the collector system equivalent and pad-mounted step-up transformer are often neglected, and the wind farm is modeled as multiple generators on the station transformer level. The reason is that often times wind farms are not owned by transmission operator and the only measurements that transmission operators have available are on the points of connection of wind farm with the transmission system and the status of incoming lines from wind farms. Figure 7.2a illustrates typical example. However, such a model is not accurate for system studies and the dynamic model from planning model (WECC basecase) cannot be coupled with it. For that reason Peak Reliability undertook significant efforts with its member utilities through West-wide System Model and Basecase Reconciliation Task Force (WBRTF) to remodel the wind farms in EMS so that models from EMS and WECC basecase are matched one to one. Figure 7.2b illustrates the remodeled EMS model. That way EMS model and WECC basecase model match one to one.

Controllable shunt reactive devices like mechanically switched capacitor and reactors, static VAR compensators, STATCOMS, or thyristor switched shunt capacitors and reactors are modeled as static VAR devices (SVD) or switched shunts in planning models. Such devices are grouped together and modeled as individual blocks within an SVD in planning models, if they regulate the same terminal [2, 9].

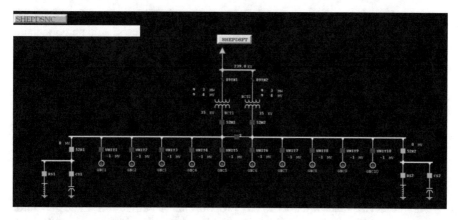

Fig. 7.2a Typical EMS model of wind farm. Collector system equivalent and pad-mounted step-up transformer are neglected

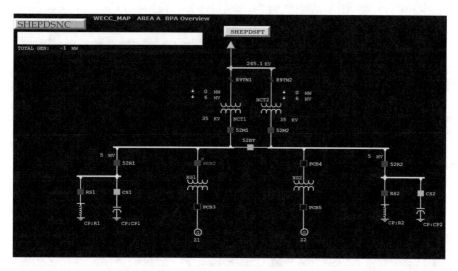

Fig. 7.2b Wind farm from Fig. 7.2a remodeled so that EMS model matches WECC basecase and MVWG recommendations

Power flow algorithms in commercial software's like PSLF and PSS/E are designed to enable or disable individual blocks to emulate control actions and in switching sequence of individual devices. In the operations model, each capacitor or reactor bank is modeled explicitly at different nodes and not grouped together as an SVD. In addition, power flow solution techniques used in real-time tools adjust shunts or reactor banks based upon sensitivity or priority list. This difference in modeling philosophies leads to difference in simulation results when using online and offline tools for the same operations model. To reconcile this difference and facilitate use of node-breaker or hybrid model in planning software products, either a good automatic grouping mechanism is required to group such devices in a consistent manner and emulate their actions or new solution techniques are required which would eliminate the need for grouping individual elements into SVD and instead use sensitivity or priority based schemes. A hybrid model would need to support the modeling methodologies and algorithms used in planning and operations. Further, power flow algorithms would have to be enhanced to adjust shunts or reactors using the breakers connected to them. In addition, large 500 kV substations often have automatic voltage control (AVC) which coordinates capacitor switching in substation. Generic AVC models would need to be developed which would work for both planning and operation models.

Station service loads are not directly proportional to output of a single generator in a power plant but rather to the output of entire power plant in many cases depending upon plant configuration. This is often the case if the different generators share the same GSU or in combined cycle power plant. PSLF enables the users to associate station service load with individual generators or group of generators. However, more work is needed to determine the best approach to model station service loads in all the planning models and commercial software packages and to link them to either individual generators or entire power plant.

In the operation model, voltage schedules are typically set to the estimated or measured voltages. Planning models need detailed voltage schedules to emulate how voltage control devices would respond to this. Accurate voltage schedules need to be added in the operations model.

Typically three-winding transformers are modeled as three two-winding transformers for legacy reasons in the operations models like WSM. This could be attributed to the lack of a good three-winding transformer models in the past in real-time operations tools or attributed to modeling practices to be compatible with data formats which could only model three-winding transformers as three two-winding transformers. Even though technologies have changed and modern software packages like PSLF support three-winding transformer models, the use of old modeling practices is still prevalent as seen in WSM. The planning community on the other hand uses three-winding transformer models to properly represent such transformers in planning models like WECC basecases. To reconcile this modeling difference, automated tools are required to convert the legacy three-winding transformers to latest three-winding transformer models when importing the operations model for planning studies.

Planning software like PSLF allows specification of equipment data on its own nameplate ratings. Use of nameplate data eliminates the possibility of errors associated with converting data to different bases and is very useful to add engineering checks to detect suspect data. Due to legacy reasons, some of the data in WSM and other operations model are still specified on 100 MVA base. This not only leads to discrepancy between operations and planning models but also prevents engineering checks to detect suspect modeling errors. For example, transformer impedance typically ranges from 8 to 14% on the nameplate rating. However without information of the nameplate rating, it becomes difficult to check for such data errors in the operations model.

7.5   Future Changes Recommended in Models

To help industry gradually migrate to a unified planning and operations model, several changes are required in the existing planning as well as operations model. This section describes the changes which are required in the existing planning and real-time operations model to help prepare the planning community to the use of the operations model.

Substation data should be populated in the planning models like WECC, MVWG, or other planning basecases. Presence of substation data would be useful to merge operations and planning models or to incrementally read node-breaker operations model on a substation basis in bus-branch-based planning model. The ability to import operations data in a planning model for selected substations will encourage planners to use and get accustomed to the node-breaker model for portions of the system and understand its benefits. In addition, it would also encourage planners to use the hybrid model by modeling the study regions of interest with detail node-breaker representation and rest of the system with bus-branch model.

Station service loads are currently modeled as loads with ID of "SS" in planning models like WECC basecases. However these loads are not associated with any generators. WSM enables station service load to be associated with a given generator. To better align the planning and operations models in WECC, station service loads should be associated with generators in current planning models with the exception of combined cycle and other special configurations. Automated tools in GE PSLF help to directly link existing station service loads to generators. Although regulatory authorities like NERC emphasize on the modeling station service loads in MOD standards [10], they are not found in all the planning models. Planning groups like MVWG or standards like ENTSO E CIM (IEC 61970 CIM14) and others should advocate use of the station service load models when representing large generators. Reliability coordinators should update the data submission policies to enforce the modeling of station service loads for large generators in a consistent manner which makes it easier to associate station service loads to generator or group of generators.

Data providers and working groups which assemble planning models should be encouraged to submit balancing authority area information in addition to control areas, if the balancing authority area is different from its control area. There are 38 balancing authority areas in WSM and 21 control areas in WECC planning models. Ability to represent control area and balancing authority area would eliminate the need to reconcile any differences by redefining control areas in planning models based upon balancing areas in operations model or vice versa. This would also enable use of accurate AGC representation in planning models.

In order to facilitate use of same equipment model in operations and planning, real-time models like WSM should be updated to enable user to model equipment data on its own nameplate rating rather than on 100 MVA base.

To facilitate new techniques to estimate station service loads due to lack of measurements in real-time models like WSM, information of turbine type, unit, or fuel type should be added in operations model. This would be important to identify different types of units like coal, nuclear, gas, hydro, and others and estimate their station service loads. Such information is already available in market models or production simulation planning models like WECC TEPPC database.

7.6 Bringing Node-Breaker Model in Offline Tools

One of the obstacles for acceptance of node-breaker model by the system planning community was the unavailability of the capability in traditional study software to use node-breaker model. In order to overcome this barrier, BPA, Peak Reliability, and WECC worked closely with different software vendors to implement a full topology model in traditional study software so that the node-breaker model can be used outside of state estimator platforms in full topology format [3–10]. Currently, PowerWorld, GE PSLF, and V&R Energy have this capability implemented. BPA, WECC, and Peak Reliability have agreed on a format so that the abovementioned tools use the same format of node-breaker model exported from the state estimator system in full topology. The subsections below provide a brief description of the node-breaker capability for each of these software platforms.

7.6.1 GE PSLF

WECC and Peak Reliability worked closely together with General Electric Energy Consulting (GE) to develop node-breaker capability in GE PSLF (GE Positive Sequence Load Flow) Program. PSLF is the most widely used software for planning and model validation in the western interconnection. PSLF has been enhanced to support node-breaker, bus-branch, or the hybrid model (parts of system are represented with node-breaker model, and parts of system are represented with bus-branch model).

Switching devices like breakers, disconnects, fuses, links, ZBR, and others can be modeled as breakers in PSLF. PSLF includes built-in topology processing which is seamless to the user, when using the bus-branch, node-breaker, or hybrid model. Advanced features to automatically use switching devices, when opening or closing a branch or clearing faults, are directly available in PSLF. PSLF includes an algorithm to automatically detect the breakers which would need to be switched in order to close or open a branch.

It is often difficult to model stuck breaker schemes easily in planning models due to lack of information about the breakers and their locations. System planners need detailed one-line diagram to correctly simulate the stuck breaker schemes. PSLF includes the built-in option to emulate stuck breaker schemes when using the node-breaker or hybrid model.

Different tiers of one-line graphics are available in PSLF which makes it efficient and easy to traverse the network topology. Without such advanced features, it often becomes challenging to visualize the network topology due to presence of large numbers of switching devices. PSLF includes the option to import State Estimator mismatches as pseudo loads when importing state estimator model. This makes it easy to run sensitivities to evaluate the impact of the state estimator mismatches when using operations cases.

GE/PSLF also uses labels instead bus numbers for identification of power system assets and for matching steady-state model to dynamic model. BPA, Peak Reliability and WECC developed unified naming convention for labels to avoid any discrepancies in the label naming [9]. Currently, there is an effort underway to populate WECC basecase with labels in order to create permanent link in between WSM and WECC basecases. That way system planners will still have bus numbers which they are used to, and, at the same time, they will get acquainted to labels that are created using EMS names. Introducing labels into the basecases will help to further identify differences in between two models and eventually eliminate them. Power system assets can be referred using traditional bus-based identifier (bus number or bus name + base voltage) or EMS labels in PSLF. EMS labels are stored as long IDs in PSLF for each power system asset. The scripting language in PSLF (known as EPCL) has been enhanced to support EMS labels.

PSLF is the first software to be successfully used in the western interconnection to validate dynamic models using WSM. GE, WECC, and Peak Reliability jointly developed a new framework which makes it possible to directly link WSM and planning models. The new architecture makes it possible to use WSM model to validate planning dynamic model for model validation studies and to perform frequent disturbance analysis to meet regulatory requirements such as NERC MOD B 033. A number of recent disturbances in WECC were validating using the node-breaker model in PSLF by importing WSM case and planning data. The results of one such disturbance-based model validation are presented later in this chapter.

All three software use WSM (West-wide System Model) cases provided in comma-delimited format (*.csv). Full non-consolidated WSM model consisting of over 115,600 nodes or 16,000 buses is used for all computations.

Another important capability of offline power flow programs in conjunction with full topology model is their capability to visualize substations and to switch to bus-branch views when needed. Typically, on EMS side, a lot of efforts are invested in substation display design. To avoid duplication of the work, EMS displays can be transferred offline for better visualization improving communication between operation study engineers and dispatchers.

7.6.2   PowerWorld

PowerWorld has the most experience using the state estimator platform used by ISO-New England, TVA, BPA, and Peak Reliability; however they have implemented this ability for importing data from a few other state estimator platforms as well. The original impetus for this work came from working with ISO-New England starting in 2006 [8].

Within the PowerWorld Simulator, the full node-breaker model may be directly read in, and simulations run from these models. Starting with the traditional bus-branch data structures, the primary additional information is a specification with each transmission branch indicating what type of a branch it is: *line, transformer, breaker, disconnect, load break disconnect*, zero-impedance branches (*ZBRs*), *fuse, ground disconnect*, or *series capacitor/reactor* [10]. This information can then be handled appropriately internally to treat nodes connected by switching devices or ZBRs as a single electrical point. Other additional but highly useful information, such as the specification of substation records in the model, is also typically available in the EMS environment so this is read in as well.

With only this additional information, the software can then perform any necessary integrated topology processing as needed to perform numerically robust solutions. The various software tools such as contingency analysis, transient stability, available transfer capability, PV and QV curves, Optimal Power Flow, and sensitivity analysis will internally perform topology processing when appropriate during their respective solution processes.

With the additional information available, other features can be implemented to make performing engineering analysis easier. For example, at first look, contingency definitions can become much more complex in the node-breaker model because taking a line outage involves opening several breakers (often 2 at each end of the line). To reduce this burden, software features are available that allow a contingency definition to be defined simply as "open breakers necessary to isolate this device," and the software will interrogate the network topology and automatically determine what switching devices to open (PowerWorld only allows a branch designated as a *Breaker* to do this.)

Another vitally important change when switching to using the EMS node-breaker based model is to switch to a method of identifying devices in the model, which is not based on the node or bus identifiers. For example, in traditional bus-branch models, a transmission line is identified by the buses (typically the bus numbers) to which the line is connected. This is problematic as EMS models have a data structures that do not make bus identifiers a central part of their identification system. Instead each device has a string (or several possible strings) that uniquely identify the device. The PowerWorld Simulator data structures have permitted string "labels" to identify devices since 2001; however this had not been extensively used until working with ISO-New England starting in 2006 [8]. This work has continued with others including the Bonneville Power Administration and Peak Reliability through today [8–13].

7.6.3 V&R Energy

The node-breaker model in V&R Energy's POM Suite/ROSE software is utilized to perform real-time computations as well as bridge real-time and offline (e.g., planning) analyses.

7.6.3.1 POM Suite/ROSE Analysis Framework

POM Suite/ROSE has three different frameworks for using node-breaker model for real-time and offline analysis:

1. Reading State Estimator case in node-breaker model and saving it as a branch-bus model [11, 12]
2. Creating a "hybrid" model when part of the model is node-breaker and part in bus-branch based on the planning case.
3. Inserting real-time data into planning model:

   – Inserting state estimator case into a planning case
   – Inserting historical real-time measurements into a planning case

   This section presents framework (1) above. For this framework, analysis includes:

- Reading of State Estimator (SE) case in node-breaker model
- Performing necessary topology processing
- Performing steady-state analysis

   – Voltage stability analysis, AC contingency analysis, RAS implementation, determining automatic remedial actions, or any other load flow computation

- Using PMU data for Voltage Stability Analysis
- Performing transient and small-signal stability analysis
- Automatic conversion of the node-breaker model to bus-branch model format

7.6.3.2 ROSE Architecture for Voltage Stability Analysis at Peak Reliability

Peak-ROSE uses a node-breaker model to perform Voltage Stability Analysis at Peak Reliability, [11]. Peak-ROSE works in real-time and offline modes. ROSE architecture in real-time mode is shown in Fig. 7.3.

State Estimator (SE) cases used for Peak-ROSE analysis are WSM (West-wide System Model) cases provided in comma-delimited format (*.csv). They were transferred from Peak Reliability's EMS system to Peak-ROSE software every 5 min. Full non-consolidated WSM model consisting of over 115,600 nodes or 16,000 buses is used for all computations; see Fig. 7.3.

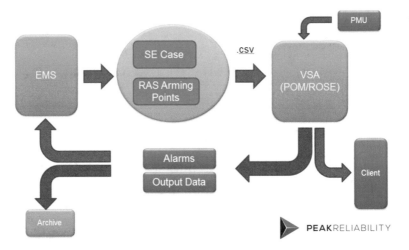

Fig. 7.3 ROSE real-time architecture at Peak Reliability

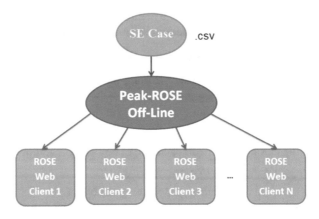

Fig. 7.4 ROSE offline architecture at Peak Reliability

As Peak-ROSE reads the export file and contingency lists, it performs topology processing and, as a result, adds new buses to the model. These additional buses are numbered such that the new number always includes the number of the original bus, before it was split. Every time topology processing is performed, Peak-ROSE generates a file that lists buses that were split and how they were split by contingencies.

Peak-ROSE architecture in offline (e.g., study) mode is shown in Fig. 7.4. The input to Peak-ROSE offline is WSM cases.

After a WSM case is loaded, it is displayed in the form of ROSE data tables and one-line diagrams which are automatically built.

In offline mode, WSM cases may be converted to bus-branch model and automatically saved as Siemens PTI *.raw files in ver. 33 for further offline analysis. Breakers are modeled as zero-impedance lines in the bus-branch model.

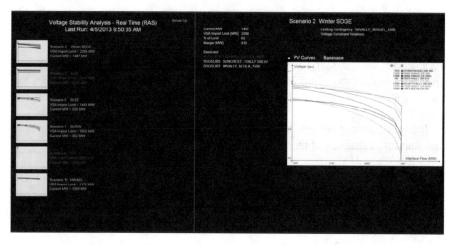

Fig. 7.5   Performing PV curve analysis using node-breaker model

7.6.3.3   Using Node-Breaker Model for Voltage Stability Assessment

Node-breaker model is used for Voltage Stability Analysis (VSA) as follows:

- Determining interface limits
- Performing PV curve analysis (see Fig. 7.5):

  - Stresses the system in terms of load increase and power transfer and computes post-contingency voltages
  - Running multiple power flow computations to determine the stability margin

- Performing VQ curve analysis:

  - Adding a fictitious condenser bus
  - V is independent variable; Q injection is dependent
  - Providing a set of scheduled voltages at a bus

Contingencies that are applied during VSA are breaker-oriented contingencies from RTCA.

7.6.3.4   Using Node-Breaker Model for RAS Modeling

Node-breaker model is used in Peak-ROSE for RAS modeling [13]. RAS actions are breaker-oriented. RAS are automatically triggered provided that certain triggering conditions are met and arming status is enabled (where available). Breaker-oriented contingencies may be executed with or without RAS.

RAS actions implemented in Peak-ROSE are:

- Event based
- Condition based
- Combination of event based and condition based

7.6.3.5 Using Node-Breaker Model for Determining Automatic Corrective Actions

Peak-ROSE automatically develops corrective remedial actions to improve power system transfer capability. Remedial actions include switchable shunts, ULTC, phase shifters, generation MW and MVAR redispatch, load shedding, and line switching.

Remedial actions are determined to alleviate post-contingency violations of voltage stability, voltage, and thermal limits.

7.6.3.6 Advantages of Using Node-Breaker Model

Advantages of using node-breaker model during real-time and offline Voltage Stability Analysis include:

- Accurate representation of the real-time system (e.g., use of an actual State Estimator snapshot)
- Ability to perform real-time contingency analysis
- Accurate modeling of RAS
- Ability to automatically save WSM model as a bus-branch model

7.7 Validation (Cross Comparison of Two Models)

While it might not be possible to make the two models identical, results from simulation of actual system events should illustrate consistency between the two models. Ensuring consistency with system events assures system planners and operators that results of system studies are trustworthy. WECC staff, Peak Reliability, WECC Modeling and Validation Work Group, and WECC Joint Synchronized Information Subcommittee have worked together to make this happen within the Western Interconnection. Ensuring consistency between real-time and offline models can lead to dynamic assessment of voltage stability and transient stability limits instead of using seasonal limits. It would improve reliability and unlock available transmission capacity locked within seasonal limits. This paper illustrates some of the results achieved within WECC.

Ensuring consistency in between two models requires benchmarking models against each other and against measurements for actual system events. The validations include comparison of frequency, voltages, active, and reactive power responses for major transmission paths and buses during system events. This paragraph describes the methodology and shows some of obtained results from a specific system event that occurred in WECC in the spring 2014. The event consists of a 500-kV outage followed by generation drop to prevent overload, shunt capacitor, and reactor switching for voltage control and insertion of the brake. Generation drop was around 2,600 MW. There was also an unplanned tripping of another unit about 40 s later. Details are presented in reference [15].

7.7.1 Methodology

Following system events, MVWG request the WSM snapshot prior to the event. The operational information was mapped on the 2014 planning case (WECC conventional basecase). The event sequence was then replicated using both models, the 2014 planning case (bus-branch) and the WSM directly in full topology format (node-breaker). Both dynamic simulations were performed using GE/PSLF. Results were then compared against the actual measurements (PMUs).

7.7.2 Frequency Response

Figure 7.6 shows very good agreement between the simulated and actual frequency response for the given event.

7.7.3 Voltage Response

Figures 7.7 and 7.8 show the simulated and actual voltage response traces superposed for the given event.

7.7.4 Active Power Response

Figure 7.9 shows the simulated and actual MW response for some major WECC transmission corridors.

In general, both planning and operating models had very good agreement with the actual event, building the confidence in the study tools used in the West.

7.7.5 Discussion

Result of dynamic simulation using WSM and planning model match reasonably well. Some differences in results are due to following:

- Planning model (WECC basecase):
  - Dynamic model library is 100% mapped to BCS.
  - Composite load model is used with BCS.
  - It is difficult to adjust BCS to pre-contingent condition to match 100% WSM.

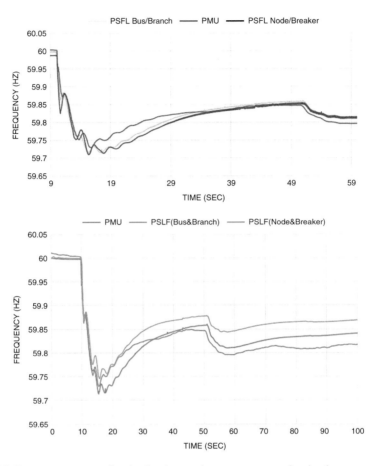

Fig. 7.6 Frequency response for simulated event (measurement vs. planning basecase vs. real-time WSM

- WSM (node-breaker):

  - Dynamic model is mapped for 94% of MW rating of units.
  - ZIP load model is used in conjunction with WSM.
  - WSM is adjusted based on SCADA measurements.
  - HVDC dynamic missing.

- PMU:

  - There is difference in between PMUs and SCADA measurements for certain PMUs;

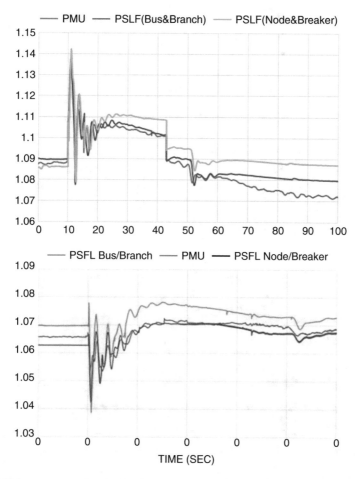

Fig. 7.7 Voltage response for simulated event (measurement vs. planning basecase vs. real-time WSM

7.8   Impact of Real-Time Model on MODE-33 Standard

The NERC Steady-State and Dynamic System Model Validation Standard, MOD-033-1, was created to establish consistent validation requirements to facilitate the collection of accurate data and building of planning models to analyze the reliability of the interconnected transmission system. One of the requirements in this standard is that each planning coordinator shall implement a documented data validation process. WECC created the System Model Validation Task Force (SMVTF) under the WECC Modeling and Validation Work Group (MVWG) to facilitate the MOD-033-1 validation process and to enhance the model validation.

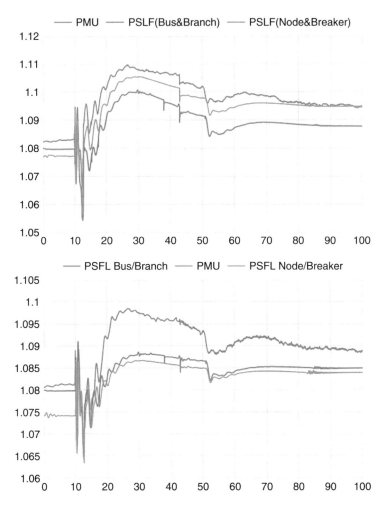

Fig. 7.8 Voltage response for simulated event (measurement vs. planning basecase vs. real-time WSM

This section is meant to provide basic information on how WECC staff and Peak RC prepare the study cases for model validation from SE case and perform steady-state and dynamic system model validation. The cases may be leveraged for use by its members, as applicable, to meet the MOD-033-1 compliance.

Here is a brief overview of MOD-033-1: requirements (for the complete requirements, see the NERC MOD-033-1 standard):

R1. Each PC shall implement a documented data validation process that includes the following attributes:

   1.1. Comparison of the performance of the PC's system in a planning power flow model against actual system behavior represented by State Estimator (SE) case or other real-time data sources

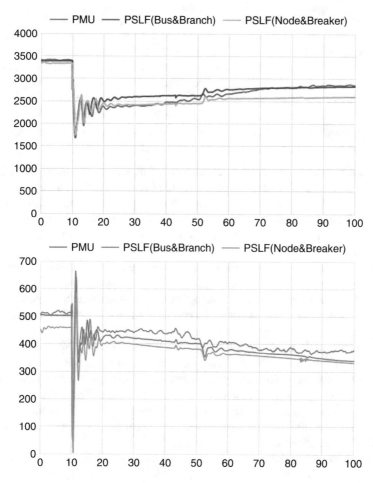

Fig. 7.9 Active power response for one major WECC transmission path for simulated event (measurement vs. planning basecase vs. real-time WSM

    1.2. Comparison of the performance of the PC's system in a planning dynamic
        model against actual system response
    1.3. Guidelines the PC will use to determine unacceptable differences in perfor-
        mance under 1.1 and 1.2
    1.4. Guidelines to resolve unacceptable differences identified under 1.3

R2: Each RC and TOP shall provide actual system data necessary to the PC to perform validation under Requirement R1 within 30 calendar days of a written
request.

Validation of the planning power flow and dynamic models is to be performed at
least once every 24 months.

The focus of MOD-033-1 is the comparison of the performance of the PC's portion of the existing system for steady-state and dynamic response for a local event. Additionally, it is specified in the standard that a dynamic local event could also be a subset of a larger disturbance involving large areas of the grid. NERC's main emphasis is the utilization of local disturbances for the evaluation of the model; however, there are numerous advantages in the use of large disturbance events within WECC, if available and relevant to the PC's validation of its system model. Following arguments emphasize importance and advantages of using large interconnection wide disturbance instead of a local one:

- Dynamic system model validation requires full knowledge of which units are online within the interconnection during the event used for validation.
- Interconnection dynamic response including frequency response (initial and primary) depends on all units within a system, and the planner must know which units are online in the remote parts of the system in order to validate the simulated frequency response.
- After a disturbance occurrence, real and reactive power flows on major transmission lines and paths are directly affected by the response of many generating unit. If the status and outputs of online generating units in areas remote from the actual disturbance are not known, the results of the event simulation may significantly differ from the actual measurements.
- Correct dynamic modeling of generators is essential to properly simulate its behavior. Dynamic oscillatory behavior of the system depends on available rotational masses and excitation system response of the generators within the system. For example, after a system disturbance, the generating units in California may oscillate against generating units in Alberta. Hence, unless the generating units in Alberta are represented accurately, the resultant simulation response may significantly differ from the measurements.
- Usage of large interconnection events for system validation has the benefit that all affected PCs within WECC may be able to use the same power flow basecase and dynamic model. Using the system common case may significantly reduce time needed for basecase preparation and potentially could enhance the model validation process. In the case that a system event does not cause a significant impact on a PC footprint, the PC may choose to evaluate local events that could provide better validation of its power flow and dynamic models.

The adjustment of the WECC basecase utilizing exclusively local events may lead to an over-tuning of the models in an attempt to match the simulation to the field measurements. For instance, if all PCs modify the modeling parameters within their area based only on local events, the sum result may lead to the incorrect validation of the WECC-wide system model, particularly the dynamic response.

There are essentially two components of planning model validation for the MOD-033-1 standard: steady-state (R1.1.1) and dynamics (R1.1.2). Both require adjustments of the WECC power flow basecase, either to the selected date and time or to pre-contingency event conditions. While the steady-state and dynamic system models can be validated separately, it may be more logical and efficient to use the same event and power flow case for validation of both (R1.1.1 and R.1.1.2). Note that an accurate steady-state model is needed for dynamic validation but a steady-state model validation does not require having a system event. Both of these topics are tightly linked but will be discussed separately.

This guideline is not meant to be perceived as the only acceptable document to be used for model validation. Additional information is provided in references [14, 15].

The first step in MOD-033-1 model validation is to select an event against which system response will be validated. Large system event occurs infrequently and unplanned, and the one to be used for model validation must be selected carefully. An example of a useful large event for system model validation is the loss of PDCI and associated RAS that include multiple generation tripping. This type of event and corresponding impacts is observed and recorded widely within the interconnection. Other useful events are transmission line faults followed by the loss of a large amount of generation. SMVTF will choose and prepare at least two WECC-wide system event cases annually. If time permits, and should other interesting event(s) occur, SMVTF may decide to prepare additional study cases. Another potential case that may be selected is a case for steady-state validation only, such as heavy winter peak or heavy summer peak scenario or choosing disturbance that eventually occurred during system conditions close to heavy winter or peak summer conditions.

Some events may not be suitable for MOD-033-1 validation purpose. These events may include, for example, asymmetric events that include highly unbalanced flows such as single-pole reclosing or an event that occurred at the top of the hour when generating units are ramping up or down. In study simulation, during initialization process, we assume that all generating units are static with fixed outputs, but over the course of the simulation progress, where the time frame typically lasts 60–120 s, some of the units may ramp up or down, and they would need additional modeling efforts to simulate. Such effort is unlikely to add value to model validation.

Once a system event is selected, the data (i.e., from state estimator, SCADA, and PMU) for the system and the time duration being simulated should be acquired. The data can be requested through the Reliability Coordinator (RC) and/or TOPs (per MOD-033-1 R2). The RC in WECC is Peak Reliability (Peak RC). Peak RC can provide a snapshot from their State Estimator (SE) data prior to and immediately after the event. Peak RC will use PMU data available from different part of BES to validate general dynamic response. Individual PCs will use their own data for validation of their portion of the system dynamic response.

To enhance the process and data accuracy, Peak RC staff will simulate the selected event using WSM directly and compare the simulation results to available PMU data. The following are two reasons for this process:

- To validate WSM itself for Peak Reliability
- To make sure that WSM snapshot provided for model validation represents event accurately since this snapshot is used as modeling inputs to the WECC power flow basecase for pre-contingency operational conditions

The process of WSM validation is relatively quick since the WSM case is a representation of the pre-event conditions adjusted by real-time SCADA measurements. Only event sequence needs to be investigated, prepared, and then included in the study. Since the WSM case uses the same dynamic model as the WECC power flow case, all modeling issues found in this process will be reported to the WECC staff.

After WSM validation is performed, the WECC power flow case needs to be adjusted and prepared based on the pre-contingency operating conditions. This process will be performed by the WECC staff, and it may require a few weeks to prepare the power flow study case. This is the most time-consuming part of the process.

For the preparation of the event basecase, the WECC steady-state planning model must be modified with generation dispatch, topology, and load changes based on the real-time data noted above in order to achieve a close match to actual system condition for the selected time. Reference [14] from the NERC MWG document provides more details on this process. There could be some limitations on the part where the WECC staff can modify the power flow study case for the overall WECC-wide footprint. Additional refinement of the power flow study case may be required by the PC for their own planning area.

The main intent for validating a steady-state power flow model is to compare the pre-disturbance measurement (e.g., bus voltages, real and reactive power flow on system elements and paths, generation dispatch, phase shifter settings, LTC tap positions, etc.) to the power flow solution from the WECC study case that is adjusted to the pre-disturbance operating conditions. The desired outcome would be a close match of the results obtained between the power flow simulation and the real-time measured data.

If the results are not a good match, based on engineering judgment, it is necessary to investigate the cause(s) of the discrepancies. Therefore, it is recommended to adjust voltages by allowing LTC taps, SVCs, and generators to adjust automatically based on measured conditions (AVR selection for power flow solution) and then to compare simulated tap positions and MVAR values to the actual values. This process is helpful to pinpoint the issues and to correct transformer tap positions, controlled points, and transformer impedances. Note that there will be differences between the WECC power flow case and WSM (state estimator snapshot) study case. The difference is mainly due to mismatches that are introduced to the SE power flow case during the process of state estimations. The mismatches appear as small MW and MVAR loads that are added to the study case.

In order to create an event case using a WECC basecase, the following will need to be adjusted in the WECC basecase, generation dispatch, voltage, load dispatch, and transmission line status and adjusting the flows on the phase-shifting transformers to match the WSM case.

In mapping the WSM[1] to the WECC basecase, start on the outer rim of the WECC footprint such as Alberta. Set the generation dispatch from the WSM into

---

[1] Instead of using WSM, PC may use SCADA data. Advantage of using WSM is in fact that WSM is already adjusted case that represents state of the system for given pre-disturbance event operating conditions for all WECC footprint.

the WECC basecase. This can be done by running a script that will use the WSM generator ID that will match the same ID in the WECC basecases. Next adjust the load dispatch to match the net MW interchange in or out of Alberta. The load dispatch is typically a little different due to different losses between the WSM and WECC basecases. Repeat these steps for all areas in the WECC basecases.

Setting the voltage in the WECC planning case to match the WSM case is a manual process. Starting with the area that has been affected by the disturbance, take the actual voltage of a bus from the WSM and adjust the schedule voltage parameter in the WECC basecase.

Adjusting the transmission line status is another manual process. In the WSM case under the SECDD record, you can display a field called DST and filter for values of 0. These are lines that are out of the service in the real-time case (WSM) and their statuses need to be adjusted in WECC basecase. Set the phase-shifting transformers to match the same flow that the WSM is showing across it. Once an acceptable steady-state model has been developed, the next step is to create a good dynamic data file. To begin, we will need to select the dynamic data file that is available with the approved WECC power flow basecase that was selected in Sect. 5.4. For the process of adjusting generation dispatch, load modeling, etc., in the preparation of the steady-state model, it may be necessary to have additional steady-state model adjustments as well as dynamic model adjustments. For example, individual generator outputs may need to be lowered in the steady-state power flow case, or generator real power capability parameter may need to be incremented in the dynamic model to achieve a clean dynamic initialization result. There are multiple reasons: such as (1) no plant load modeled in WSM; (2) state estimation mismatches that are not presented in basecase but they exist in WSM as additional positive/negative loads; (3) unable to map some generators one to one in between two cases. Additionally, new composite load models may need to be modeled if they are missing in the original dynamic data. It is important that the adjusted dynamic data file is initialized properly with the steady-state power flow model. In other words, generators must initialize within governor limits, missing dynamic models must be created appropriately, and initialization warning/error messages must be addressed. After the power flow case and dynamic data are prepared, a few transient runs should be performed using the new dynamic data. A no-disturbance simulation should produce flat lines; a ring-down simulation (insertion of the Chief Joseph Braking Resistor) should produce traces that initially oscillate but damp out acceptably. In addition, a few additional disturbances should be simulated (such as the double Palo Verde generation loss) with acceptable transient results. All netted generating units in the dynamic model in a specific PC's footprint should be addressed and corrected by the PC.

The next step is to create an accurate sequence of events and switching file. The TOPs of the system where the event happened should have the most accurate time sequence data. The switching sequence is created based upon the sequence of event data, such as from DFRs, relays, and other information such as SCADA or dispatcher's logs. Sequence component currents and voltages are recorded by relays.

The following example illustrates the above procedure.

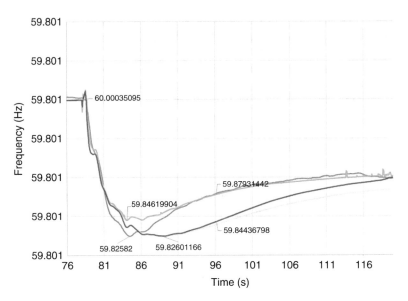

Fig. 7.10 GE/PSLF simulated system frequencies vs. PMU data on August 08, 2017 event

In August 8, 2017, event (PSLF vs PMU), 1634 MW of generation was tripped by Colstrip RAS action following 500 kV Line 1 and 500 kV Line 2 tripping. The following graphs on Figs. 7.10 and 7.11 show PMU measurement vs. WECC planning model simulation with basecase power flow adjustments by the WSM pre-disturbance SE case. WECC basecase (planning model) is used for MOD-33 compliance. This example also illustrates cross-validation of two interconnection models, planning (bus-branch) and operation (node-breaker). Simulations are performed using GE PSLF software.

7.9 Executive Summary

Utilities use different offline software tools for operation planning studies. Usually, the regulatory or regional entities such as FERC, NERC, WECC, etc. rely on offline analytical tools and planning models to perform system reliability risk assessment and event analysis. It requires tremendous efforts for planning engineers to replicate real-time system operating scenarios or system event cases in a planning basecase closely and accurately.

To overcome the gap between real-time WSM and WECC planning models, Peak RC worked with WECC and BPA to develop a mechanism to bridge two different models as follows:

- Create a master translation table for mapping nodes in the WSM with buses in WECC planning basecases and cross-mapping individual units in between.
- Insert EMS long ID for major network equipment in the WSM into WECC planning basecases.

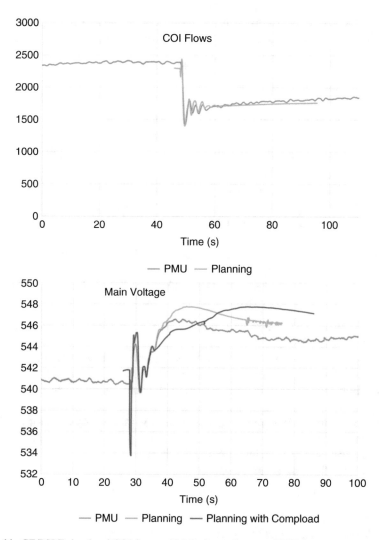

Fig. 7.11 GE/PSLF simulated COI flow and Malin bus voltage vs. PMU data

- Enable using WECC planning model in the *.dyd file to run dynamic or transient simulation on a WSM SE savecases.
- Export WSM SE snapshot cases into either node-breaker model files in *.csv format or PTI v30/v34 bus-branch model files in *.raw format.

It shall be mentioned that Peak RC collaborated with the software vendors of planning study tools, i.e., GE/PSLF, PowerWorld, and V&R Energy to consolidate their WSM node-breaker model import formats into one common *.csv format importable for all three vendor products. This work significantly reduced planning engineers of WECC member entities to run offline operation studies using the WSM export cases.

WECC has developed a procedure to set up system model validation cases from WSM SE snapshot cases. It consists of the following steps:

A. **Set up power flow case in WECC planning basecase**

1. Pick event to be studied
2. Pick starting case
3. Extract generation outputs from Peak RC's WSM
4. WSM-WECC Basecase Master Mapping file

   (a) Mapped units
   (b) Unmapped units
       WECC staff create a script to adjust base load flags for event cases

       - If the turbine types are 14, 21, 22, 23, 24, 31, and 40, sets flag to 2
       - If Pgen is greater than 78% of Pmax unless for hydro of 95%, sets flag to 2
       - If unit falls in between 30 and 78% or 30% and 95% for hydro, sets flag to 0

5. Transfer WSM generation outputs to planning case for each balancing authority (BA).
6. Adjust loads in planning case to match interchange from WSM case for each BA.
7. Adjust phase-shifting transformers in planning case to match flows in WSM case.
8. Adjust topology in planning case to match system conditions for the day of the event.
9. Adjust voltage schedules on high kV buses.
10. Check flows on major paths.

B. **Setting up dynamics for event case**

1. Get most recent Master Dynamic File (MDF).
2. Read Dynamics File in PSLF.

   (a) Run scripts to net generation that is on that has an output that is less than 3% of Pmax.
   (b) Clean up missing models.
   (c) Fix initialization errors.

       - Typically these are hydro plants that are tuned to lower MWcaps.

   (d) Create composite load model for season and time.
   (e) Run a no disturbance test to correct any issues.
   (f) Run event sequence and compare results to PMU data.

After those joint efforts, data sharing and model conversion between the WSM and WECC basecase models can be achieved successfully. The remaining effort is focused on automatic setting up of WECC basecase power flow and dynamic files from real-time WSM or equivalent EMS SE model solution. Once it's achieved, WECC planning model can be updated and set up with real-time EMS SE solution and/or ICCP measurements to support non-EMS analytical tools for real-time assessment with no or minimal user configuration.

References

1. "WECC Data preparation manual" by WECC Systems Review Working Group. (available on-line) http://www.wecc.biz/committees/StandingCommittees/PCC/TSS/SRWG/Shared%20 Documents/Forms/AllItems.aspx
2. "MMWG Procedure Manual V10" by Eastern Interconnection Working Group.
3. Dmitry Kosterev and Donald Davies, "System Model Validation Studies in WECC".
4. FERC and NERC joint report on "Arizona-Southern California Outages on Sept 8, 2011".
5. "NERC MOD 33" standard under development by North American Reliability Council.
6. Kincic, S. et al. (2012, July). Impact of massive deployment of PMUs on reliability coordination and reporting – PES GM San Diego.
7. Overholt, P., Kostrev, D., Eto, J., Yang, S., & Lesieutre, B. (2014, May/June). Improving reliability through better models. *IEEE Power & Energy*, *12*, 3
8. Ramanathan, R., & Tuck, B. (2013). *BPA's experience of implementing node breaker model for power system operations studies*, UPEC 2013.
9. Ramanathan, R., Tuck, B., Kincic, S., Davis, D., & Zhang, H. (2015). *Standardization of equipment naming convention for Node/Breaker model*. To be submitted for PES GM 2015, Denver, Co.
10. http://www.powerworld.com/files/PowerWorld-Integrated-Topology-Processing1.pdf (White Paper on PowerWorld's website written in August 2009).
11. Marzinzik, C. M., Grijalva, S., & Weber, J. D. (2009, January 5–8). Experience using planning software to solve real-time systems. *System Sciences*, 2009. HICSS '09. 42nd Hawaii international conference
12. Marzinzik, C. M., Weber, J. D., & Davis, C. M. (2012, Febuary 24–25). *Designing a power system planning tool around real-time models*. Power and energy conference at Illinois (PECI), 2012 IEEE.
13. Ramanathan, R.; Tuck, B.; O'Brien, J. (2013, July, 21–25). BPA's experience of implementing remedial action schemes in power flow for operation studies," Power and Energy Society General Meeting (PES), 2013 IEEE.
14. Procedure for validation of Powerflow and dynamic cases, NERC http://www.nerc.com/ comm/PC/Model%20Validation%20Working%20Group%20MVWG/Model_Validation_ Procedures_2011_12.pdf update to the newer version.
15. A New Framework to Facilitate the Use of Node-Breaker Operations Model for Validation of Planning Dynamic Models in WECC- PES GM 2016, Boston, MA.

Chapter 8
EMS System Architectures, Cybersecurity, and ICCP Implementation

Hongming Zhang

Acronyms

BES	Bulk Electric System
DTS	Dispatcher Training Simulator
eLSE	EPG's enhanced Linear State Estimator
EMS	Energy Management System
EPG	Electric Power Group
FERC	Federal Energy Regulatory Commission
GE	General Electric
ICCP	Inter-Control Center Communications Protocol
LOV	Peak RC's Loveland Control Center
NERC	North American Electric Reliability Corporation
OAG	GE's Open Access Gateway, i.e., ICCP application
OMS	Washington State University's Oscillation Monitoring System
PP	GE's PhasorPoint product
PI	OSIsoft Plant Information product
PEAK	Peak Reliability or Peak RC (i.e., Reliability Coordinator)
PowerTech	PowerTech Research Lab Inc.
RC	Reliability Coordinator
RCSO	Reliability Coordinator System Operator
ROSE	Region Operation Security Existence
RTDMS	EPG's Real-Time Dynamics Monitoring System
RT-VSA	Real-Time Voltage Stability Analysis
SCADA	Supervisory Control and Data Acquisition
TSAT	Transient Stability Analysis Tool
VAN	Peak RC's Vancouver Control Center
V&R	V&R Energy Research Laboratory
WECC	Western Electricity Coordinating Council
WSM	West-wide System Model
WSU	Washington State University

© Springer Nature Switzerland AG 2021
H. Zhang et al., *Advanced Power Applications for System Reliability
Monitoring*, Power Systems, https://doi.org/10.1007/978-3-030-44544-7_8

8.1 EMS System Architectures

8.1.1 EMS Server Configuration

Peak RC maintained two geographically separated data centers in Loveland, CO, and Vancouver, WA, respectively. Each data center had a cluster of servers consisting of a primary and secondary server. In the event that the connection to the primary server is lost, the secondary server capabilities allowed it to take control and essentially act as the "primary server."

For a resilient communication service, a BA/TOP needs to provide secondary network paths that become instantly operational so that the failure of a primary path will not result in a ICCP communication failure.

There were three separate EMS server networks configured at the Peak RC control rooms:

- Production EMS (Prod/EMS domain), Test EMS (Test/TEMS domain), and Development EMS (Dev/DEMS Domain. ICCP data flow was sent from Prod EMS to Test EMS for every 30 s and then sent from Test EMS to Dev EMS for every minute.
- Dev EMS servers were used for internal software development and vendor software patches testing (first-pass test). Dispatcher Training System (DTS) servers were also deployed onto the Dev environment (i.e., DEMS domain) so that the DTS was able to support new software testing against extreme system operation conditions, i.e., islanding, blackout, and so on.
- The Test EMS was a full replication of Production EMS environment, primarily used for EMS model update testing and new software release production integration test (second-pass test). Per specific request, Reliability Coordination System Operator (RCSO) and/or Real-time Operation Engineer (ROE) could be granted access to Test EMS servers for end user acceptance testing and verification on new software features/display changes prior to production deployment (Fig. 8.1).

8.1.2 EMS System Integration

The Peak RC Prod EMS was seamlessly integrated with many other software products:

- Montana Tech's Modal Analysis Software (MAS) integrated into GE PhasorPoint (PP)
- OSIsoft Plant Information (PI) for data historian
- PowerTech's Transient Security Assessment Tool (TSAT)
- V&R Energy's Region of Stability Existence (ROSE) engine-based Voltage Security Assessment (VSA)

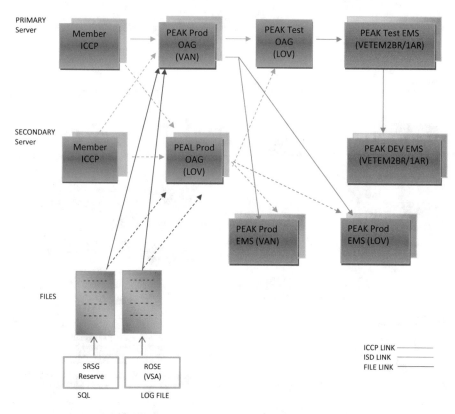

Fig. 8.1 Peak RC EMS system architecture

In the Peak RC's Test and/or Dev EMS environment, several new software products were deployed for engineering analysis and validation testing between 2017 and 2019:

- Washington State University's Oscillation Monitoring System (OMS) tool suite
- EPG's enhanced Linear State Estimator (eLSE)
- EPG's Real-Time Dynamics Monitoring System (RTDMS)

The diagram in Fig. 8.2. illustrates the Peak RC's EMS system integration design, which was implemented for its engineering lab environment successfully in 2017. This EMS integration design was proven very effective. It provided Peak RC engineers, interns, and vendor developers a sophisticated software test bed with minimal cost and least deployment effort. The engineering lab was used productively and extensively for over 3 years before Peak RC shut down the whole DEMS environment on November 1, 2019.

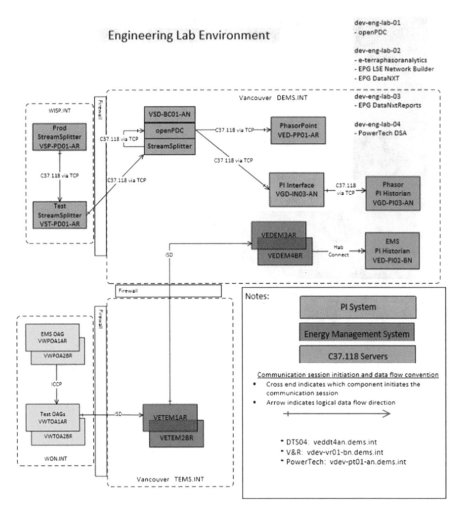

Fig. 8.2 Peak RC engineering lab EMS system integration diagram

8.2   Cybersecurity Enforcement in the EMS Environment

8.2.1   Terminology

SNMP Simple Network Management Protocol is an Internet Standard protocol for collecting and organizing information about managed devices on IP networks and for modifying that information to change device behavior.

SIEM Security Information and Event Management (SIEM) combines security information management (SIM) and security event management (SEM). SIEM provides real-time analysis of security alerts generated by applications and network hardware.

AoRs Area of Responsibilities. In SCADA and Network Applications, user Access Permission can be defined upon area of responsibility. For example, users of one entity are assigned with specific AoR that restricts them to access the EMS displays and application solution results exclusively for their own footprint.

8.2.2   EMS Cybersecurity Outline

NERC and FERC have been focused on cybersecurity standard development and compliance enforcement for the last few decades [1–3]. Leading EMS software vendors have worked with utility customers to implement advanced control room functions in compliance with the NERC CIP standards [4–8].

Peak RC adopted and complied with the NERC critical infrastructure protection ("CIP") reliability standards version 5 ("CIP Version 5 Standards") in its Prod and Test EMS environment through its RC operation business cycles. Peak RC met the NERC CIP requirements on the EMS environment by:

- Built-in vendor software cybersecurity solutions and rigorous EMS software deployment testing and maintenance procedures.
- Managing cybersecurity and operations contingency consistently.
- Adopted the system security architecture design criteria.

  – System hardening
  – Network security design
  – Single sign-on
  – Smart card
  – Change management and control process
  – Support anti-virus software
  – Support content inspection

- Access control.

  – Authentication
  – Authorization

- Data security.

  – Secure data in transit – IEC62351.

    Bump-in-the-wire (BITW) – SSL/TLS
    SSH/SFTP
    HTTPS
    SOAP and RESTFul
    Kerberos encryption

  – Secure data in storage.

    Use the encryption libraries provided by operating system, the application
    server containers, and RDBMS to protect data in files and memories.

- Common logging service (CLS) for logging and system auditing.

  – Security Information and Event Management (SIEM)
  – Simple Network Management Protocol (SNMP) agents

- Prompt vulnerability response and security patches deployment.

  – Heartbeat monitor for all real-time applications identified critical for RC
    operational awareness.
  – Implement specific operation procedures for EMS, ICCP, and/or PI system
    failover, as well as loss of other real-time assessment tools.
  – 7×24 on shift operation engineer support plus on call support from IT and
    network applications teams.
  – Fast track change management for urgent software fixing and EMS model
    update patches.

- Data archiving in compliance with the NERC CIP standards.

  – Record SCADA data in HDR files continuously.
  – Autosave EMS SE basecase snapshot file for every 5 min.
  – Backup RT-VSA data files after each run (~5 min).
  – Backup online TSAT data files after each run (2–10 min).
  – Archived files were periodically moved into the data drives, which were
    accessible to authorized users.

- Specify the network security zones.

  – Zone 1: General office areas
  – Zone 2: Control rooms
  – Zone 3: Data centers

- Must work with firewalls and IDS/IPS.

  – Set up firewalls between any two different networks/domains:

    Corporate network
    Production/EMS network
    Test/TEMS network
    Dev/DEMS Network

  – Secured FTGs were configured to transfer files between corporate drive and
    local desktops or laptops bi-directionally.

- Secured FTGs were configured to transfer files between DEMS and TEMS bi-directionally.
- Secured FTGs were configured to transfer non-executable files between TEMS and EMS bi-directionally.
- Autonotification for any illegal file transfer operations

8.2.3 Access Control

Peak RC fully implemented GE/Alstom's EMS access control features described in Fig. 8.3. It includes both authentication and authorization processes.

8.2.4 Logging and Auditing Based on SIEM Integration (Fig. 8.4)

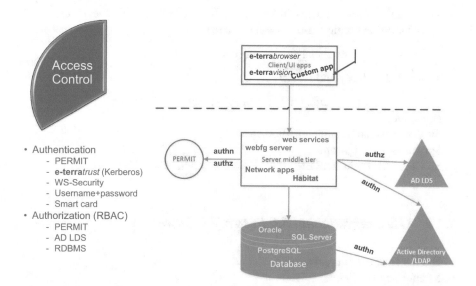

Fig. 8.3 GE/EMS built-in access control diagram

SIEM Integration

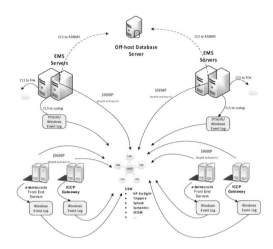

- Common Logging Service
  - Windows Event Log and syslog support
  - Consistent message format across e-terra product lines
  - Pick and choose your messages sent to SIEM
- SNMP Agents

Fig. 8.4 SIEM integration for logging and auditing

8.2.5 Authentication and Authorization

- Support centralized user account management.
  - Managing user accounts in one centralized location. This means creating/ updating/removing a user account in one place.
- Utilize standard user management services.
  - Active Directory and LDAP servers
- Provide a unified authentication and authorization framework for NMS applications.
- Support two-factor authentication.
- Support single sign-on.
- Support both traditional EMS/SCADA applications and web services.

8.2.6 GE e-terrahabitat 5.8 Security Features

- Habitat runs under **local service** account
- Minimizing privileges
- Ease of password management (local service has no password)
- Habitat runs by users who have limited privileges
- Multiple habitat instances run as Windows service
- Manage users and computers in Active Directory

- Manage privileges with GPOs
- Eliminate shared accounts
- Tighten the umask so world has no rwx access after installation

8.3   Peak RC ICCP Application

8.3.1   Terminology

- **Client:** An ICCP user who request services or objects owned by another ICCP user acting as a server.
- **CNP** (Communications Network Processor): The machine running the OAG software.
- **Data Item (DI):** A data element to be exchanged, this can be analog or status values.
- **Data Set:** A collection of data items to be exchanged or forwarded to a client.
- **e-terra*comm*:** GE/ALSTOM's product that consists of communication server software and a set of protocol link applications to interconnect utilities using a range of communications protocols.
- **FileLink:** Process of the OAG system that implements data transfers between a host (local computer being communicated to using ISD) and the CNP.
- **ICCP** (Inter-Control Center Communications Protocol): Supports the exchange of operational data between EMS/SCADA systems.
- **ICCP Client (ICCPCL):** The role of an ICCP node that acquires data from an ICCP server. The client defines the data to be received and the conditions under which the data is to be received and enables and disables the reception of that data.
- **ICCP Node:** The computer connected to the data exchange network that performs ICCP client functions, ICCP server functions, or both.
- **ICCP Server:** The role of an ICCP node that makes data available to an ICCP client. The ICCP server responds to requests for data exchange definitions, requests to enable data exchange, and requests to disable data exchange. The ICCP server sends data to a requesting ICCP client based on the conditions of transfer defined by the ICCP client.
- **ISD** (Inter-Site Data): A registered protocol developed by GE/ALSTOM to support the exchange of data among GE/ALSTOM SCADA sites.
- **ISDLINK:** An OAG link application using ISD to communicate with a host running an ALSTOM **e-terra*scada*** subsystem.
- **LINK:** A virtual connection to a remote communications partner. The number of links that are modeled corresponds to the number of simultaneous connections with which you wish to communicate onto that partner.
- **OAG:** Open Access Gateway (GE/ALSTOM ICCP software).

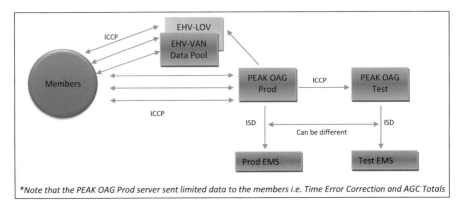

Fig. 8.5 Peak RC OAG/ICCP data path

- **OAGMODEL:** The **e-terra***comm*'s modeling and validation application. All modeling is done in the OAGMOM database.
- **OAGSERVE:** The database serving task of the **e-terra***comm*.
- **PATH:** Designates the actual connection instances that are possible for a LINK. When multiple PATH records occur under a LINK record, the link application will make a connection on only one of the PATHs at a time.
- **Remote:** Describes a remote communications partner – an entity with which you wish to exchange data, such as another control center or a local EMS system.
- **Server:** An ICCP user that is the source of data and provides services for accessing the data.

8.3.2   Peak RC ICCP Implementation Overview

Inter-Control Center Communications Protocol (ICCP) allows for data exchange over wide-area network (WAN) between utility control centers. Peak RC, however, has chosen to use the WON (WECC Operations Network). ICCP provides for the exchange of real-time power system information including status and measured values. As illustrated in Fig. 8.5, the Peak RC Open Access Gate (OAG, GE's terminology for ICCP) application consists of both client and server software. The client software connects to other members on the network to request point, while the server software responds to client request by returning the requested.

8.3.3   Type of Communication (ICCPLINK vs. ISDLINK)

ICCPLINK is the industry standard for master to master communications, e.g., external ICCP is transported to the Peak RC OAG via the ICCPLINK. Quality codes, such as manual set and telemetry failure, are transmitted along with the data.

ISDLINK is a GE (formerly ALSTOM) protocol that facilitates the communication of inter-site data between SCADA systems, e.g., communication between the Peak RC Production OAG and the Peak RC Production EMS.

8.3.4   GE's Open Access Gateway (OAG) Application

8.3.4.1   Introduction

All OAG modeling is done within the *OAGMOM* database, and with each WSM model build, the OAGMOM has to be updated to enable receiving and sending ICCP measurements between two control centers.

OAGMODEL is the GE's application used to support database modeling. A valid OAGMOM is imported into a different application, the *OAGSERVE*. The OAGSERVE is for operational/online use, and its use is limited when it comes to updating the OAG database.

Navigation between the modeling displays is provided by various means available on the menu bar below.

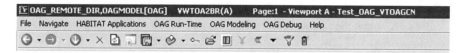

HABITAT Applications provides the Process Manager, Configuration Control, Process Manager, etc.

OAG Run-Time provides access to the displays and modifications that can be done while OAG is running.

OAG Modeling pull down menu provides access to all modeling displays.

***Some display names have the same name but different functionality based on which pull-down menu they are retrieved from. In this document such displays will be followed by the source in parenthesis. ***

8.3.4.2   OAGSERVE Master (OAG Run-Time)

This display on Fig. 8.6 shows information for the database currently "online" and allows retrieval of a savecase into an OAGSERVE clone.

8.3.4.3   Remote Directory (OAG Run-Time)

This lists all the remote sites Peak RC has established communications with via ICCP; additionally this list has the local EMS system which enables data to be exchanged between the Loveland (_EMSL) and Vancouver servers (_EMS).

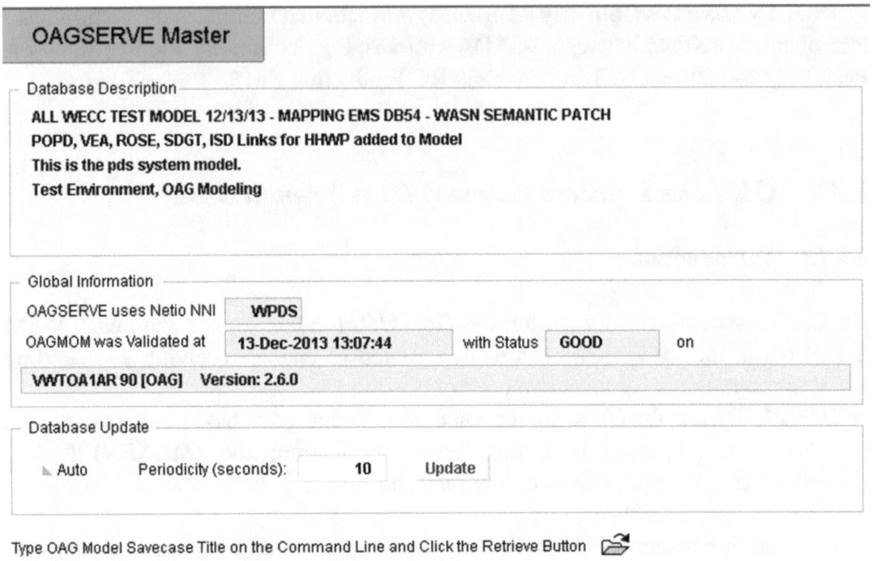

OAGSERVE Master

Database Description

ALL WECC TEST MODEL 12/13/13 - MAPPING EMS DB54 - WASN SEMANTIC PATCH
POPD, VEA, ROSE, SDGT, ISD Links for HHWP added to Model
This is the pds system model.
Test Environment, OAG Modeling

Global Information

OAGSERVE uses Netio NNI   **WPDS**

OAGMOM was Validated at   **13-Dec-2013 13:07:44**   with Status   **GOOD**   on

VWTOA1AR 90 [OAG]   **Version: 2.6.0**

Database Update

⊾ Auto   Periodicity (seconds):   **10**   Update

Type OAG Model Savecase Title on the Command Line and Click the Retrieve Button 🖻

Fig. 8.6 OAGSERVE Master Display

Peak RC maintains two separate data centers (VAN, LOV), and as such, all remote partners will have three remote sites, for example:

WECC: ICCPLINK links member ICCP to Peak RC Prod OAG.
WECC_EMS: ISDLINK to VAN site.
WECC_EMSL: ISDLINK to LOV site.

The status of an ICCP link shows either "Up/Normal" in green or "Down/Bad" in red color (Fig. 8.7).

When one of the remotes is selected, a **status of topology** display is called and positioned onto the communication topology for that remote. The display describes the quality and qualities of the connection between OAG and remote partners. The hierarchy of this topology has record type:

Remote: Remote communications partner – the entity with which you exchange data or a local EMS system.
Link: Virtual connection to entity with which you exchange data. This can be either ICCPLINK or ISDLINK.
Path: Actual connection instances that are possible for a link. A LINK record can have multiple PATH records, but the link application will make connection on only one of the PATHS at a time. The diagram below shows four paths available under the Peak link; however three paths have the connection status Con: **DOWN** and only one path with Con: **UP** (Fig. 8.8).

Status of Topology Directory

AESO	AVA	AZPS	BCTC	BHPL	BPA
CAISO	CECD	CFE	CHPD	CSU	DOPD
EHV	EPE	EWEB	FEUS	GCPD	IID
IPC	LDWP	NVE	NWE	PACE	PACW
PGAE	PGE	PNM	PRPA	PSCO	PSE
POPD	SCE	SCL	SCPD	SDGE	SDGT
SMUD	SRP	SRSG	SWTC	TEP	TID
TPWR	TSGT	ROSA	ROSP	VEA	WACM
WALC	WASN	WAUW	WECC	AESO_EMS	AVA_EMS
AZPS_EMS	BCTC_EMS	BPA_EMS	BHPL_EMS	CASO_EMS	CECD_EMS
CFE_EMS	CHPD_EMS	CSU_EMS	DOPD_EMS	EPE_EMS	EWEB_EMS
FEUS_EMS	GCPD_EMS	HRWP_EMS	IID_EMS	IPC_EMS	LDWP_EMS
NVE_EMS	NWE_EMS	PACE_EMS	PACW_EMS	PGAE_EMS	PGE_EMS
PNM_EMS	PRPA_EMS	PSCO_EMS	PSE_EMS	POPD_EMS	ROSA_EMS
ROSP_EMS	SCE_EMS	SCL_EMS	SCPD_EMS	SDGE_EMS	SMUD_EMS
SRP_EMS	SRSG_EMS	TEP_EMS	SWTC_EMS	TID_EMS	TPWR_EMS
TSGT_EMS	VEA_EMS	WACM_EMS	WALC_EMS	WASN_EMS	WAUW_EMS
WECC_EMS	AESO_EMSL	AVA_EMSL	AZPS_EMSL	BCTC_EMSL	BPA_EMSL
BHPL_EMSL	CASO_EMSL	CECD_EMSL	CFE_EMSL	CHPD_EMSL	CSU_EMSL
DOPD_EMSL	EPE_EMSL	EWEB_EMSL	FEUS_EMSL	GCPD_EMSL	HRWP_EMSL

Fig. 8.7 Remote Directory topology

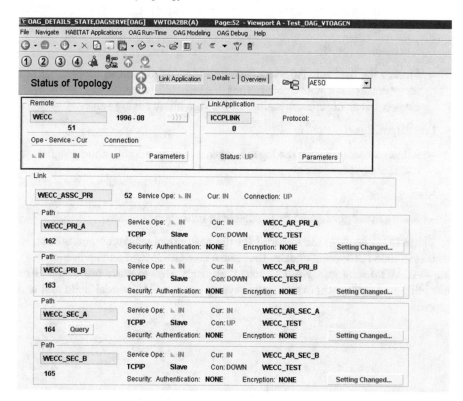

Fig. 8.8 Status of OAGSERVE topology

Other OAG displays for ICCP update and OAG database upload are as follows:

- Remote Directory
  - List of already modeled remotes.
- Link applications
  - Models the link applications used (ICCPLINK, ISDLINK, etc.) These form the link between OAGSERVE (data server) and a remote partner.
- Data items
  - Type: Models all the DI exchanged
  - Source: Specifies from which remote a DI comes from (ICCP scope)
- DI permission
  - Models the remote permissions and defines which sites can access a given DI
- Data sets (definition)
  - Models data sets
- Data sets (contents directory)
  - Lists existing data sets uses to access their contents (i.e., which DI are in these data sets)
- Data semantics conversion
  - Models the data semantics conversions. This is modeled to support defined representations of status, conversions to alternate representations, and conversions between different meanings of the sign bit on analog values.
- Log dispositions
  - Allows selecting where the log messages issued by the validation program go (database or file)
- Message log
  - Shows messages logged by the database validation program

Those OAG displays are intended for the new model validation and upload, as well as limited ICCP point update. To create a new OAGMOM and update an existing OAGMOM with massive ICCP changes, users need to make use of the GE's modeling tool suite, **e-terra**source, or develop custom OAG model build tool and process.

8.3.5   Peak RC ICCP/OAG Model Build Process

8.3.5.1   WSM Data Maintenance Access DB

Peak RC developed an offline WSM Data Maintenance Access database (DB) tool described in Fig. 8.9, to manage all WSM-relevant ICCP (both inbound and outbound) points, network equipment limits, and special measurement (SPMEAS) tables containing dynamic path limits, RAS data, AGC signals, etc. This inhouse-developed Access DB facilitated the Peak RC's Modeling team to update ICCP points and network limits for monthly EMS model update.

The WSM Data Maintenance Access DB comprises a complete list of SCADA mapping for the existing ICCP points and populates associated with network equipment limits. The Access DB was built in many scripts to automate data sanity check, user-defined calculations, and creating files for the OAG model build process. For each monthly EMS model update between 2009 and 2019, the Peak RC's Modeling team continually updated a number of ICCP points, SCADA mapping, and network limits in the Access DB upon new model change submittals from Peak RC's member utilities.

Figure 8.10 shows an example of WSM SCADA mapping table. It includes the following data fields:

- SUBSTN: Substation name.
- DEVTYP: Device type.

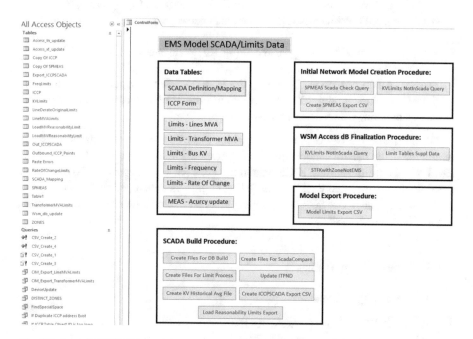

Fig. 8.9 Peak RC WSM Data Maintenance Access DB overview

Fig. 8.10 Peak RC WSM SCADA mapping table view

- DEVICE: Device ID.
- ID: SCADA Measurement ID, i.e., MW/MV (analog), STTS (status), RAS, etc.
- ICCP ObjectID: Primary ICCP key for data exchange between the sender and the receiver.
- Zone: Company ID sending or receiving ICCP measurements to or from Peak RC.
- ITPND_Analog: A node ID to map the SCADA telemetry point to the node in the SE network model. Correctly assigning the ITPND enables the SE solved values matching the SCADA measurements properly.

All entries in the SCADA mapping tables must be consistent with the existing and/or expected network model changes represented in the Peak RC's EMS and the EMS of the entities sending ICCP data to the Peak RC control centers. To eliminate possible data errors, the Peak RC Modeling engineers developed many scripts and procedures to check out SCADA mapping errors and clean them up before launching an OAG model update.

8.3.5.2 Prepare Input Files

Main input data files include:

- Online PROD savecase (OAGSERVE)
- OAG Model savecase online (OAGMODEL)

- WSM Data Maintenance Access DB
- Online SCADA savecase

There are eight steps for the modelers to prepare the input files for OAG model build:

1. Create WSM_Data_Maintenance_Pre[DBxx].accdb in the EMS Model Test environment for Pre-DBxx OAG Model.
2. Generate the Access_ICCP list from the Access DB.
3. Check for differences or errors between OAGSERVE (OAG Input.csv) and the Access database (Access_ICCPList).
4. Run "Duplicate ICCP Query" check and fix the duplicate points if any:

   (a) Correct this by deleting in **SCADA Mapping** or **ICCP** Table, as appropriate.

5. Prepare Pre [DBxx] _ICCP_Compare.xlsx spreadsheet.
6. Complete creation of WSM_Data_Maintenance_Pre-DBxx.accdb.

   (a) Append new SCADA Mapping points to WSM_Data_Maintenance_ Pre[DBxx].accdb:

      (i) Copy and paste all "NEWPOINT" records from Pre[DBxx]_ICCP_ Compare.xlsx, *NewScadaMapping* tab → WSM_Data_Maintenance_ Pre[DBxx].accdb, *SCADA_Mapping* table

   (b) Append new ICCP points to WSM_Data_Maintenance_Pre[DBxx].accdb.

7. Generate the Pre-DBxx ICCPSCADA List by the custom VB+ scripts.
8. Compact and repair WSM_Data_Maintenance_Pre [DBxx].accdb.

8.3.5.3 Create the Pre-DB OAG Model Case for the New Database [DBxx]

- Import online and newly added ICCP.
- Combination of online and upcoming DB. An example of combining online and upcoming ICCP point lists in the Access DBxx is given in Fig. 8.11.
- Records include:

   - ICCPs deleted for upcoming DB
   - ICCPs added for upcoming DB
   - Combination of online and upcoming DB

- Append new ICCP points to frozen WSM Data Maintenance DB. As an example, illustrated on Fig. 8.12, user must Append records with unique DINAME, i.e., Substn$Devtyp$Device$ID.
- Create OAGMOM/OAGModel savecases using the final [DByy] OAGModel case as the starting savecase.

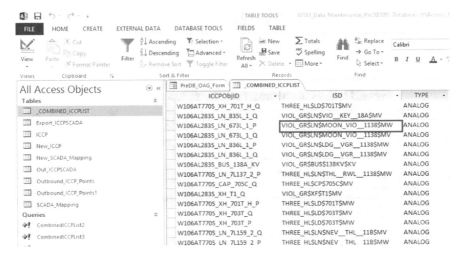

Fig. 8.11 Combining online and upcoming ICCP points in the Access DBxx

SUBSTN	DEVTYP	DEVICE	ID	ICCPObjID	ITPND_ANALOG	ZONE	ZTPND_ANALOG	ALARM_CAT	ACURCY
NEWPOINT	LN	NEWPOINT0001	MV	W040_03015048	201	PGE			MBR2
NEWPOINT	LN	NEWPOINT0002	MW	W040_03015047	201	PGE			MBR2
NEWPOINT	D	NEWPOINT0003	MW	W042_BINGEN_MW	615	PACW			MINJ
NEWPOINT	DSC	NEWPOINT0004	STTS	W001_P125562		PACE		SW	
NEWPOINT	DSC	NEWPOINT0005	STTS	W001_P125558		PACE		SW	
NEWPOINT	DSC	NEWPOINT0006	STTS	W001_P125559		PACE		SW	
NEWPOINT	DSC	NEWPOINT0007	STTS	W001_P125560		PACE		SW	
NEWPOINT	DSC	NEWPOINT0008	STTS	W001_P125553		PACE		SW	
NEWPOINT	DSC	NEWPOINT0009	STTS	W001_P125554		PACE		SW	
NEWPOINT	DSC	NEWPOINT0010	STTS	W001_P125561		PACE		SW	
NEWPOINT	DSC	NEWPOINT0011	STTS	W001_P125563		PACE		SW	
NEWPOINT	PCB	NEWPOINT0012	STTS	W001_P125458		PACE		P1	
NEWPOINT	PCB	NEWPOINT0013	STTS	W001_P125462		PACE		P1	

Fig. 8.12 Append new records with unique DINAME

- Run Perl script oag_model_create.pl, and create OAGModel\OAGMOM savecase.

  If new remotes (ICCP Link) were added, write a *.rio script to assign read permission for all outbound points.

8.3.5.4   Load Pre-[DBxx] Savecase on Test OAG Server (TOAG) for Validation

Validate Pre-OAG DBxx in TOAG twice and ensure no errors in the validation logs. These are the new oagmodel_oagmom_PreDBxx savecases (VAN and LOV).

8.3.5.5   Perform Pre-[DBxx] Validated Savecase Checks

8.3.5.5.1   Obtain the Following Savecases from Prod Servers

(a) Validated pre-[DBxx] for both VAN and LOV (Pre[DBxx] OAGMODEL/
    OAGMOM)
(b) Online Production savecase ([DByy] OAGSERVE/OAGMOM)
    Note: This is the savecase that Dev provides in the very beginning.
(c) Test OAG (TOAG OAGMODEL/OAGMOM)

8.3.5.5.2   Perform Initial OAG Savecase Checks Before Pre-DBxx
Propagation

(a) Run perl script: oag_model_compare.pl.
(b) ICCP-SCADA translation.

   – SCADA ID: Substn$Devtyp$Device$ID
   – DINAME; Substn$Devtyp$Device$ID

(c) Zone/remote.

   – OAG uses "zone" to assign permissions and data sets.

8.3.5.5.3   Compare Pre-[DBxx] (LOV) vs. Pre-[DBxx] (VAN)

(a) Run perl script: oag_model_compare.pl in the OAG build work dir.

   – Compare Online Production savecase [DByy] vs. Pre-[DBxx] (LOV or
     VAN)
   – **Verification:** Outcome is new/added points and data set differences.
   – **Sanity Check:** Number of points added equals sum of new SCADA map-
     ping and new ICCP points in the PreDBxx_ICCP_Compare spreadsheet.
     There should not be any deleted points or mapping.

8.3.5.6   Data Quality Check After Pre-DBxx Load on Production

1. Perform initial garbage data check.

   (a) Navigate to the OAG build work dir.
       Run perl script: find_oag_garbage.pl [oag_savecase] [scada_savecase.

8.3.5.7   Create Final OAGMODEL/OAGMOM Savecase

- Frozen WSM Data Maintenance.

    – Correct invalid data.
    – Final ICCPSCADA.

- Repeat OAG model build process.
- Repeat data quality checks.

    – Ensure points mapped in OAG are mapped in SCADA.

8.3.5.8   Perform Final Savecase Checks

1. Perform final OAG and savecase checks before current database model propagation using validated cases.

    (a) Run perl script: oag_model_compare.pl.

2. These are the final OAGMODEL\OAGMOM savecases to load for the final
   OAGMOM.

8.3.5.9   Post-production Upload Garbage Checks

1. After model goes into production and EMS failover, perform final garbage data
   check, and create ICCP garbage/Invalid ICCP list.
2. Verify permissions and outbound points received by entities correctly.

8.3.6   Online ICCP Database Update

Upon completion of model changes, the following steps can be taken to implement
changes to OAGSERVE. It is assumed backup savecases of OAGMODEL\
OAGMOM and OAGSERVE\OAGMOM were created prior to editing
OAGMODEL.

1. Validate model.

    - Navigate to OAG Modeling>OAGMODEL Master.
    - Edit the DB description.
    - Run database validation; status should be GOOD. If validation results in
      warnings or fails, review logs and fix.

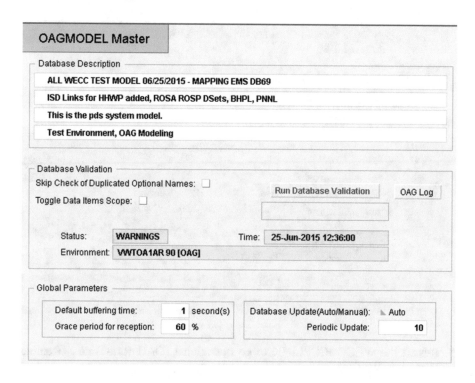

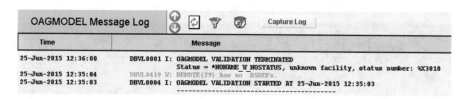

1. Save OAGMODEL/OAGMOM savecase.
2. Retrieve OAGMODEL savecase in OAGSERVE clone.

   - Stop process via ProcMan: FileLink, ISDLINK, ICCPLINK, and OAGSERVE.
   - Navigate to OAG Run-time > OAGSERVE Master.
   - Enter saved OAGMODEL savecase on command line.
   - Use icon 📂 to retrieve the OAGMODEL savecase.
   - Restart processes via ProcMan: FileLink,ISDLINK,ICCPLINK, and OAGSERVE.

OAGSERVE Master

Database Description
ALL WECC TEST MODEL 06/25/2015 - MAPPING EMS DB69
ISD Links for HHWP added, ROSA ROSP DSets, BHPL, PNNL
This is the pds system model.
Test Environment, OAG Modeling

Global Information
OAGSERVE uses Netio NNI   WPDS
OAGMOM was Validated at   25-Jun-2015 12:36:00   with Status   WARNINGS   on
VWTOA1AR 90 [OAG]   Version: 2.6.0

Database Update
⊾ Auto   Periodicity (seconds):   10   Update

Type OAG Model Savecase Title on the Command Line and Click the Retrieve Button 🖝

3. Restart the WECC ICCPLINK to go operational (OAG Run-Time > Remote Directory>WECC).

   • Toggle back the Remote IN service.

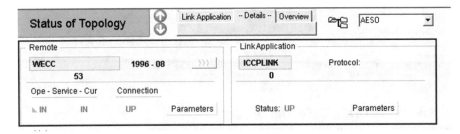

4. Verify model changes on OAG Run-Time.

8.4   Summary

Peak RC a built highly redundant and reliable EMS environment with GE's built-in cybersecurity reinforcement features and internal CIP compliance process, including rigorous change management procedures for vendor software patches deployment and EMS model update onto the production. Peak RC implemented the ICCP application for receiving over 150,000 SCADA measurements for every 2–10 s from Western entities for wide-area real-time assessment successfully between 2009 and 2019. To manage a large amount of ICCP points, Peak RC developed the WSM Data Maintenance DB in Access database format and automated OAG/ICCP model build process to achieve 111 times successful build of the WSM EMS model update, including OAG/ICCP models.

References

1. NERC CIP Standards online available: https://www.nerc.com/pa/Stand/Pages/CIPStandards.aspx
2. NERC Cyber Attack Task Force Final Report (2012.) online available: https://www.yumpu.com/en/document/read/11702317/cyber-attack-task-force-final-report-nerc
3. FERC Cyber Security Focused Areas A-4. Staff Presentation online available: https://www.ferc.gov/industries/electric/indus-act/reliability/cybersecurity/11-21-19-A-4-presentation.pdf
4. GE Advanced EMS Platform online available: https://www.ge.com/digital/sites/default/files/download_assets/advanced-ems-platform-from-ge-digital.pdf
5. GE/Alstom Grid Solution Network Management North American User Group Conference online available: https://www.gegridsolutions.com/alstomenergy/grid/grid/ug/Network-Management-Solutions/index.html
6. ABB Cyber Security Network Management online available: https://new.abb.com/process-automation/process-automation-service/advanced-digital-services/abb-ability-cyber-security-services/cyber-security-network-management
7. Siemens Cyber Security in Systems and Solutions online available: https://new.siemens.com/global/en/products/energy/energy-automation-and-smart-grid/grid-security/system-security.html
8. PNNL Secure ICCP Final Report online available: https://www.pnnl.gov/main/publications/external/technical_reports/PNNL-26729.pdf

Chapter 9
Conclusion

Hongming Zhang

Acronyms

AESO	Alberta Electric System Operator
ATC	Available Transfer Capability
BC Hydro	British Columbia Hydro and Power Authority
BA	Balancing authority
BPA	Bonneville Power Administration
CAISO	California Independent System Operator
DER	Distributed Energy Resources
DTS	Dispatcher Training System
EMS	Energy Management System
ERCOT	Electric Reliability Coordinating Council of Texas
FERC	Federal Energy Regulatory Commission
IROL	Interconnection Reliability Operating Limit
NERC	North American Electric Reliability Corporation
NWPP	Northwest Power Pool
PEAK	Peak Reliability or Peak RC
PMU	Phasor Measurement Unit
RAS	Remedial Action Scheme
RC	Reliability Coordinator
RTCA	Real-Time Contingency Analysis
SOL	System operating limit
SPP	Southwest Power Pool
SSAT	Small Signal Analysis Tool
TOP	Transmission operator
TSAT	Transient Security Assessment Tool
TTC	Total Transfer Capability
UDSA	Universal Data Sharing Agreement
UFLS	Under-frequency load shedding

© Springer Nature Switzerland AG 2021
H. Zhang et al., *Advanced Power Applications for System Reliability Monitoring*, Power Systems, https://doi.org/10.1007/978-3-030-44544-7_9

RT-VSA	Real-Time Voltage Stability Assessment
WECC	Western Electricity Coordinating Council
WI	Western Interconnection
WISP	Western Interconnection Synchrophasor Program
WSM	West-wide System Model

9.1 Introduction

In closing this book, the Western Interconnection is a unique grid. It is geographically diverse with scattered populations served by thousands of miles of transmission lines. Peak Reliability, the balancing authorities (BA), and transmission operators (TOP) have stepped up over the years to solve problems that can arise in such a unique system. Peak RC shut down the operation systems in December 3, 2019, and closed the doors in December 13, 2019, following smooth transitions of its Reliability Coordination services to RC West of CAISO, Western Reliability of SPP, and BC Hydro, respectively.

During the past 10 years of operation, Peak RC made a good partnership with western entities, which has resulted in enhanced reliability for the populations served. And at the request of our stakeholders, Peak RC has developed innovative tools and processes, and many will remain effective after Peak RC closed, continuing on at RC West, SPP, and other utilities. This chapter starts with a debrief of Peak RC function development and wind down and then summarizes highlights of Peak RC's achievements of technology innovations and industry leadership. The following sections will discuss the possibility of extending Peak RC's proven work to resolve emerging challenges of grid operations under high penetration of renewable energy resources.

9.2 Life Cycle of the Organization

9.2.1 January 2009

WECC consolidates Pacific Northwest Security Coordinator, Rocky Mountain Desert RC, and California Mexico RC into a single RC.

9.2.2 June 2013

WECC members approve bifurcation into Regional Entity (WECC) and a Reliability Coordinator (Peak Reliability).

9.2.3 January 1, 2014

Peak Reliability is officially launched, with control centers in Vancouver, Washington, and Loveland, Colorado, providing a wide-area view.

9.2.4 December 7, 2017

Responding to potential changes to its RC Area footprint, Peak RC explores a range of options to advance reliability while reducing overall costs.

9.2.5 July 18, 2018

Peak RC's solutions don't gain traction, and Peak announces it will cease operations at the end of 2019.

9.2.6 August 6, 2018

Peak RC releases results of its stakeholder comment process, affirming that the majority of Peak RC's funders support the wind down of the organization and the transition of RC services from Peak to alternative providers at the end of 2019.

9.2.7 September 18, 2018

Board of Directors ratifies decision to close Peak RC by Dec. 31, 2019. Peak begins work on the organizational wind down.

9.2.8 December 3, 2019

Peak RC ceases operation as the Reliability Coordinator.

9.2.9 December 13, 2019

Peak RC closes its doors.

9.3   Highlights of Peak's Reliability Services

Peak RC's Area included

> all or parts of 14 western states, British Columbia and the
> northern portion of Baja California, Mexico.

74million

Peak was the single, unbiased entity focused exclusively on reliability for the more than
74 million residential customers in the West.

1.6million

Peak's Reliability Area was **1.6 million square miles**, with **110,129 miles**

very 10 seconds, data was collected from more than **62 TOPs and 37
individual BAs** which equates to

more than **177,000** real-

time data points.

111   Interconnection-wide model updates

performed by Peak to the West-wide System Model.

Every 5 minutes

Peak ran real-time contingency analysis which included close to
300 RAS schemes and/or Protection Relays significantly
impacting BES in the Western Interconnection.

There were $\mathbf{12{,}556}$ participants in Peak's Dispatcher Training Simulation (DTS) exercises.

9.4 Achievements

With its Interconnection-wide view and singular reliability focus, Peak RC created a reliable and efficient operational platform for the Western Interconnection. Its employees also participated in many national reliability improvement efforts.

Many of the tools and processes developed by Peak RC that the Interconnection has benefited from over the years include the following.

9.4.1 West-wide System Model (WSM)

WSM represents full topology of the system and has been one of the most valuable accomplishments of Peak RC. Prior to WSM, Western Interconnection system models were bus-branch models (i.e., not representing full topology of the system) or regional models, thus not representing the entire system.

9.4.2 Dispatcher Training Simulator (DTS)

Through DTS, Peak RC conducted Interconnection-wide training exercises that gave BAs and TOPs a real-time replication of the system for the entire Western Interconnection and allowed operators to practice their response to system disruptions and restoration based on realistic conditions and events but in a simulated environment.

9.4.3 Enhanced Curtailment Calculator (ECC)

ECC allows entities to reliably maximize the assets on the grid, using real-time information to provide detailed real-time situational awareness of the variables contributing to transmission system operating limit (SOL) exceedances over a given Western Interconnection Path or facility.

9.4.4 Real-Time Voltage Stability Analysis (RT-VSA)

This tool monitors multiple voltage stability Interconnection Reliability Operating Limits (IROLs) in the Western Interconnection.

9.4.5   Hosted Advanced Applications (HAA)

HAA is a Cloud-based service that Peak RC provided to the TOPs so they could gain access to some of the same situational awareness tools as Peak. Also, when a TOP's primary real-time assessments tools were not available, it could rely on HAA to perform real-time assessments.

9.4.6   WECC Interchange Tool (WIT)

This tool facilitates BA compliance with NERC Standard BAL-004-WECC-3 requirements.

9.4.7   Reliability Messaging Tool

RMT is a Web-based system that conveys important real-time updates related to electrical system operation, improving situational awareness and wide-area monitoring of the Bulk Electric System.

9.4.8   Transient Security Assessment Tool (TSAT)

This tool monitors expected system transient response for frequency, voltage, and flow and provides accurate situational awareness for system operators and engineers necessary before the event actually occurs. Peak RC also made several improvements to its real-time model and mapped the real-time model to the dynamic data so state estimator cases could be directly studied in transient domain.

9.4.9   Western Interconnection Synchrophasor Program (WISP)

Peak RC coordinated the implementation of WISP that saw the installation of Synchrophasor technology across the Interconnection and also developed several tools based on Synchrophasor data, such as mode meter and linear state estimator.

9.4.10   Forced Oscillation Detection Source Locator Tool (FODSL)

In collaboration with Washington State University (WSU), Peak RC developed a tool that can locate the source of forced oscillations in the system. Peak has transferred the intellectual rights of this tool to WSU so that it can continue to provide benefit to futureRCs as they seek benefit from the implementation of Phasor Measurement Units (PMU) in the Western Interconnection.

9.4.11   Operating Plan Database Coordination

Peak RC over the years has created a database of operating plans coordinated among TOPs for operational reliability issues related to system outages. Peak RC has provided the entire database to future RCs for their future use and benefit so that this knowledge is not lost and they don't have to re-create those operating plans from scratch.

9.4.12   Daily Operational Excellence Metrics

Peak RC developed a process where various operational excellence metrics were measured on a daily basis and processes/tools were improved when issues were identified.

9.4.13   RC Workbook

Peak RC internally developed a tool that provided situational awareness information to the system operators at a single application, e.g., RTCA results, RAS information, outages, and forecast data.

9.4.14   Universal Data Sharing Agreement (UDSA)

Peak RC coordinated a UDSA among all funding members so that necessary operational data could be shared among all BAs and TOPs for improved situational awareness. Prior to UDSA being in place, data sharing among BAs and TOPs was restricted.

9.4.15   Remedial Action Scheme (RAS) Modeling

Peak RC not only modeled all of the remedial action schemes into its operational tools but also modeled non-RAS automatic schemes that could impact the system, for accurate situational awareness for system operators and engineers.

9.4.16   System Operating Limit Methodology

Peak RC developed an Interconnection-wide SOL methodology which not only improved reliability but also eliminated the need to take unnecessary mitigation actions when situational awareness tools did not show problems.

9.4.17   Outage Coordination

Peak RC developed an Interconnection-wide outage coordination process that provided BAs and TOPs the appropriate timelines to coordinate mitigation actions to support reliability and at the same time take their required maintenance outages, as per their schedules, in a more coordinated fashion.

9.4.18   Seasonal Planning Process

Peak RC developed an Interconnection-wide seasonal coordination process that improved upon the existing seasonal planning processes that BAs and TOPs were using, so that expected system conditions could be studied prior to the season and appropriate operating plans could be coordinated among various sub-regional groups.

9.4.19   Dynamic Ratings Application

Peak RC implemented a dynamic ratings application in its Energy Management System (EMS) that allowed TOPs to use temperature-based ratings for their transmission facilities, thus allowing the use of maximum allowable transmission capacity.

9.5 Recognitions

Peak RC employees have worked with the new Reliability Coordinators (RC) and others to ensure a smooth transition of RC operations. Our employees have been involved in certification teams for the RCs, worked with the RC transition team and the Member Advisory Committee (MAC), and attended all WECC RC and transition team meetings, committed to a successful transition. Through what could have been a difficult process, employees have received accolades for their professionalism and dedication, doing what's best for reliability and what's right for the Western Interconnection. The RC West Oversight Committee, the Western Electricity Leaders (WEIL), and SPP's Western Reliability Committee have all provided their acknowledgment for a job well done (see Appendices 1, 2, and 3). Most employees already have new jobs lined up with utility companies in the Western Interconnection, a mark of how respected they are as experts in the industry.

The wind down of operations completed smoothly and was cut off on schedule. It also was accomplished at lower cost than anticipated.

9.6 Emerging Challenges and Recommendations for Grid Operation Monitoring

9.6.1 Grid Operations with High Renewable Energy Penetration

In the last decade, many renewable and inverter-based generation resources, such as solar PV and wind farm units, were continuously integrated into the WECC system. The renewable energy portion keeps growing rapidly. In the last few years, the Peak RC's EMS recorded that the renewable energy generations from solar PV and wind farm plants reached 24% of the WECC system total generation and exceeded 50% of the CAISO total online generations for a short time, respectively.

The penetration rate of overall renewable generations is expected to continuously increase for the next decades. The state of California is obligated to meet its 50% renewable resource mandate by 2030 or 100% clean electricity by 2045. Arizona Public Service (APS), the largest electricity utility in the state Arizona, announced the goal of achieving 100% carbon-free electricity by 2050. The goal includes a nearer-term 2030 target of achieving a resource mix that is 65% percent clean energy.

As the portion of renewable generation in BA's or RC's online generation capacity total continuously increases, it becomes imperative to manage contingency reserve in a reliable and economic manner. In this regard, the Reserve Sharing program under North West Power Pool (NWPP) has demonstrated significant cost savings through coordinated planning and operation among participating entities in the WECC footprint over the last decades. By sharing contingency reserve, participants are entitled not only to use their own "internal" reserve resources but to call on other participants for assistance if internal reserve does not fully cover a contingency or disturbance. If the case of NWPP is extended to a higher operating authority and a larger system scale, such as RC-to-RC or Interconnection-to-Interconnection, each RC or interconnection will have more cost-effective and less custom interruptible means to manage emerging challenges with reduced system reserves and uncontrolled frequency dip in a power system of high renewable penetration.

Some factors such as HVDC control mode, unit control setting, and PSS settings across the impacted power systems for oscillation modes are significant.

A large-scale oscillation could impact multiple RCs' footprints under an interconnection such as EI, WI, and ENTSOE. Prompt coordination between the TSOs or RCs plays a vital role in the mitigation of the inter-area oscillation issues.

Dynamic evaluations on system behavior are becoming more and more necessary. A full system model, initialized from pre-disturbance SE basecase solution, is essential for system model validation and post-event analysis.

High penetration of renewable energy resources is changing western system operating paradigms. The sustained declining of traditional generators with a large inertia mass requires RCs to monitor system frequency response sufficiency in real time and assure for compliance of NERC's Interconnection Frequency Response Obligation (IFRO) persistently.

To manage the emerging challenges of BES operations effectively, the RCs may consider using real-time analytical tools and technology innovations to enhance their situational awareness capabilities in the following areas.

9.6.2 Loss of Massive Wind Generations

On August 13, 2019, ERCOT experienced an operation emergency condition driven by the hot weather.

Electricity demand in ERCOT system hit an all-time high of 74,531 megawatts as people blasted their air conditioners on that Monday afternoon and totaled 74,310 megawatts at 4:34 pm local time, August 13 (Tuesday), according to ERCOT. At one point on Tuesday afternoon, the region had just 2121 megawatts left in power reserves, less than 3% of total demand on the system. Lackluster breezes contributed to the higher prices significantly. As one can see from Fig. 9.1, the total wind power generation of ERCOT on August 13, 2019, has plunged from the peak by nearly 75% at around 4:00 pm.

Loss of massive wind generation resulted in the ERCOT system being operated in an extreme low operating margin. The ERCOT Operator had to issue an emergency alert, calling on all power plants to ramp up and asking customers to conserve. The event underscores how dependent the region's power grid has become on wind farms, which now make up about a 25% of the gen capacity in Texas.

The low operating margin risk of ERCOT on the week of August 12, 2019, reminds ones to think about how to model such massive renewable generation loss

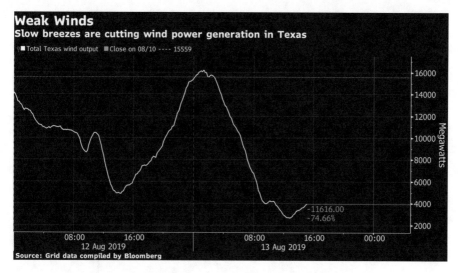

Weak Winds
Slow breezes are cutting wind power generation in Texas

Fig. 9.1 ERCOT wind power generation variations on August 13, 2019

contingencies in RTCA and perform real-time assessment on risk of cascading effect due to weather conditions. This is one of the emerging challenges for future grid operations. There is a big gap in the existing EMS tool suite in modeling and monitoring such massive of wind generation in real time.

9.6.3   Interaction Between Renewable Energy Generations and RAS

9.6.3.1   Wind Generation vs. Load Demand

It is worth noting that wind ramps and load ramps do not necessarily coincide with each other. Figure 9.2 shows load profile (upper trace) and wind profile (lower trace) for one specific day. In that specific example, it can be seen that during morning load pick up wind generation ramps down, putting additional stress on the system operators and increasing need for balancing the system.

There are a large number of RAS within WECC and the Pacific Northwest. The role of RAS is to automatically mitigate emergency system conditions such as path overloads. They are triggered automatically, based on preprogrammed thresholds and operating conditions. Many RAS include a generation drop as part of their mitigation scheme that can relieve line or path loading very quickly. When generation is dropped as part of a RAS, AGC is automatically blocked to avoid picking up dropped generation. If the cause of the RAS triggering is not immediately mitigated, the generation dispatcher needs to contact the scheduler and request an e-Tag curtailment prior to returning the AGC to service. Once the e-Tag is curtailed, the

Fig. 9.2 Negative
correlation between
demand and wind
generation

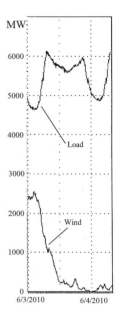

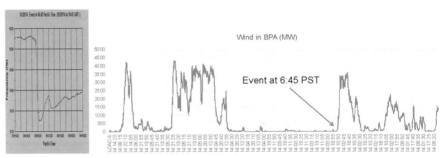

Fig. 9.3 (**a**) System frequency at the January 29, 2014, event; (**b**) BPA wind generation total in MW

set points for the units are changed, and the AGC is placed back in service. Since
wind displaces conventional units, RAS effectiveness is potentially changed. Adding
more wind online displaces more conventional units. At some point there will be no
RAS-associated generation to drop if there is an overabundance of wind generation
online. The effectiveness of the single RAS depends on quantity of wind generation
online (full wind vs. no wind). Additional studies need to be performed to assess the
effectiveness of RAS for different wind conditions. The consequence of failing to
adjust remedial actions is potentially dropping too much or too little generation.

Figure 9.3 (a) and (b) shows another example of the WECC PDCI RAS tripping event.

On January 29, 2014, a system event occurred in the Western Interconnection. In
particular,

- John Day-Grizzly #1 500 kV line relayed with #2 line out of service for maintenance.
- PATH65/PDCI tie RAS was initiated to drop 2200 MW generation; the system frequency declined from 59.993 to 59.75 Hz.
- SRP Springerville Unit 4 tripped, rejecting 400 MW at the same time.
- At the event time, total wind generation in BPA was close to 0 (very luckily). Just after a couple of hours, the BPA wind output went up to nearly 3000 MW.

If the system event had occurred while BPA wind generation was at the maximum output level, the system frequency could be hit by a big dip and might be recovered much more slowly.

9.6.3.2 Impact on Remedial Action Schemes (RAS)

9.6.3.2.1 Impact on System Studies

Large amounts of wind variability, displacement of conventional generation due to wind, the wheeling of regulation reserve through the transmission system, and reactive power issues cause stress on the transmission system. Consequently, more frequent operation studies are needed. One of the main challenges that transmission system operation faces is correct system operating limit (SOL) and Available Transfer Capability (ATC) assessment. While thermal limits are generally well known, they usually apply to short transmission lines (up to 50 mi.). For the longer lines, however, voltage stability and transient stability limits are typically the determining factors for power transfer capacity.

These limits are much more difficult to assess accurately and are generally set very conservatively. They are identified by planning and operation studies. In the environment where there is a high wind penetration, it is increasingly important to rely on studies initialized from real-time snapshots to perform more frequent reliability assessment of the power system.

Routinely, system operation studies and reliability analyses are executed for operations planning to ensure that the system can operate reliably, to prepare for scheduled outages or in case of the system events (N-1, N-1-1, N-k). Operational studies are usually performed a few days in advance based on static cases from the planning model. Such a case is adjusted to resemble, as much as possible, the current system conditions or to reflect predicted system conditions. This procedure can be time-consuming and yield inaccurate results. Because of wind volatility and inaccuracy of long-term wind forecasts, it is more difficult to plan for operations. Near-real-time studies become a necessity for ATC evaluation. Such studies are often based on cases provided from State Estimator (SE) within the EMS platform.

The SE provides the platform for Real-Time Contingency Analysis (RTCA), Real-Time Voltage Stability, and Real-Time Transient Stability Assessment. The SE also can automatically generate save cases after each run to facilitate postmortem analysis if needed. Following are a few examples of modern studies based on a real-time snapshot similar to those that are carried by WECC Reliability Coordinators.

9.6.3.2.2   Near-Real-Time Study

When system conditions warrant (predicted wind gusts, dynamic schedule during heavy loading conditions, etc.) or prior to a scheduled outage, a snapshot from the SE should be used as a starting point for power flow analysis. After adjusting the wind farm output according to the wind forecast or dynamic schedule, one can run the power flow again. The two obtained solutions should be compared for differences in flows and voltages to assess any potential reliability issues. After that, a contingency analysis can be performed together with voltage stability and transient stability analyses.

9.6.3.2.3   Impact on Contingency Analysis

System operation is especially concerned with the impact of the most severe contingencies on the system. Usually, the most severe contingencies are the loss of a large nuclear unit, a DC link, or a major transmission line. A large amount of geographically concentrated wind generation can lead to the loss of a large number of wind farms in short time period due to excessive wind. For example, for a Vestas V82-1.65 MW wind turbine, cutout speeds are 10 min for 44.7 mi/h, 1 min for 53.68 mi/h, and 1 s for 71.58 mi/h. Such system conditions are usually not studied adequately, and their impact on the system is largely unknown.

9.6.4   Recommendation 1: Monitor System Inertia Response by Real-Time TSAT

Peak RC collaborated with Powertech to enhance the TSAT software feature for calculating system inertia and IFRO Measure A-C on WECC system or individual balancing authority (BA) level. Figure 9.4 shows a snapshot of Area Frequency Response Summary table calculated against Palo Verde unit 1 tripping contingency. Real-time calculation of system inertia and frequency response measures will be essential for monitoring the impact of high penetration of renewable generation in the future. The RCs/ISOs and/or NWPP may use real-time calculated system inertia and IFRO measures to examine sufficiency of the system spinning reserve levels and determine the new requirements of spinning reserves on individual BA level or a reserve sharing group levels.

No.	Name	Inertia	A-B	A-C	C-B
1	AESO	55652.16	102.177	83.025	1.231
2	APS	32395.16	26.730	36.635	0.730
3	AVA	7949.87	1.970	2.217	0.888
4	AVRN	0.00	7.042	6.761	1.042
5	BANC	10800.65	10.409	12.213	0.852
6	BC-HYD	53549.15	56.376	48.623	1.160
7	BPA	63144.30	111.720	98.500	1.134

Area Frequency Response Summary - APS0U005 (PALO VERDE UNIT 1)

Fig. 9.4 WSM-TSAT calculated area frequency response summary

Online TSAT can be used to perform real-time assessment of IROLs/SOLs including transient RAS and dynamic components, which are unable to model in RTCA accurately. Peak RC used online TSAT solutions to back up the RT-VSA solved IROLs/SOLs for about 2 years. It helped the RCSOs to determine whether a VSA limit drop was a true concern or a false alarm several times effectively.

9.6.5   Recommendation 2: Enhance RTCA with Cascading and Look Ahead Analysis

Real-time monitoring on creditable contingency exceedances or limit violations with inclusive RAS/SPS protection interactions is one of the most critical control room functions in accordance with the NERC standards/requirements. Accurate RAS/SPS models, dynamic rating limits, and cascading outage scenario assessment in real time are the weakest areas in the existing RTCA application at control rooms.

The Western RCs shall enhance their RTCA tools with cascading analysis and look ahead CA capabilities for fast and accurate assessment on potential blackout scenarios. This was one of the lessons learned from the September 8, 2011, Pacific Southwest blackout event.

RTCA and look ahead CA applications shall

New challenges for RTCA to monitor [N-k]/MUC creditable contingencies and loss of massive renewable generations due to weather conditions are analyzed by leveraging ERCOT and BPA wind generation cases that adversely impact the system operation reliability. Software and visualization enhancements for complex RAS modeling, predictive RTCA, automatic cascading outage analysis, and the framework integrating RTCA with TSAT for transient/dynamic RAS evaluation can be achieved from Peak RC's experience described in Chaps. 2 and 4.

9.6.6   Recommendation 3: Enable RTCA for [N-k] Contingencies Monitoring

Peak RC modeled 9000+ contingencies and 515 RAS/SPS/non-RAS schemes in RTCA, including 500+ [N-2] and/or multiple contingencies (MUC) (some are always creditable, but most are conditionally creditable). Currently, only a small portion of those [N-2]/MUC contingencies are enabled in the Western RCs' RTCA for real-time screening.

Some of [N-2]/MUC CTGs involve transient RAS and UFLS models, which are unable to model in RTCA correctly. It takes more effort to validate impact of MUCs. In reality, [N-2] or MUC outages occurred sometimes.

Regarding [N-2]/MUC monitoring in RTCA, there will be some special scenarios for real-time assessment. Figure 9.5 gives an example of future scenarios in this regard. Once the WECC system is split by WECC-1 RAS operation, RTCA won't be able to report risk of insufficient spinning reserves and inertia response support by traditional generation resources in the Southern Island.

It's observed that RTCA has limitation on solving [N-k]/MUC contingencies properly, as many of the [N-k]/MUC contingencies are protected by the transient RAS and/or under-frequency load shedding (UFLS) devices in reality. To resolve the problems, Peak RC was able to integrate online TSAT with EMS/RTCA to model the required RAS and UFLS components properly for real-time assessment. The framework of TSAT-RTCA integration is illustrated in Fig. 9.6 for the Western RC's reference.

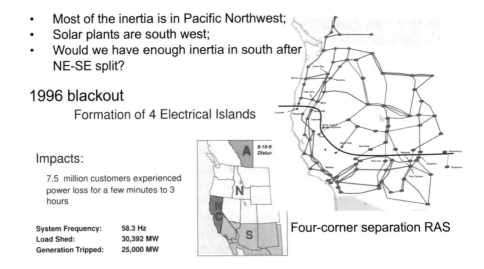

- Most of the inertia is in Pacific Northwest;
- Solar plants are south west;
- Would we have enough inertia in south after NE-SE split?

1996 blackout
Formation of 4 Electrical Islands

Impacts:

7.5 million customers experienced power loss for a few minutes to 3 hours

System Frequency:	58.3 Hz
Load Shed:	30,392 MW
Generation Tripped:	25,000 MW

Four-corner separation RAS

Fig. 9.5 WECC system separation against high renewable penetration (future)

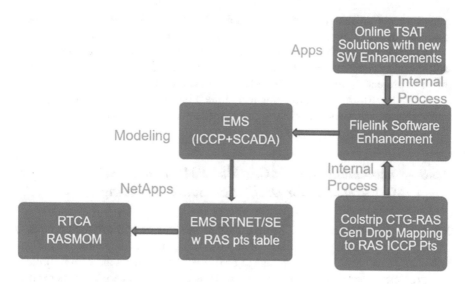

Fig. 9.6 Framework of integrating TSAT with EMS for ATR RAS monitoring

9.6.7 Recommendation 4: Evaluate IROLs/SOLs in Real Time

As it becomes a rising trend to retire large and traditional generating units and make up the capacity by renewable generation resources, North America power grids are facing a series of challenges:

- Loss of massive renewable generations (i.e., over 50% total wind or solar generation drop in a region or an interconnection at the same time) could happen and move the system reserve margin down to a very low level.
- In case of shortage of adequate system reserves, the system frequency stability is threatened by [N-1] or [N-k] large unit tripping or RAS gen drop actions.
  Once such contingency occurs in reality, a system frequency dip will probably activate under-frequency load shedding schemes to avoid a system frequency collapse.
- To overcome massive under-frequency load shedding, the balancing authorities (BAs) in North America should collaborate with each other to create a larger contingency reserve pool for sharing resources under normal and emergency conditions with minimize risk and cost on the average.
- To monitor such a large contingency reserve pool, system operators need to run Real-Time Voltage Stability Analysis tool to evaluate resource deliverability and identify reactive sufficiency at major load centers so that both large power transfer and specific contingency reserves can be delivered from a group of committed source(s) to a sink region without violation of SOLs.

Peak RC implemented V&R ROSE RT-VSA tool in RC control room through close collaboration with vendor and entities. After extensive validation test and enhancements on the software and models, the ROSE RT-VSA tool was proven adequate for providing near-real-time assessment on multiple IROLs.

The RT-VSA tool can be modified to be used for TTC and ATC calculation, as well as real system reactive sufficiency assessment in the future. Peak RC performed the pilot projects to expand the RT-VSA use cases in this regard. Those use cases require high-performance computing technique if the numbers of TTC/ATC simulation scenarios and load centers for reactive sufficiency assessment in real time.

9.6.8   Recommendation 5: Combine PMU and SCADA Measurements for Oscillation Source Locating

It remains a challenge for the Western RCs to monitor system oscillation modes, identify root cause of low oscillation damping events, and develop effective mitigation plans in real time, as the coverage of PMU measurements on the BES is limited. The problems become more challenging when a dominant system mode is excited by a forced oscillation at the close frequency.

Use synchronized measurements across the interconnection during grid disturbances or abnormalities to baseline the oscillatory performance of the interconnection. The following data will be required for oscillation event analysis and simulation:

- Synchronized PMU measurements collected for multiple system events
- Historical Mode Meter Results exportable from PI
- Online TSAT Archive cases captured before the events
- System Topology Changes stored in EMS SE auto savecases and/or PI
- AGC control measurements
- WECC Path/Cutplane Flow and SOL changes

To clearly understand resonant oscillation phenomena and estimate mode damping accurately for development of correct mitigation plans, we would recommend to (1) use the high-fidelity Interconnection-wide operational model for oscillation modal analysis under contingencies using TSAT and SSAT; (2) use both PMU and ICCP measurements for forced oscillation source unit locating; and (3) develop high-performance online Fast Sub-Space Identification (FSSI) tools for deployment in control room to decompose an inter-area mode and a forced oscillation automatically quickly.

9.6.9   Recommendation 6: Use Both EMS and PMU Data for Cybersecurity Assurance

PMU provides high-resolution measurements of system dynamics. FNET/GridEye visualization software enables control rooms to capture the system disturbance locations and frequency oscillation evolution from PMU measurements quickly. For

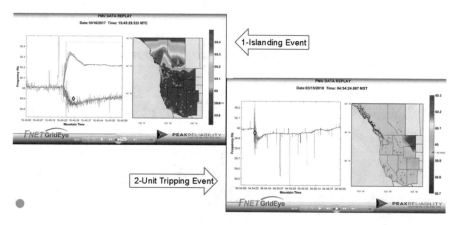

Fig. 9.7   FNET/GridEye visualizations on two system events: islanding and loss of one large unit

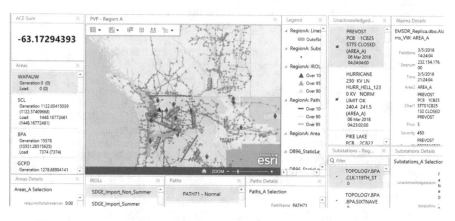

Fig. 9.8   Integrating and visualizing both EMS and PMU application data on PI-ESRI graphic view

example, from Fig. 9.7, one can clearly observe system islanding condition and large unit trip event, respectively.

By leveraging OSIsoft PI-ESRI product, Peak RC managed to visualize both EMS data (i.e., SCADA ICCP measurements, SE, and RTCA solution results) and PMU downsampled measurements on a geographic view shown on Fig. 9.8. Because EMS and PMU are two independent data sources, in case one of the data source encounters an issue (such as ICCP link is down or a cyber-attack), the other data source can still provide real-time situational awareness to the control rooms and help ones to detect the cyber-attack and recover the corrupted data correctly.

Appendices

Appendix 1: CAISO-RC West Resolution of Appreciation

Resolution of Appreciation
Peak Reliability
October 23, 2019

WHEREAS, Peak Reliability (and its predecessor, Western Electricity Coordinating Council) has faithfully acted as the highest level of authority responsible for the reliable operation of the Bulk Electric System in the Western Interconnection for more than 10 years,

WHEREAS, Peak Reliability has been an essential partner in the transition of reliability coordinator services to RC West, and

WHEREAS, Peak Reliability has continuously demonstrated its commitment to reliability,

BE IT RESOLVED, that the RC West Oversight Committee formally recognizes Peak Reliability and its staff for the professional and thorough support of the transition of RC Services to RC West and the continued dedication to the reliability of the Western Interconnection. Thank you for "assuring the wide area view"!

Michelle Cathcart
Chair

Steve Cobb
Vice Chair

Appendix 2: Western Reliability Executive Committee Resolution

SPP *Southwest Power Pool*

Western Reliability Executive Committee

RESOLUTION

Whereas fourteen entities in the Western Interconnection will transfer reliability coordination (RC) services from Peak Reliability to Southwest Power Pool, Inc. (SPP) on December 3, 2019; and

Whereas Peak Reliability has reliably served and supported the Western Interconnection and electric utility industry as a whole since 2014; and

Whereas SPP has received significant support from Peak Reliability for the past two years during the reliability coordination transition process; and

Whereas with Peak's participation and support, the SPP West RC project has progressed efficiently with recognition through the North American Reliability Corporation (NERC) certification process; and

Whereas on December 3, 2019, SPP will begin reliability coordination services in the Western Interconnection following the outstanding model the Peak Reliability has provided since 2014; then

Be it now therefore resolved that the Western Reliability Executive Committee unanimously recognizes Peak Reliability for their devoted and conscientious service and we express our deepest gratitude for assisting in the RC transition. SPP and their Reliability Coordinator customers thank Peak for working together on this effort, enabling SPP to keep the lights on – today and in the future! Peak's outstanding efforts will have an everlasting impact on the Western Interconnection Reliability.

Resolved this twenty-fifth day of September, Two thousand and nineteen.

Keith Carman
Chair, Western Reliability Executive
Committee

C.J. Brown
Staff Secretary

Appendix 3: Western Electric Industry Leaders (WEIL) Recognition Letter

October 11, 2019

Mr. John Procario
Chair, Board of Directors

Ms. Marie Jordan
President and CEO

Peak Reliability
7600 NE 41st Street, Suite 201
Vancouver, WA 98662

SUBJECT: Recognition of Peak Team During RC Transition

Dear Mr. Procario and Ms. Jordan:

The Western Electric Industry Leaders (WEIL) would like to extend our sincere appreciation to Peak Reliability for your facilitation of the transition of Reliability Coordinator (RC) services to the new providers.

Peak has worked both professionally and collaboratively in the transfer of data and processes from Peak to the new RC providers. The team members have been solution-oriented, with a focus on the continued reliability of the Western Interconnection. In addition, Peak has managed the wind-down effort in a professional and proactive manner that is anticipated to be well within budget.

WEIL recognizes the difficulties that a transition like this can present to both the organization and its employees. We appreciate Peak's efforts in successfully managing this transition and request that you extend this appreciation to the entire Peak team.

Sincerely,

        Elliot E. Mainzer
        Administrator and Chief Executive Officer
        Bonneville Power Administration

Appendix 4: The Final WSM Factsheet

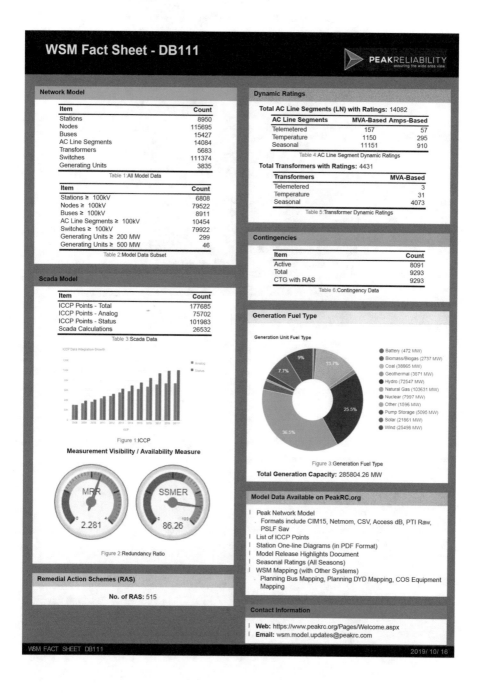

WSM Fact Sheet - DB111

PEAKRELIABILITY
assuring the wide area view

Network Model

Item	Count
Stations	8950
Nodes	115695
Buses	15427
AC Line Segments	14084
Transformers	5683
Switches	111374
Generating Units	3835

Table 1:All Model Data

Item	Count
Stations ≥ 100kV	6808
Nodes ≥ 100kV	79522
Buses ≥ 100kV	8911
AC Line Segments ≥ 100kV	10454
Switches ≥ 100kV	79922
Generating Units ≥ 200 MW	299
Generating Units ≥ 500 MW	46

Table 2:Model Data Subset

Scada Model

Item	Count
ICCP Points - Total	177685
ICCP Points - Analog	75702
ICCP Points - Status	101983
Scada Calculations	26532

Table 3:Scada Data

Figure 1:ICCP

Measurement Visibility / Availability Measure

MRR 2.281

SSMER 86.26

Figure 2:Redundancy Ratio

Remedial Action Schemes (RAS)

No. of RAS: 515

Dynamic Ratings

Total AC Line Segments (LN) with Ratings: 14082

AC Line Segments	MVA-Based	Amps-Based
Telemetered	157	57
Temperature	1150	295
Seasonal	11151	910

Table 4:AC Line Segment Dynamic Ratings

Total Transformers with Ratings: 4431

Transformers	MVA-Based
Telemetered	3
Temperature	31
Seasonal	4073

Table 5:Transformer Dynamic Ratings

Contingencies

Item	Count
Active	8091
Total	9293
CTG with RAS	9293

Table 6:Contingency Data

Generation Fuel Type

Generation Unit Fuel Type

- Battery (472 MW)
- Biomass/Biogas (2737 MW)
- Coal (38865 MW)
- Geothermal (3671 MW)
- Hydro (72547 MW)
- Natural Gas (103631 MW)
- Nuclear (7997 MW)
- Other (1596 MW)
- Pump Storage (5095 MW)
- Solar (21861 MW)
- Wind (25498 MW)

Figure 3:Generation Fuel Type

Total Generation Capacity: 285804.26 MW

Model Data Available on PeakRC.org

- Peak Network Model
  - Formats include CIM15, Netmom, CSV, Access dB, PTI Raw, PSLF Sav
- List of ICCP Points
- Station One-line Diagrams (in PDF Format)
- Model Release Highlights Document
- Seasonal Ratings (All Seasons)
- WSM Mapping (with Other Systems)
  - Planning Bus Mapping, Planning DYD Mapping, COS Equipment Mapping

Contact Information

- **Web:** https://www.peakrc.org/Pages/Welcome.aspx
- **Email:** wsm.model.updates@peakrc.com

WSM FACT SHEET DB111

2019/ 10/ 16

Index

© Springer Nature Switzerland AG 2021
H. Zhang et al., *Advanced Power Applications for System Reliability
Monitoring*, Power Systems, https://doi.org/10.1007/978-3-030-44544-7